Memoir 27

A Color Illustrated Guide To

Carbonate Rock Constituents, Textures, Cements, and Porosities

Peter A. Scholle

U.S. Geological Survey

Published by
The American Association of Petroleum Geologists
with the support of
The American Association of Petroleum Geologists Foundation
Tulsa, Oklahoma, U.S.A., 1978

Published February 1978
Third Printing, March 1981

Library of Congress Catalog Card No. 77-94380
ISBN: 0-89181-303-9

Printed by
Rodgers Litho
Tulsa, Oklahoma

Table of Contents

Introduction

The purpose of this book is to make available to geologists who may not be specialists in carbonate petrography a volume which illustrates the major grains, textures, cements, and porosity types found in carbonate rocks.

Successful hydrocarbon exploration in carbonate rocks is a difficult and complex problem. Carbonate rocks not only have complicated and varied depositional patterns, but also are subject to extensive post-depositional alteration which may radically alter original porosity and permeability relationships. Because these *diagenetic* changes can occur at many stages of burial and(or) uplift, the timing of such events relative to hydrocarbon migration and structural deformation is important. Petrography commonly is the most valuable tool which can be used to resolve such interpretational problems in potential reservoir rocks.

This book is designed as an introduction for the explorationist or student, and is by no means a complete treatise or textbook. However, it does include a wide variety of examples of skeletal and nonskeletal grains, cements, fabrics, and porosity types. It also encompasses a number of noncarbonate grains which can occur as common accessory minerals in carbonate rocks or which may provide important biostratigraphic or paleoenvironmental information in carbonate rocks. With this guide, people with little formal training in petrography should be able to examine thin sections or peels under the microscope and interpret the main rock constituents and their depositional and diagenetic history.

Carbonate petrography, because it involves very few minerals (mainly calcite with subordinate dolomite, quartz, and clay), is primarily a qualitative art. One must learn to recognize the distinguishing characteristics of skeletal grains of various ages. Different directions of sectioning will produce radically different appearances, as will different stages of diagenetic alteration. There are no simple diagnostic tests (such as measuring birefringence, optic figure, etc.) which will absolutely identify a bryozoan, for example. It is simply a question of experience. Comparison of grains in thin sections with photographs of identified grains in this and other books will quickly allow geologists to correctly identify the majority of the rock-forming grains in their samples. A selected bibliography is provided to permit the interested reader to pursue details only briefly covered in this book and, particularly, to supplement the interpretive aspects of petrographic work.

A particular attempt has been made to illustrate typical rather than spectacular examples of many grains and fabrics. For example, grains which were originally composed of aragonite normally undergo wholesale diagenetic alteration which commonly obliterates primary structural features. Therefore, such grains are shown in their extensively altered state because this preservation is typical. A table of original mineralogies is provided to help the reader anticipate preservation problems. Furthermore, examples have been included from rocks of Cambrian to Holocene age because of the enormous evolutionary changes in organisms (and, therefore, carbonate fabrics) through time.

In terms of the overall costs of hydrocarbon exploration today, the financial investment needed for petrographic work is virtually insignificant. A basic polarizing microscope can be purchased currently for $1,000 to $5,000 depending on optical quality, accessories, and other factors. Thin sections can be purchased for $2 to $5 each from a number of commercial labs. Acetate peels (see technique section of the bibliography) can be made in any office in minutes from polished rock slabs, and can provide a remarkable amount of information. Outcrop samples, conventional cores, sidewall cores, and cuttings samples all can be examined microscopically although textural information decreases with decreasing sample size. Even the investment of time involved in petrographic work need not be great relative to the potential for problem solving. Few other techniques are as valuable and accurate for the identification of preserved, destroyed, or created porosity, or the prediction of depositional and diagenetic trends.

Research conducted over the past 25 years has outlined many principles of deposition and diagenesis in carbonate sediments. Facies models have been established for modern (as well as ancient) reefs and other bank-margin deposits, for tidal-flat and sabkha sedimentation, for basinal deposition, and for other environments. Diagenetic studies have pointed out the influence of syndepositional marine cementation, early freshwater diagenesis, and later subsurface compaction-dissolution phenomena. This work has clearly shown that, although carbonate depositional and diagenetic patterns may be complex, there is commonly a large volume of information recorded in the rocks which can be used to decipher this record. For example, virtually all carbonate grains are produced either directly by, or through the indirect influence of, organisms. Unlike clastic terrigenous particles, carbonate grains are rarely transported far from their original site of formation. Thus, the identification of a biologically produced suite of grains will often reveal considerable information about the environment of deposition. Likewise, marine, freshwater, and late subsurface cements each have distinctive, recognizable fabrics.

Petrography, when used in close conjunction with well-log analysis, seismic interpretation, regional geology, and other studies, can be an invaluable tool for applying these recently developed principles of carbonate sedimentology to ancient rocks. Furthermore, it is best applied by the explorationist who is deeply involved in techniques other than petrography, for he is in the best position to ask the right questions—questions petrography may be able to answer. That is the goal of this volume.

Explanation of Captions

Each photograph in this manual has a description in standard format. The first two lines give the stratigraphic unit (including geologic age) and state or country of origin. This is followed by a description of the photograph. The last line of the caption defines the type of lighting used and the scale of the photograph. The following code is used for lighting:

X.N.—crossed nicols (cross-polarized light)
P.X.N.—partly crossed nicols
G.P.—gypsum plate (Quartz Red I plate) inserted
R.L.—reflected light

The absence of any symbol indicates that transmitted light with uncrossed polarizers was used. All scales are given as a certain number of millimeters (or micrometers μm, for scanning electron microscopy, SEM) for a scale bar of uniform length (1.25 cm or 0.5 inches) for all photographs. Thus, a figure of 0.38 mm indicates that a length of 1.25 cm on the printed picture is equivalent to 0.38 mm on the original specimen.

Acknowledgements

An earlier version of this book was compiled at Cities Service Oil Company Research Laboratory and much of that original version is included here. I would like to express my great appreciation to Cities Service for their kind permission, even encouragement, to publish this volume. During the seven years of its existence, this book (in various unpublished editions) has been reviewed or commented upon by hundreds of persons and it has grown largely through the contribution of samples by persons too numerous to mention.

Thanks also go to R. M. Forester, R. B. Halley, and N. J. Silberling who reviewed copies of this book; J. P. Bradbury, R. N. Ginsburg, and W. J. Meyers who kindly contributed photographs; and L. R. Tomlinson and J. L. Hennessy who greatly aided in the editorial preparation of this volume.

Finally, I would like to express my appreciation to the three petrographers who spent many hours looking down a microscope with me and whose teaching dedication made this volume possible: R. G. C. Bathurst, A. G. Fischer, and R. L. Folk.

35mm Slides Available

A set of 100 selected photographs from this book are available in 35 mm slides. The set is designed to aid instructors who might use the text for teaching purposes. To order the set, or to request a free brochure, contact: AAPG Publications, P.O. Box 979, Tulsa, Oklahoma 74101.

A similar set of slides for clastic rocks keyed to AAPG Memoir 28, "A Color Illustrated Guide to Constituents, Textures, Cements, and Porosities of Sandstones and Associated Rocks," also is available from AAPG.

Glossary: Carbonate Petrography

(Source references given only for carbonate rock classification)

Allochems an inclusive term for carbonate grains or particles, as contrasted to carbonate mud matrix and clear calcite cement; includes fossils, oolites, pellets, etc. (Folk, 1959, 1962).

Anhedral a single crystal or crystal fabric which does not show well-defined typical crystallographic forms.

Bahamite granular limestone resembling the Holocene deposits in the Bahamas; composed largely of pellet-like aggregates of carbonate mud.

Beach Rock a friable to well-cemented rock consisting of calcareous sand cemented by calcium carbonate crusts precipitated in the intertidal zone. Generally found as thin beds dipping seaward at less than 15°.

Biochemical deposited by chemical processes under biologic influence. For example, the removal of CO_2 from sea water by aquatic plants may cause calcium carbonate to precipitate.

Biolithite limestone made up of organic structures growing in place and forming a coherent, resistant mass during growth (Folk, 1959, 1962).

Biomicrite limestone composed of skeletal grains in a micrite matrix (Folk, 1959, 1962).

Biosparite limestone composed of skeletal grains with sparry calcite cement (Folk, 1959, 1962).

Bladed in reference to sparry calcite cement, defined as including crystals with a length-to-width ratio between 1.5:1 and 6:1.

Borings openings created in relatively rigid rock, shell, or other material by boring organisms. The rigid host substrate is the feature which distinguishes borings from burrows; the latter form in unconsolidated sediment.

Boundstone carbonate rock showing signs of grains being bound (by organisms) during deposition (Dunham, 1962).

Calcarenite limestone composed predominantly of sand-sized calcium carbonate grains (carbonate sand).

Calcilutite limestone composed of lithified calcareous mud (lime mud).

Calcirudite limestone composed predominantly of calcium carbonate fragments larger than sand size (carbonate conglomerate).

Calcispheres silt- or sand-sized spheres of clear (sparry) calcite, some with a discernible wall and some without . Probably of diverse origin, algal spores being a likely major variety.

Calclithite a rock formed chiefly of carbonate clasts derived from older, lithified limestone, generally external to the contemporaneous depositional system. Commonly formed along downthrown sides of fault scarps.

Caliche (calcrete) surficial material such as sand, gravel, or cobbles cemented by calcium carbonate in arid climates as a result of evaporative concentration of $CaCO_3$ in surface pore waters. Often characterized by crusts, pisolites, reverse grading, autofracturing, and microstalactitic textures.

Cavernous Porosity a pore system characterized by large openings or caverns. Such porosity is too large to be identified in normal subsurface cores but is recognizable during drilling by large drops (0.5 m or greater) of the drill bit.

Chalk a limestone which consists predominantly of the remains of calcareous nannoplankton (especially coccoliths) and microplankton (especially Foraminifera). Chalks are often considered to be soft and friable, although diagenetic alteration can lead to complete lithification of chalks.

Clastic as used by most sedimentary petrologists, composed of particles that have been mechanically transported, at least locally. Specifically includes limestones made up of fossils or other allochems that have been moved by waves or currents. (Note that most facies mappers use clastic for terrigenous rocks and not limestones).

Coated Grains a general term for grains with coatings or rims of calcium carbonate; includes oolites and superficial oolites, pisolites, and algal-coated grains.

Detrital used in different ways by different authors and hence undefinable out of context. Sometimes synonymous with clastic, sometimes with terrigenous, and sometimes restricted to rocks composed of broken older rocks.

Diagenesis changes in sediments or sedimentary rocks which occur after deposition but excluding processes involving high enough temperature and pressure to be called metamorphism.

Dismicrite disturbed micrite; a carbonate mud that contains stringers or "eyes" of sparry calcite resulting from filling of burrows, slump or shrinkage cracks, or other partial disruption on the sea floor (Folk, 1959, 1962).

Eogenetic occurring during the time interval between final deposition and burial of the newly deposited sediment or rock below the depth of significant influence by processes that either operate from the surface or depend for their effectiveness on proximity to the surface.

Equant in reference to sparry calcite cement it is defined as including crystals with a length to width ratio of less than 1.5:1.

Euhedral refers to a single crystal or crystal fabric which shows well-defined typical crystallographic forms.

Extraclast a detrital grain of lithified carbonate sediment (lithiclast) derived from outside the depositional area of current sedimentation. The rock composed of these grains would be a calclithite. See also *intraclast.*

Fenestrae (fenestral fabric) primary or penecontemporaneous gaps in rock framework larger than grain-supported interstices. Commonly associated with algal mats and can result from shrinkage, gas formation, organic decay, etc.

Fibrous in reference to sparry calcite cement it is defined as including crystals with length-to-width ratios greater than 6:1.

Geopetal Structure any internal structure or organization of a rock indicating original orientation such as top and bottom of strata.

Grains (1) solid particles whose physical limits may encompass many crystalline entities. Distinctions as between *coarse grained* and *coarsely crystalline* are not always observed but are fundamental; (2) the friable aggregates of silt-sized carbonate crystals that are forming today on the Bahaman Platform from the partial cementation of crystals in contact with each other.

Grain-supported refers to the fabric of a rock in which the grains (allochems) are in contact with each other, even though they may have a mud (micrite) matrix.

Grainstone carbonate rock composed of grains with no carbonate mud in the interstices (Dunham, 1962).

Grapestone sometimes used for aggregates of silt-sized carbonate crystals (*grains*, def. 2), but more properly applied to grape-like clusters of such aggregates.

Interstices technically, voids; but used mostly for areas that were voids in the initial sediment, though they are now filled.

Intraclast a fragment of penecontemporaneous, commonly weakly consolidated, carbonate sediment that has been eroded from sea bottom and redeposited nearby.

Isopachous (grainskin) a descriptive term for cement which has formed as a uniform-thickness coating around grains. Common in most submarine cements.

Lithographic refers to extremely fine and uniform carbonate usually with smooth conchoidal fracture.

Lumps in modern sediments, irregular composite aggregates of silt- or sand-sized carbonate crystals that are cemented together at points of contact: in ancient carbonates, similar-appearing grains composed of carbonate mud.

Meniscus refers to a carbonate cement type formed during vadose diagenesis where cement crystals form only at or near grain contacts in the positions a water meniscus would occupy.

Mesogenetic occurring during the time interval in which rocks or sediments are buried at depth below the major influence of processes directly operating from or closely related to the surface.

Micrite microcrystalline calcite; used both as a synonym for carbonate mud (or "ooze") and for a rock composed of carbonate mud (calcilutite) (Folk, 1959, 1962).

Microspar generally 5- to 15-micron sized calcite produced by recrystallization of micrite; can be as coarse as 30 microns (Folk, 1965).

Microstalactitic (pendulous or gravitational cement) a descriptive term for cements which are concentrated on the bottom sides of grains. Such textures generally form in vadose or upper intertidal areas where pores are only partially water-filled and in which water droplets can hang from the undersides of grains.

Mold a pore formed by the selective removal, normally by solution of a former individual constituent of the sediment or rock such as a shell or ooid. Often used with modifying prefixes: oomoldic, dolomodic, etc.

Mudstone in referring to carbonates, a carbonate rock composed of carbonate mud with less than 10% allochems (Dunham, 1962).

Oncolite a pisolite of algal origin, a spheroidal form of algal stromatolite showing a series of concentric (often irregular) laminations. These unattached stromatolites are produced by mechanical turning or rolling, exposing new surfaces to algal growth.

Oolite (ooids) Oolite is a rock composed of ooids. Ooids are spherical to ellipsoidal bodies, 0.25 to 2.00 mm in diameter with a nucleus and radial or concentric structure (calcareous, hematitic, siliceous, etc.) generally grown in an agitated environment.

Oomicrite a limestone composed dominantly of ooids in a matrix of micrite (Folk, 1959, 1962).

Oosparite a limestone composed dominantly of ooids in sparry calcite cement (Folk, 1959, 1962).

Ooze carbonate mud; either the original soft sediment or its consolidated equivalent.

Orthochemical a rock constituent that is a normal chemical precipitate, as contrasted to fossils, oolites, or other mechanically or biologically deposited constituents.

Packed containing sufficient grains (allochems) for the grains to be in contact and mutually supporting, as contrasted with rocks with grains "floating" in mud (Folk, 1959, 1962).

Packstone carbonate rock composed of packed grains with carbonate mud matrix (cf. grainstone and wackestone; Dunham, 1962).

Pellet a grain composed of micrite, generally lacking significant internal structure and often ovoid in shape. Many, but not all, pellets are of fecal origin.

Pelmicrite a limestone composed dominantly of peloids (or pellets) in a matrix of micrite (Folk, 1959, 1962).

Peloid an allochem formed of cryptocrystalline or microcrystalline carbonate irrespective of size or origin. This term allows reference to grains composed of micrite or microspar without the need to imply any particular mode of origin (can thus include pellets, some vague intraclasts, micritized fossils, ooids, etc.).

Pelsparite a limestone composed dominantly of peloids (or pellets) in sparry calcite cement (Folk, 1959, 1962).

Penecontemporaneous generally referring to cements or replacement textures indicating that, in the opinion of the user, the feature or mineral formed at almost the same time as the original sediment was deposited, that is, close to the sediment-air or sediment-water interface.

Phreatic the zone of water saturation below the water table. There are marine or meteoric phreatic zones.

Pisolite a small spheroidal particle with concentrically laminated internal structure and size larger than 2 mm.

Poikilotopic (poikilitic) textural term denoting a condition in which small granular crystals are irregularly scattered without common orientation in a larger crystal of another mineral (generally sand or silt grains in a single cement crystal).

Poorly Washed a rock which has sparry calcite cement but which also has one-third to one-half of all interstices filled with carbonate mud (i.e., a poorly sorted rock).

Primary Porosity porosity present in the rock or sediment immediately after final deposition.

Pseudospar a neomorphic (recrystallization) calcite fabric with average crystal size larger than 30 microns (Folk, 1965).

Radiaxial Fabric a cavity-filling spar mosaic consisting of fibrous crystals radiating away from the wall, allied to optic axes which converge away from the wall. Also characterized by curved cleavages and irregular intergranular boundaries.

Secondary Porosity any porosity created in a sediment after final deposition.

Shelter Porosity a type of primary interparticle porosity created by the sheltering effect of relatively large sedimentary particles which prevent the infilling of pore space beneath them by finer clastic particles.

Skeletal Carbonate components (or the rocks they form) derived from hard material secreted directly by organisms. A substitute for the confusing term *organic* of some older literature.

Spar (sparry calcite) calcite sufficiently coarsely crystalline to appear fairly transparent in thin section, as contrasted to dark, cloudy-appearing carbonate mud or micrite (Folk, 1959, 1962).

Sparse allochems sufficiently scarce so as to be separated from each other by carbonate mud; less than 50% of the rock (Folk, 1959, 1962).

Spastolith refers to an ooid which has been deformed by shearing the concentric laminations away from the nucleus.

Stylolite a jagged, columnar surface in carbonate rocks which may be at any orientation relative to bedding; often associated with large amounts of insoluble material, and produced by solution and grain interpenetration.

Sucrose carbonate rocks that have appreciable intercrystal pore space and are composed dominantly of somewhat equant, uniformly sized, euhedral to subhedral crystals.

Syntaxial refers to overgrowths which are in optical continuity with their underlying grains.

Telogenetic occurring during the time interval during which long buried sediments or rocks are influenced significantly by processes associated with the formation of an unconformity.

Terrigenous derived from the land area and transported mechanically to the basin of deposition; commonly, essentially synonymous with "noncarbonate" (e.g. *terrigenous sand* vs. *carbonate sand*).

Tufa (calctufa) a sedimentary rock (limestone or silica) deposited in or around springs, lakes and rivers from emerging ground waters charged with CO_2 and $CaCO_3$ or SiO_2. Commonly builds thick calcite deposits rich in algae and higher plant material.

Umbrella Void a void (perhaps later filled with sparry calcite) produced by the presence of a large grain sheltering the underlying area by preventing infiltration of mud-sized grains. Also called shelter porosity.

Vadose designates the area above the water table (zone of aeration) in which both water and air are present in pores.

Vug a pore that is somewhat equant, is larger than 1/16 mm in diameter, and does not specifically conform in position, shape or boundary to particular fabric elements to the host rock (i.e., is not fabric selective). Generally formed by solution.

Wackestone carbonate rock composed of carbonate mud with over 10% allochems suspended in it (Dunham, 1962).

Grain Size Scales for Sediments

The grade scale most commonly used for sediments is the Wentworth scale (actually first proposed by Udden), which is a logarithmic scale in that each grade limit is twice as large as the next smaller grade limit. For more detailed work, sieves have been constructed at intervals $\sqrt[2]{2}$ and $\sqrt[4]{2}$.

The ϕ (phi) scale, devised by Krumbein, is a much more convenient way of presenting data than if the values are expressed in millimeters, and is used almost entirely in recent work.

U.S. Standard Sieve Mesh #	*Millimeters*	*Microns (μm)*	*Phi (ϕ)*	*Wentworth Size Class*	
	4096		-12		GRAVEL
	1024		-10	Boulder (-8 to -12ϕ)	
Use	256		- 8		
wire	64		- 6	Cobble (-6 to -8ϕ)	
squares	16		- 4	Pebble (-2 to -6ϕ)	
5	4		- 2		
6	3.36		- 1.75		
7	2.83		- 1.5	Granule	
8	2.38		- 1.25		
10	2.00		- 1.0		
12	1.68		- 0.75		SAND
14	1.41		- 0.5	Very coarse sand	
16	1.19		- 0.25		
18	1.00		0.0		
20	0.84		0.25		
25	0.71		0.5	Coarse sand	
30	0.59		0.75		
35	1/2 — 0.50	500	1.0		
40	0.42	420	1.25		
45	0.35	350	1.5	Medium sand	
50	0.30	300	1.75		
60	1/4 — 0.25	250	2.0		
70	0.210	210	2.25		
80	0.177	177	2.5	Fine sand	
100	0.149	149	2.75		
120	1/8 — 0.125	125	3.0		
140	0.105	105	3.25		
170	0.088	88	3.5	Very fine sand	
200	0.074	74	3.75		
230	1/16 — 0.0625	62.5	4.0		
270	0.053	53	4.25		MUD
325	0.044	44	4.5	Coarse silt	
	0.037	37	4.75		
	1/32 — 0.031	31	5.0		
Analyzed	1/64 0.0156	15.6	6.0	Medium silt	
	1/128 0.0078	7.8	7.0	Fine silt	
by	1/256 — 0.0039	3.9	8.0	Very fine silt	
Pipette	0.0020	2.0	9.0	Clay	
or	0.00098	0.98	10.0		
	0.00049	0.49	11.0		
Hydrometer	0.00024	0.24	12.0		
	0.00012	0.12	13.0		
↓ ↓	0.00006	0.06	14.0 ↓		

(From Folk, 1968)

Grain Scales: cont.

Comparison Chart For Visual Percentage Estimation (After Terry and Chilingar, 1955).

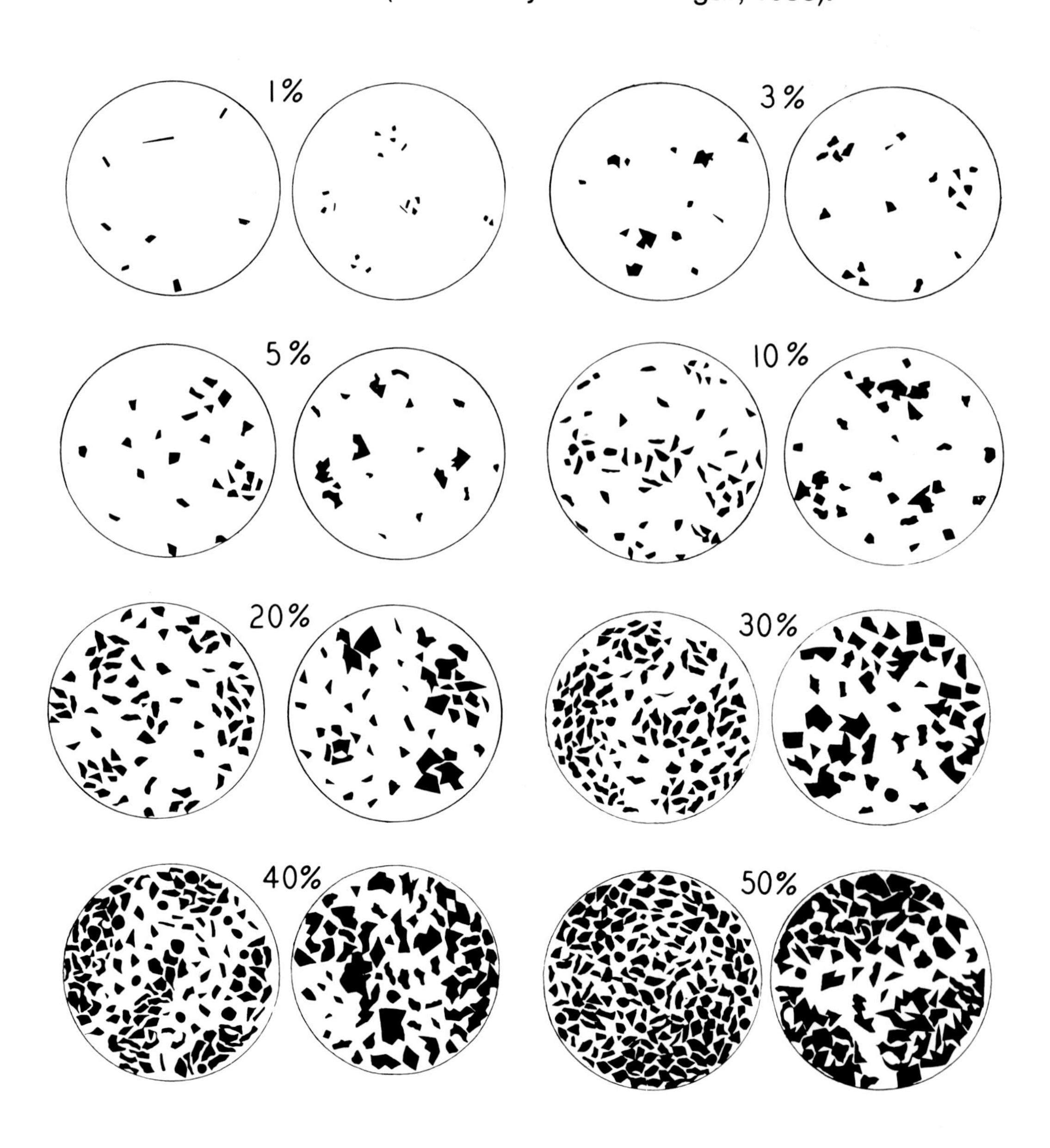

SKELETAL COMPOSITIONS

TAXON	ARAG.	CALCITE %Mg (0 5 10 15 20 25 30 35)	BOTH ARAGONITE AND CALCITE
CALCAREOUS ALGAE:			
RED		X————————X	
GREEN	X		
COCCOLITHS		X	
FORAMINIFERA:			
BENTHONIC	O	X——————X - - -X	
PLANKTONIC		X—X	
SPONGES:	O	X———X	
COELENTERATES:			
STROMATOPORIDS (A)	X	X ?	
MILLEPOROIDS	X		
RUGOSE (A)		X•••	
TABULATE (A)		X ?	
SCLERACTINIAN	X		
ALCYONARIAN	O	X———X	
BRYOZOANS:	O	X——X	O
BRACHIOPODS:		X—X	
MOLLUSKS:			
CHITONS	X		
PELECYPODS	X	X-X	X
GASTROPODS	X	X-X	X
PTEROPODS	X		
CEPHALOPODS (MOST)	X		
BELEMNOIDS & APTYCHI (A)		X	
ANNELIDS (SERPULIDS):	X	X———X	X
ARTHROPODS:			
DECAPODS		X—X	
OSTRACODES		X——X	
BARNACLES		X—X	
TRILOBITES (A)		X	
ECHINODERMS:		X————X	

X Common   O Rare   (A) Not based on modern forms

Selected Bibliography

Fossil and Grain Identification: General

Bathurst, R. G. C., 1971, Carbonate sediments and their diagenesis: New York, Elsevier, 620 p. *Although this is a general text, it offers about 75 pages of discussion, and some good thin-section and SEM photos of skeletal grains.*

Carozzi, A. V., and D. A. Textoris, 1967, Paleozoic carbonate microfacies of the eastern stable interior (U.S.A.): Leiden, E. J. Brill, 146 p. *Illustrates assemblages of grains in numerous microfacies.*

Cita, M. B., 1965, Jurassic, Cretaceous and Tertiary microfacies from the southern Alps: Leiden, E. J. Brill, 99 p. *Good photos of Mesozoic and Cenozoic fossil and grain assemblages with special emphasis on deeper-water facies.*

Cuvillier, J., 1961, Stratigraphic correlation by microfacies in western Aquitaine: Leiden, E. J. Brill, 34 p. *Photos of Jurassic to Tertiary microfaunal and macrofaunal assemblages.*

Hay, W. W., S. W. Wise, Jr., and R. D. Stieglitz, 1970, Scanning electron microscope study of fine grain size biogenic carbonate particles: Gulf Coast Assoc. Geol. Socs. Trans., v. 20, p. 287-302. *Provides some guideline for the identification of the source of very finely comminuted skeletal debris.*

Horowitz, A. S., and P. E. Potter, 1971, Introductory petrography of fossils: New York, Springer-Verlag, 302 p. *General but good quality photos of fossils, grains, and microfacies assemblages.*

Johnson, J. H., 1951, An introduction to the study of organic limestones: Colorado School Mines Quart., v. 46, no. 2, 185 p. *A very general introduction with good photos of macro- and microfossil groups. Inexpensive.*

Majewske, O. P., 1969, Recognition of invertebrate fossil fragments in rocks and thin sections: Leiden, E. J. Brill, 101 p. (plus 106 plates). *Detailed, excellent photos and determinative tables for identification of shells from structural details. Expensive.*

Milliman, J. D., 1974, Marine carbonates: New York, Springer-Verlag, 375 p. *A general text with some photos of whole and thin sectioned organisms; gives excellent summaries of geochemical data for each group.*

Moore, R. C., C. G. Lalicker, and A. G. Fischer, 1952, Invertebrate fossils: New York, McGraw-Hill, 766 p. *General text with good illustrations of internal structures of various fossil groups.*

Tasch, Paul, 1973, Paleobiology of the invertebrates: New York, John Wiley and Sons, 946 p. *A general paleontology text.*

Wilson, J. L., 1975, Carbonate facies in geologic history: New York, Springer-Verlag, 471 p. *Discusses the regional distribution of carbonate facies through time; includes 30 plates of important microfacies textures.*

Fossil Identification: Specific Groups

Treatise on Invertebrate Paleontology. *This series of books published by the Geological Society of America offers good summaries of fossil groupings, skeletal structure, and mineralogy. Most volumes have now been published and they contain extensive references. Only a few other references will therefore be included here.*

Bøggild, O. B., 1930, Shell structure of molluscs: K. Danske Vidensk. Selskabs Skrifter, 9 Raekke, Natur og Math. Afd. 2, p. 231-321. *Very detailed shell structure and mineralogy of recent and fossil mollusks. Quite accurate despite lack of X-ray data.*

Bonet, Federico, 1956, Zonificacion microfaunistica de las calizas Cretacicas del este de Mexico: Asoc. Mexicana Geologos Petroleros Bol., v. 8, p. 389-488. *Details of calcisphere, calpionellid, and foraminiferal zonation of the Cretaceous.*

Flügel, Erik, ed., 1977, Fossil Algae; recent results and developments: New York, Springer-Verlag, 375 p.

Hay, W. W., K. M. Towe, and R. C. Wright, 1963, Ultramicrostructure of some selected foraminiferal tests: Micropaleontology, v. 9, p. 171-195.

Johnson, J. H., 1961, Limestone-building algae and algal limestones: Golden, Colorado, Colorado School Mines, 297 p. *A well-illustrated introduction. Johnson also has several other books on specific algal groups of individual geologic periods also published as Colorado School of Mines Quarterlies.*

Sorauf, J. E., 1971, Microstructure in the exoskeleton of some Rugosa (Coelenterata): Jour. Paleontology, v. 45, p. 23-32.

Sorauf, J. E., 1972, Skeletal microstructure and microarchitecture in Scleractinia (Coelenterata): Palaeontology, v. 15, p. 88-107.

Tavener-Smith, R., and A. Williams, 1972, The secretion and structure of the skeleton of living and fossil Bryozoa: Royal Soc. London Philos. Trans., ser. B, v. 264 (859), p. 97-159.

Taylor, J. D., W. J. Kennedy, and A. Hall, 1969, The shell structure and mineralogy of the Bivalvia: Introduction, Nuculacea-Trigonacea: British Mus. (Nat. Hist.) Bull., Zoology, Suppl. 3, 125 p. *Includes light microscopy and SEM.*

Wise, S. W., Jr., 1970, Microarchitecture and mode of formation of nacre (mother-of-pearl) in pelecypods, gastropods, and cephalopods: Eclogae Geol. Helvetiae, v. 63, p. 775-797.

Wray, J. L., 1977, Calcareous algae: New York, Elsevier, 185 p. *A good summary of taxonomic and stratigraphic information; includes photomicrographs and some SEM's.*

Carbonate Classification

Dunham, R. J., 1962, Classification of carbonate rocks according to depositional texture, *in* W. E. Ham, ed., Classification of carbonate rocks: AAPG Memoir 1, p. 108-121.

Folk, R. L., 1962, Spectral subdivision of limestone types, *in* W. E. Ham, ed., Classification of carbonate rocks: AAPG Memoir 1, p. 62-84.

Other classifications, less frequently used, are also found in this volume.

Carbonate Cements

Bathurst, R. G. C., 1971, Carbonate sediments and their diagenesis: New York, Elsevier, 620 p. *A well-written synthesis of what is known and what isn't known about carbonate cements.*

Bricker, O. P., ed., 1971, Carbonate cements: Baltimore, Johns Hopkins Press, 376 p. *Contains a wide range of articles on cement textures and compositions from marine and nonmarine environments.*

Folk, R. L., 1974, The natural history of crystalline calcium carbonate; effect of magnesium content and salinity: Jour. Sed. Petrology, v. 44, p. 40-53. *Presents data on role of water chemistry of carbonate cement morphologies.*

James, N. P., R. N. Ginsburg, D. S. Marszalek, and P. W. Choquette, 1976, Facies and fabric specificity of early subsea cements in shallow Belize (British Honduras) reefs: Jour. Sed. Petrology, v. 46, p. 523-544.

Meyers, W. J., 1974, Carbonate cement stratigraphy of the Lake Valley Formation (Mississippian) Sacramento mountains, New Mexico: Jour. Sed. Petrology, v. 44, p. 837-861.

Carbonate Recrystallization

Bathurst, R. G. C., 1971, Carbonate sediments and their diagenesis: New York, Elsevier, 620 p. *Presents a number of alternative points of view on questions of cementation versus recrystallization.*

Folk, R. L., 1965, Some aspects of recrystallization in ancient limestones, *in* L. C. Pray, and R. C. Murray, eds., Dolomitization and limestone diagenesis, a symposium: SEPM Spec. Publ. 13, p. 14-48. *An excellent introduction to the basic concepts.*

Folk, R. L., 1973, Carbonate petrography in the post-Sorbian age, *in* R. N. Ginsburg, ed., Evolving concepts in sedimentology: Baltimore, Johns Hopkins Univ. Press, p. 118-158. *A general summary of precipitation and recrystallization.*

Purdy, E. G., 1968, Carbonate diagenesis; an environmental survey: Geol. Romana, v. 7, p. 183-228.

Sandberg, P. A., 1975, Bryozoan diagenesis; bearing on the nature of the original skeleton of rugose corals: Jour. Paleontology, v. 49, p. 587-606. *Discusses evidence for inversion of aragonite to calcite based on light microscopy and SEM.*

Porosity Classification

Choquette, P. W., and L. C. Pray, 1970, Geologic nomenclature and classification of porosity in sedimentary carbonates: AAPG Bull., v. 54, p. 207-250. *By far the best article available on this subject.*

This paper and other useful articles on the subject are contained in AAPG Reprint Series No. 5, Carbonate rocks II: Porosity and classification of reservoir rocks.

Techniques

Dickson, J. A. D., 1966, Carbonate identification and genesis as revealed by staining: Jour. Sed. Petrology, v. 36, p. 491-505.

Diebold, F. E., J. Lemish, and C. L. Hiltrop, 1963, Determination of calcite, dolomite, quartz, and clay content of carbonate rocks: Jour. Sed. Petrology, v. 33, p. 124-139.

Ellingboe, John, and James Wilson, 1964, A quantitative separation of non-carbonate minerals from carbonate minerals: Jour. Sed. Petrology, v. 34, p. 412-418.

Evamy, B. D., 1963, The application of a chemical staining technique to a study of dedolomitization: Sedimentology, v. 2, p. 164-170.

Fischer, A. G., S. Honjo, and R. W. Garrison, 1967, Electron micrographs of limestones and their nannofossils: Princeton, Princeton Univ. Press, Monographs in Geol. and Paleont. No. 1, 141 p.

Flügel, Erik, H. E. Franz, and W. F. Ott, 1968, Review on electron microscope studies of limestones, *in* G. Müller, and G. M. Friedman, eds., Recent developments in carbonate sedimentology in central Europe: New York, Springer-Verlag, p. 85-97.

Folk, R. L., 1968, Petrology of sedimentary rocks: Austin, Texas, Hemphill's, 170 p.

Friedman, G. M., 1959, Identification of carbonate minerals by staining methods: Jour. Sed. Petrology, v. 29, p. 87-97.

Goldsmith, J. R., D. L. Graf, and H. C. Heard, 1961, Lattice constants of the calcium-magnesium carbonates: Am. Mineralogist, v. 46, p. 453-457.

Goldstein, J. I., and Harvey Yakowitz, eds., 1975, Practical scanning electron microscopy; electron and microprobe analysis: New York, Plenum Press, 582 p.

Katz, A., and G. M. Friedman, 1965, The preparation of stained acetate peels for the study of carbonate rocks: Jour. Sed. Petrology, v. 35, p. 248-249.

Lane, D. W., 1962, Improved acetate peel technique: Jour. Sed. Petrology, v. 32, p. 870.

Lindholm, R. C., and R. B. Finkelman, 1972, Calcite staining: semiquantitative determination of ferrous iron: Jour. Sed. Petrology, v. 42, p. 239-242.

McCrone, A. W., 1963, Quick preparation of peel-prints for sedimentary petrography: Jour. Sed. Petrology, v. 33, p. 228-230.

McDougall, D. J., ed., 1968, Thermoluminescence of geological materials: London, Academic Press, 675 p.

Medlin, W. L., 1959, Thermoluminescent properties of calcite: Jour. Chem. Physics, v. 30, p. 451-458.

Ostrom, M. E., 1961, Separating clay minerals from carbonate rocks by using acid: Jour. Sed. Petrology, v. 31, p. 123-129.

Royse, C. F., Jr., J. S. Wadell, and L. E. Peterson, 1971, X-ray determination of calcite-dolomite; an evaluation: Jour. Sed. Petrology, v. 41, p. 483-488.

Runnells, D. D., 1970, Errors in X-ray analysis of carbonates due to solid-solution variation in composition of component minerals: Jour. Sed. Petrology, v. 40, p. 1158-1166.

Schneidermann, Nahum, and P. A. Sandberg, 1971, Calcite-aragonite differentiation by selective staining and scanning electron microscopy: Gulf Coast Assoc. Geol. Socs. Trans., v. 21, p. 349-352.

Siesser, W. G., and John Rogers, 1971, An investigation of the suitability of four methods used in routine carbonate analysis of marine sediments: Deep-Sea Research, v. 18, p. 135-139.

Smith, M. H., 1970, Identification of organic matter in thin section by staining and a study program for carbonate rocks: Jour. Sed. Petrology, v. 40, p. 1350-1351.

Weber, J. N., 1968, Quantitative mineralogical analysis of carbonate sediments; comparison of X-ray diffraction and electron probe microanalyser methods: Jour. Sed. Petrology, v. 38, p. 232-234.

Wolf, K. H., A. J. Easton, and S. Warne, 1967, Techniques of examining and analyzing carbonate skeletons, minerals, and rocks, *in* G. V. Chilingar, H. J. Bissell, and R. W. Fairbridge, eds., Carbonate rocks: New York, Elsevier, Devel. in Sedimentol. 9B, p. 253-341.

References Cited

Choquette, P. W., and L. C. Pray, 1970, Geologic nomenclature and classification of porosity in sedimentary carbonates: AAPG Bull., v. 54, p. 207-250.

Dunham, R. J., 1962, Classification of carbonate rocks according to depositional texture, *in* W. E. Ham, ed., Classification of carbonate rocks: AAPG Memoir 1, p. 108-121.

Folk, R. L., 1962, Spectral subdivision of limestone types, *in* W. E. Ham, ed., Classification of carbonate rocks: AAPG Memoir 1, p. 62-84.

—— 1965, Some aspects of recrystallization in ancient limestones, *in* L. C. Pray and R. C. Murray, eds., Dolomitization and limestone diagenesis, a symposium: SEPM Spec. Pub. 13, p. 14-48.

—— 1968, Petrology of sedimentary rocks: Austin, Texas, Hemphill's, 170 p.

—— 1974, The natural history of crystalline calcium carbonate; effect of magnesium content and salinity: Jour. Sed. Petrology, v. 44, p. 40-53.

Scholle, P. A., 1971a, Sedimentology of fine-grained deep-water carbonate turbidites, Monte Antola Flysch (Upper Cretaceous), northern Apennines, Italy: Geol. Soc. America Bull., v. 82, p. 629-658.

—— 1971b, Diagenesis of deep water carbonate turbidites, Upper Cretaceous, Monte Antola Flysch, northern Apennines, Italy: Jour. Sed. Petrology, v. 41, p. 233-250.

—— and D. J. J. Kinsman, 1974, Aragonitic and high-Mg calcite caliche from the Persian Gulf—a modern analog of the Permian of Texas and New Mexico: Jour. Sed. Petrology, v. 44, p. 904-916.

Terry, R. D., and G. V. Chilingar, 1955, Summary of "Concerning some additional aids in studying sedimentary formations," by M. S. Shvetsov: Jour. Sed. Petrology, v. 25, p. 229-234.

Skeletal Grains

Algae

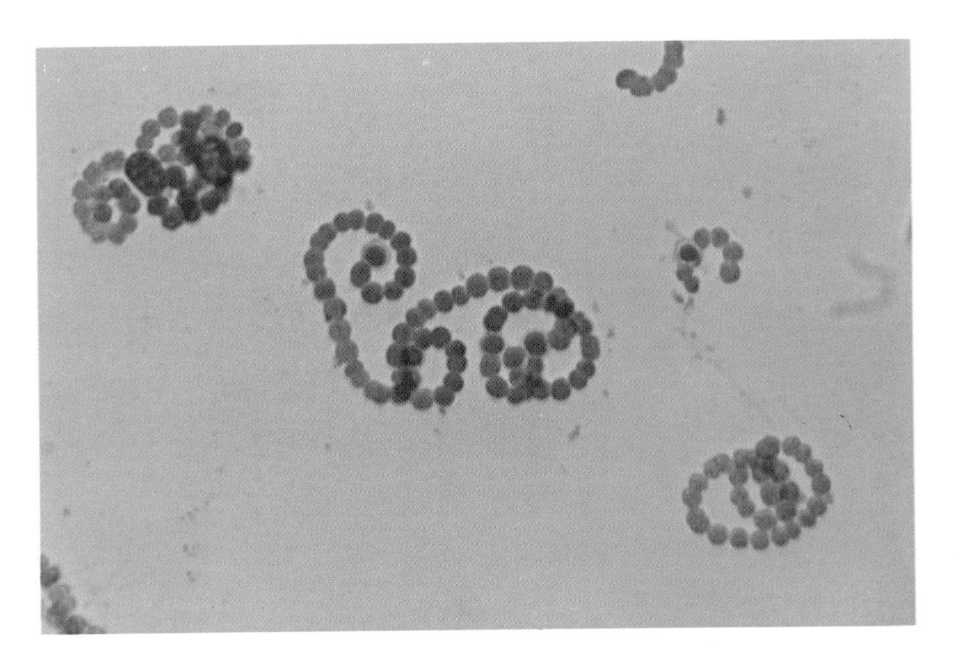

Holocene algae

Biological preparation (with stain) of *Anabaena* heterocysts. Rather typical of spherical blue-green algal bodies which link up into chains. Such algae, in combination with filamentous forms, are common in modern algal mats.

0.022 mm

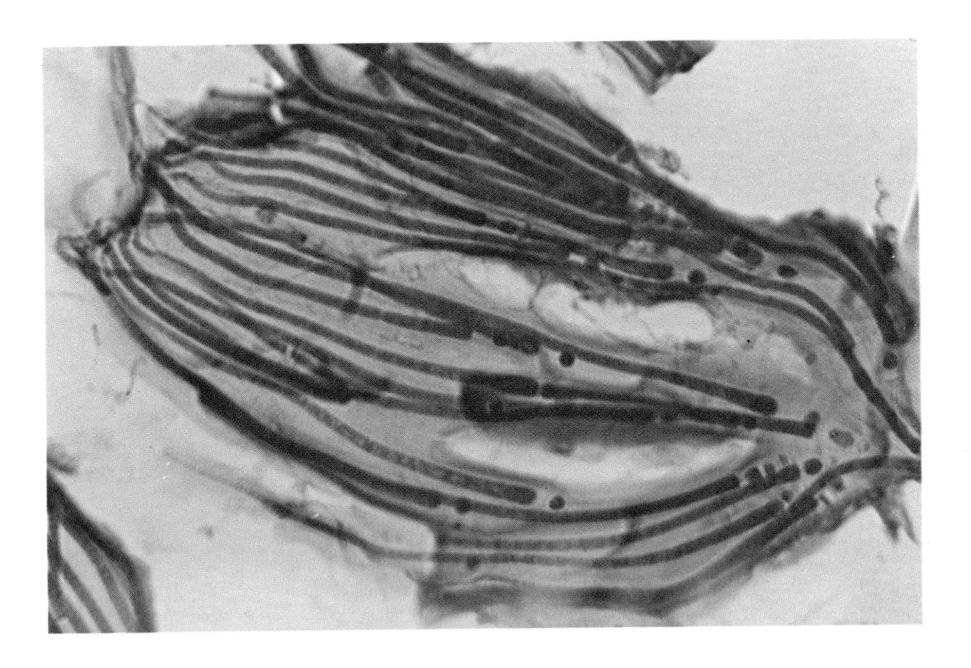

Holocene algae

Biological preparation (with stain) of *Rivularia* heterocysts. These are filamentous blue-green algae. The individual cells can be seen stained purple and the mucilaginous sheaths are stained green. Such interlocking filaments help trap particles in blue-green algal mats through a baffling effect as well as slight stickiness.

0.027 mm

Triassic Dachstein Ls., Lofer facies
Austria

A very well developed ancient example of a laminated and contorted blue-green algal mat (loferite). Dark reddish-brown color, slightly pelletal texture, and irregular lamination are characteristic of blue-green algal mats but are not always this clearly displayed.

0.64 mm

Up. Silurian part of Tonoloway-Keyser Ls.
Pennsylvania

Blue-green algal stromatolite (biolithite) showing moderate lamination, pelleting, and fenestrae. Even where algal lamination is very clear in outcrop it may be quite subtle in thin section.

0.38 mm

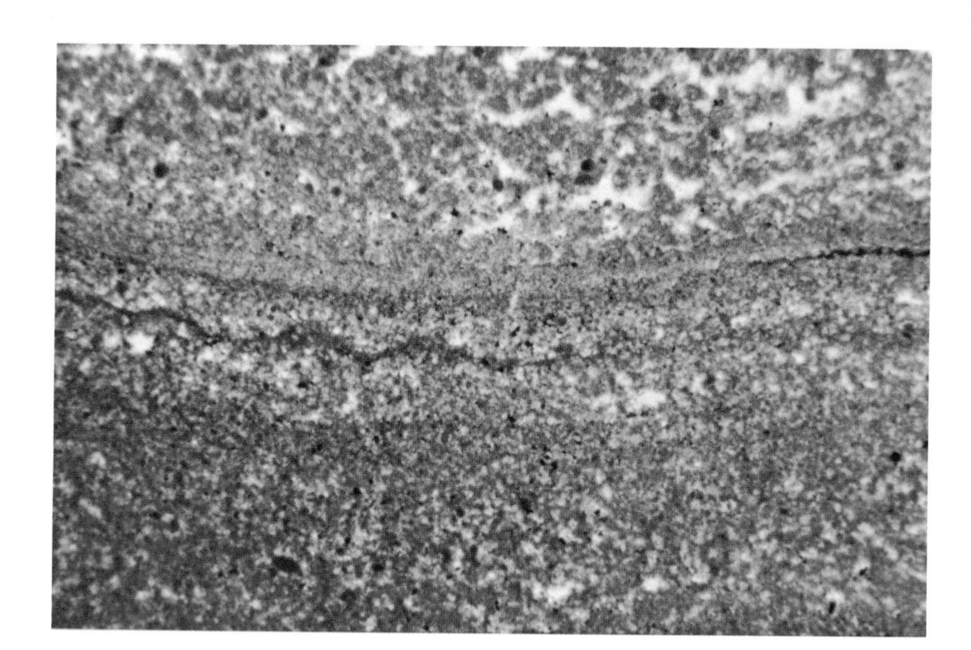

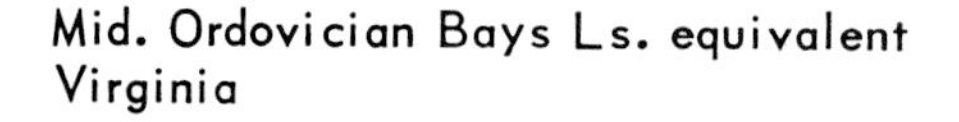
Mid. Ordovician Bays Ls. equivalent
Virginia

Burrowed blue-green algal stromatolite (biolithite) showing vague lamination, characteristic pelleting and large spar-filled vugs possibly related to burrowing or syndepositional gas generation.

0.38 mm

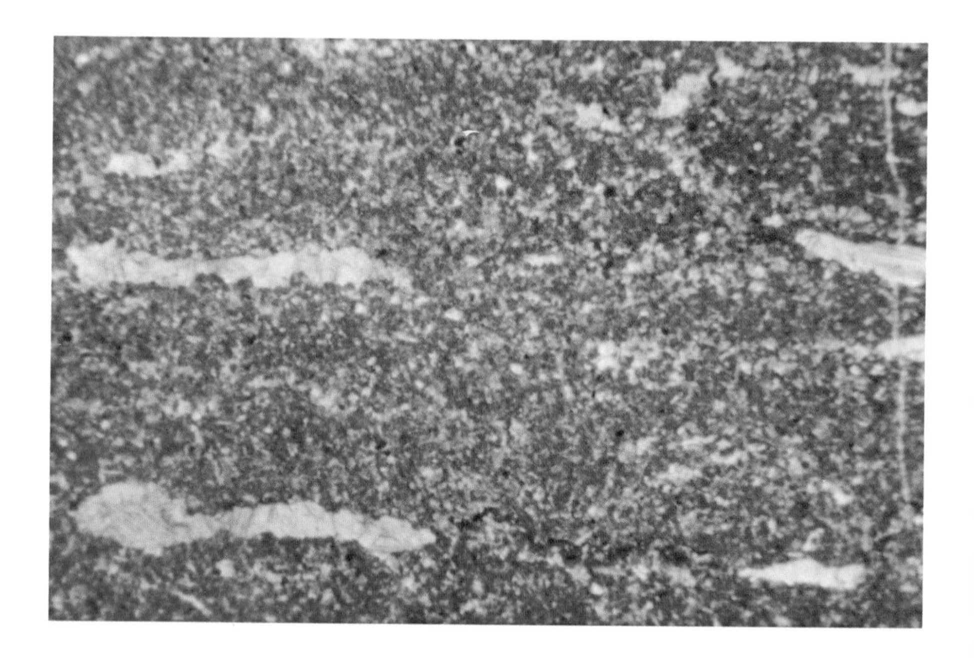

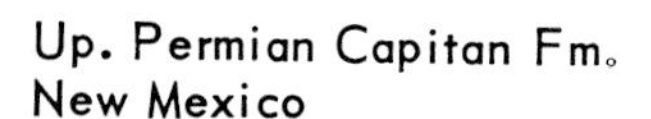
Up. Permian Capitan Fm.
New Mexico

Blue-green algae can also produce irregular encrustation of grains, as in this sponge reef. Several large, poorly preserved sponges are seen with dark, lumpy coatings (some "floating" in sparry calcite). These are often considered to be algal although some would interpret them as cement rinds.

0.64 mm

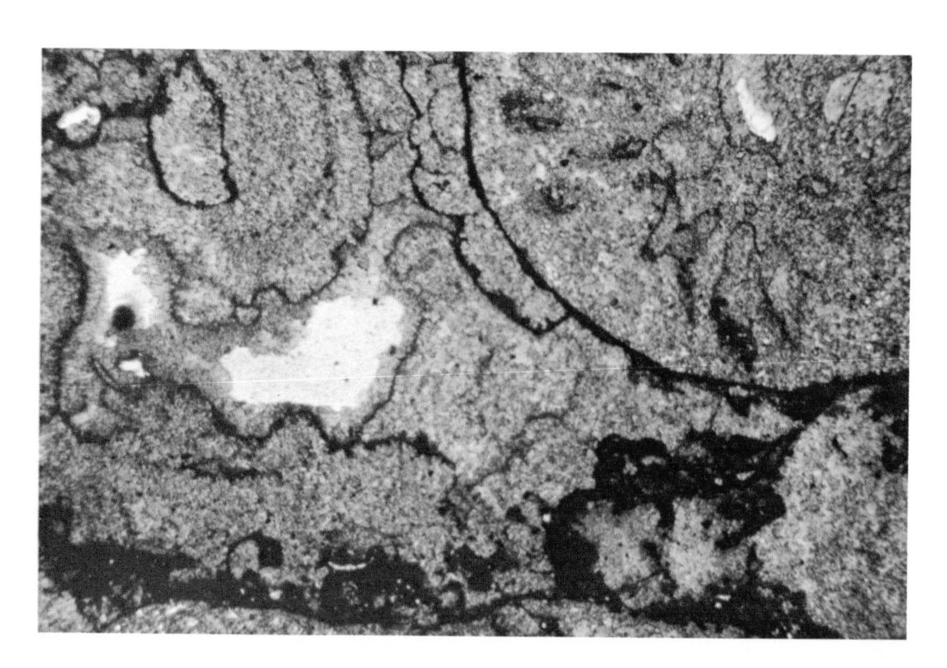

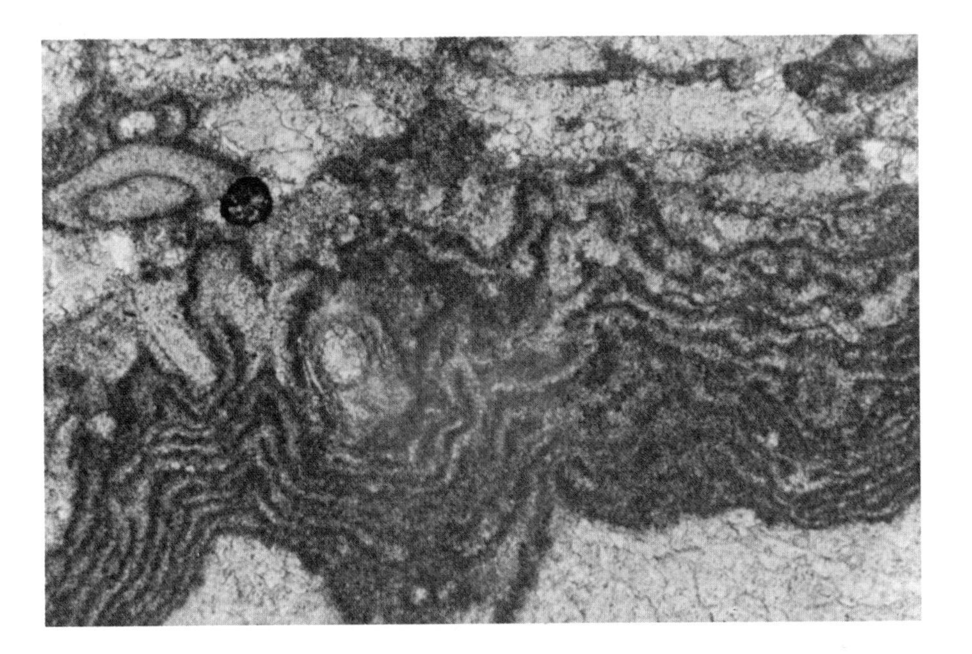

Up. Permian Capitan Fm.
New Mexico

These multiple, laminated crusts surround many grains in the Permian reef complex of New Mexico and appear to have played a major role in the lithification of the reef debris. It is presumed that these crusts are the result of blue-green algal growth.

0.19 mm

Up. Devonian Jean Marie Fm.
Canada (N.W.T.)

The dark micritic, lumpy patches are *Renalcis*, an important blue-green algal type which, in the Devonian of Canada, is restricted to the reef and very near-reef facies. It thus is an important tool in the recognition of nearby reefs.

X.N. 0.22 mm

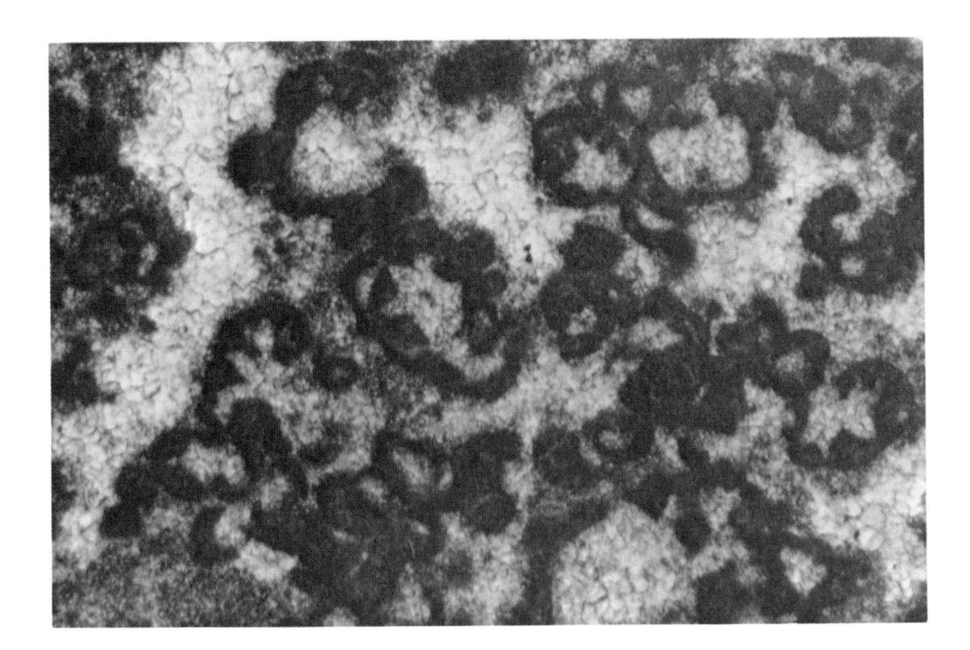

Up. Devonian Jean Marie Fm.
Canada (N.W.T.)

Another view of *Renalcis* structure illustrating the commonly scalloped outline of these blue-green algae, as well as their micritic texture.

0.25 mm

Eocene Green River Fm.
Wyoming

A head of *Chlorellopsis coloniata*, a probable blue-green algal stromatolite builder. Although the stromatolite is largely replaced by chert, one can still see laminated structure with circular, calcisphere-like bodies. These heads are common in lacustrine facies.

0.64 mm

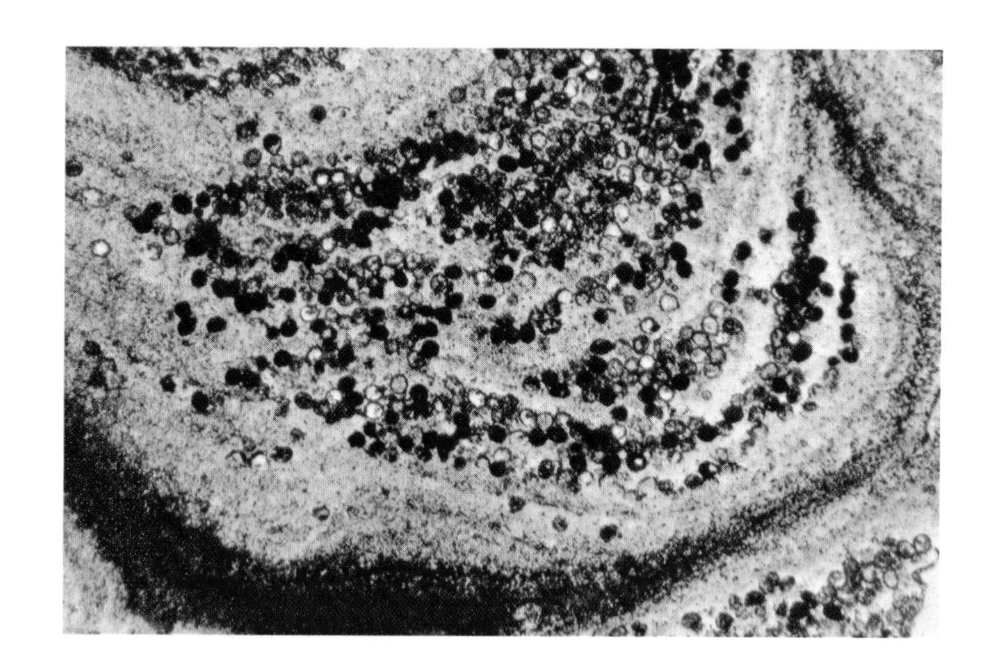

Mid. Cambrian Meagher Ls.
Montana

Closeup of an algal nodule (oncolite) showing very small, irregular tubules characteristic of the possible blue-green alga *Girvanella*. These tubes are often well preserved but can only be seen with close observation and relatively high magnification.

0.05 mm

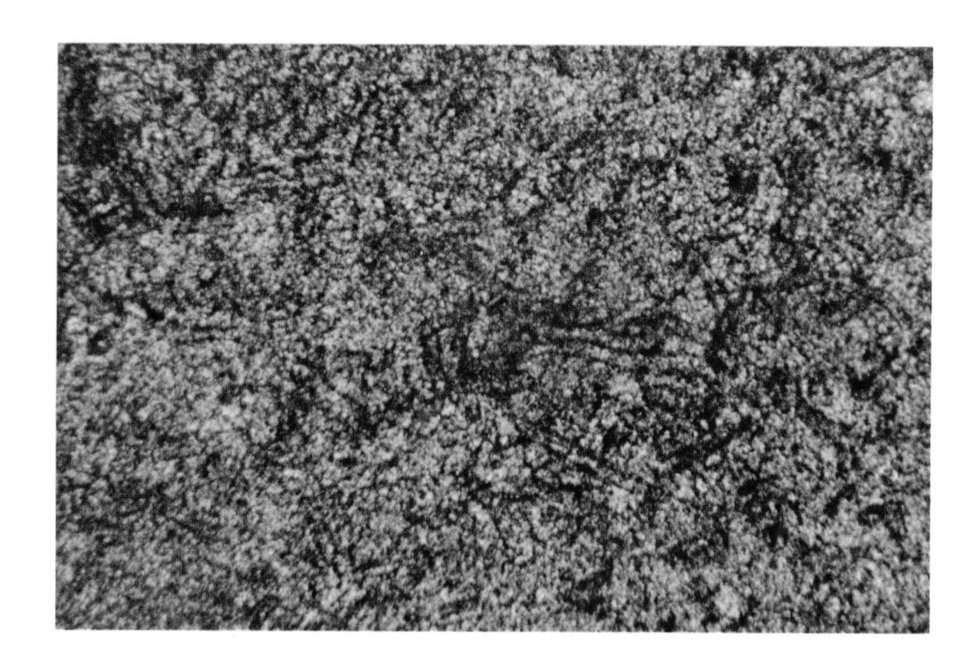

Up. Permian Capitan Ls.
New Mexico

Section of a *Tubiphytes* colony characterized by very dense, dark micritic material with broadly spaced darker banding. This is an encrustation on reef material and has been assigned by some workers to the algae and by others to the hydrozoans.

0.64 mm

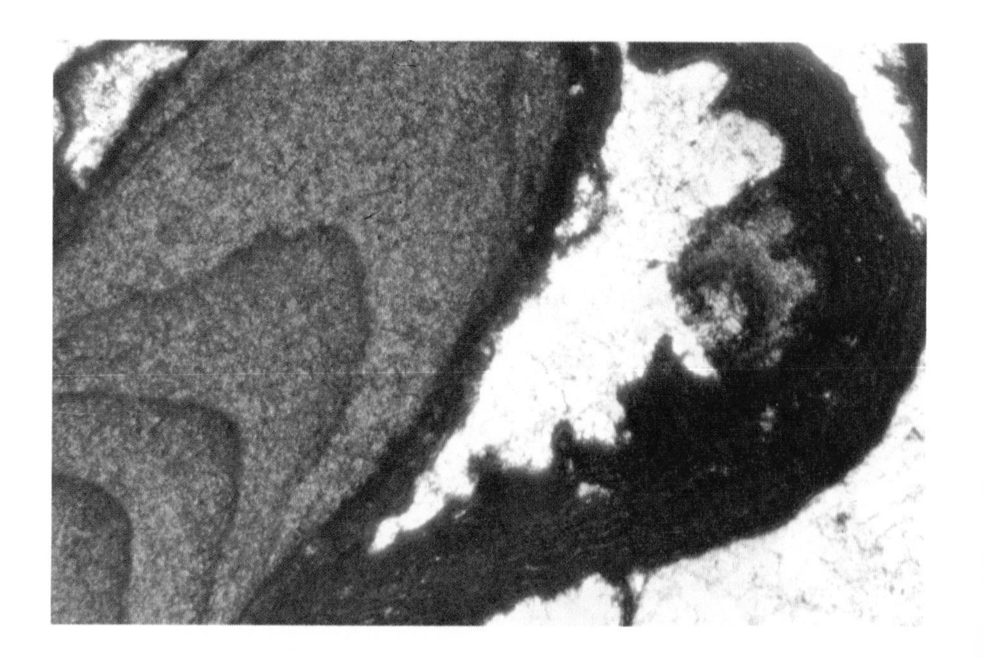

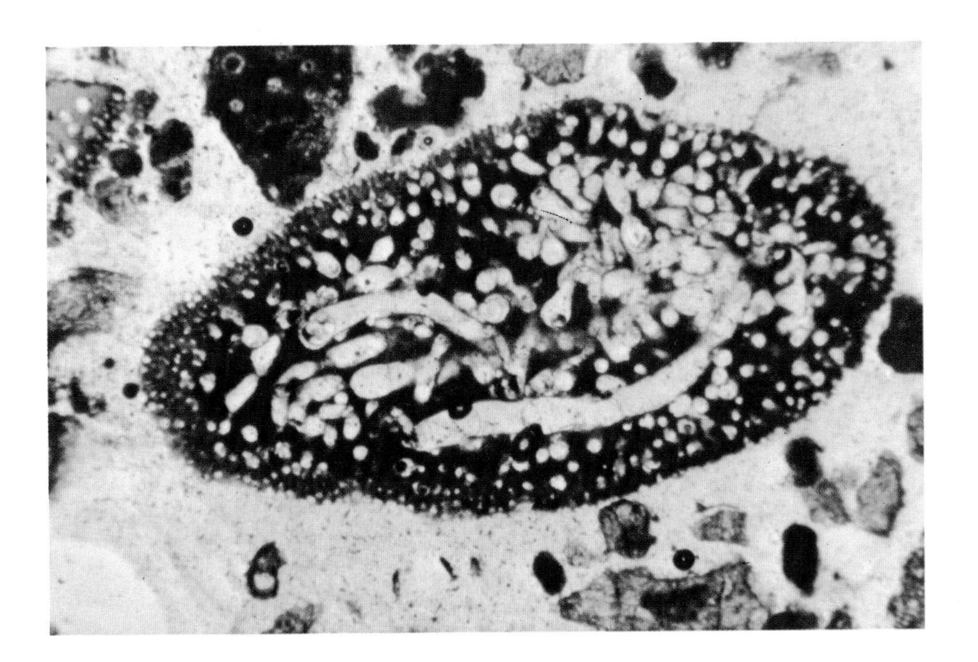

Holocene bottom sediment (Rodriquez Key)
Florida

Cross section of a Holocene Codiacian green alga (*Halimeda* sp.) showing the dark calcified plate with abundant irregular holes originally filled with organic matter. This is one of the major carbonate sand producers in modern reefs.

0.30 mm

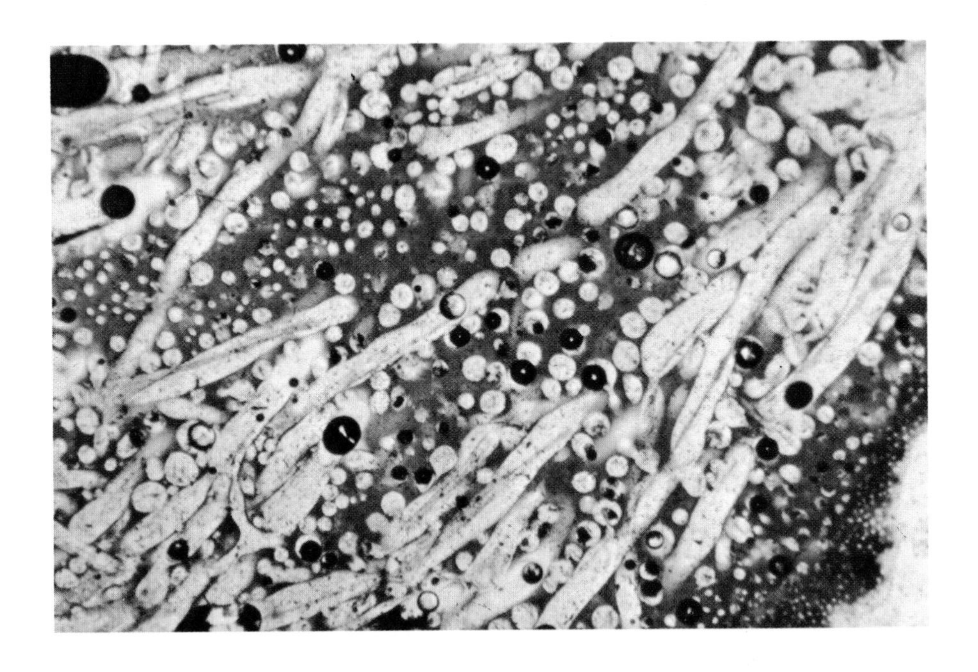

Holocene algae
Florida

Closeup of the structure of the green alga *Halimeda* showing the irregular tubes and calcified main plate. Dark blebs are air bubble images.

0.38 mm

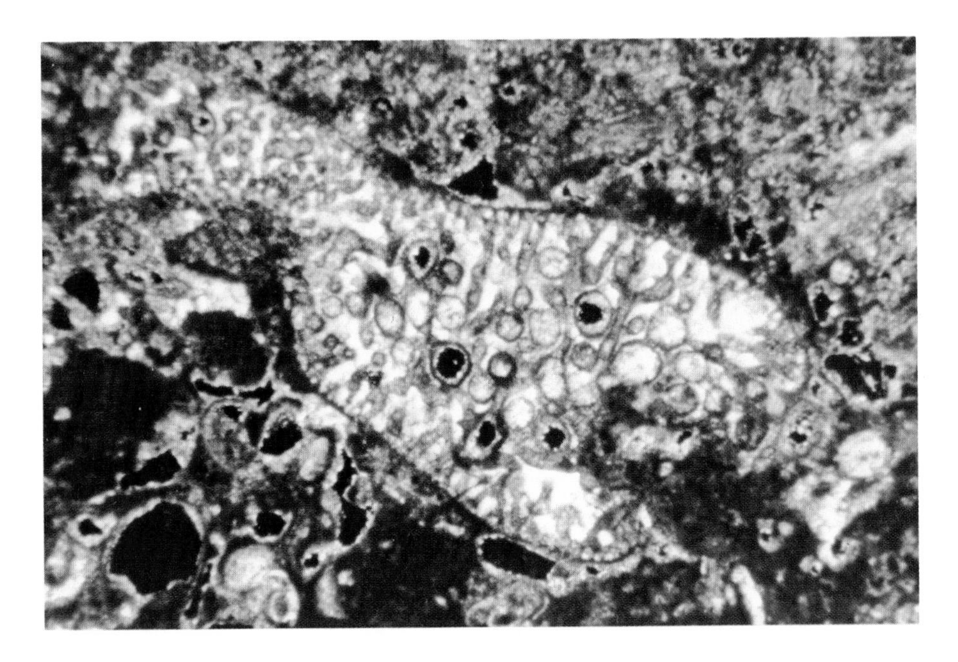

Pleistocene Key Largo Ls.
Florida

Altered grain of *Halimeda* (green alga) showing typical preservation of this type of grain. Original holes are filled with spar and most of the plate has also altered to spar. Structure shown only by micrite "envelopes" surrounding original holes.

X.N. 0.24 mm

Holocene sediment
Florida

Cross section of a *Halimeda* plate showing tubular opening, originally occupied by plant tissue, and calcified areas with abundant, interlocked, rather randomly oriented aragonite needles which make up the preservable portion of the *Halimeda* plate.

SEM Mag. 1000X 13 μm

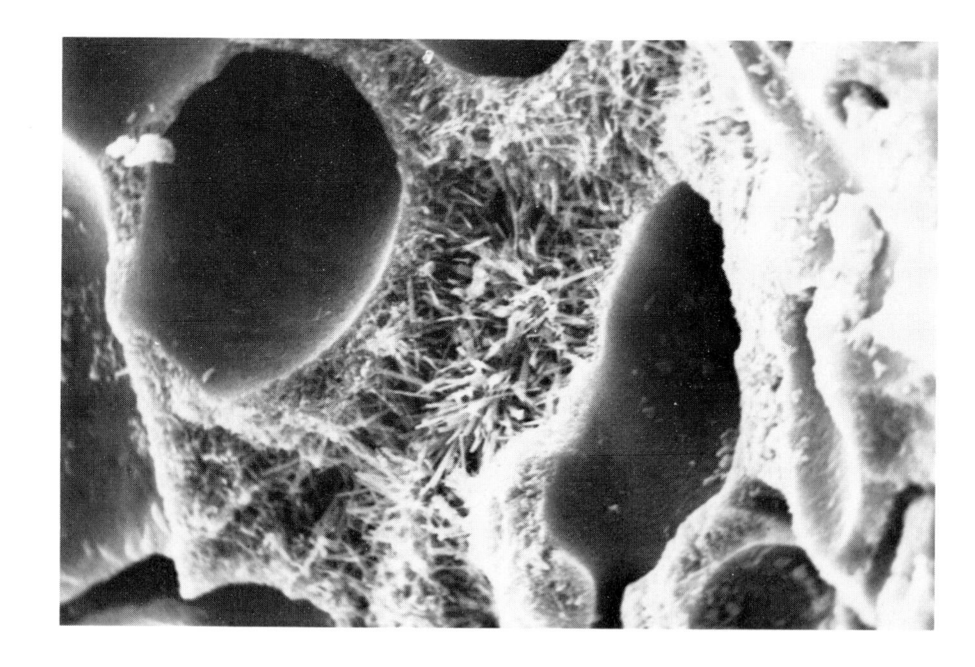

Holocene sediment
Florida

Close-up of *Halimeda* plate showing details of interlocking aragonite needles. Needles such as these are found in many species of green algae including *Penicillus*, *Udotea*, and others. When the algae decompose, the needles may be scattered and add significantly to the local production of clay-sized particles (carbonate mud).

SEM Mag. 10,000X 1.3 μm

Oligocene Suwannee Ls.
Florida

A dasyclad green alga showing typically poor preservation. Most of the lightly calcified plate structure has been destroyed and only traces of the original structure remain.

0.38 mm

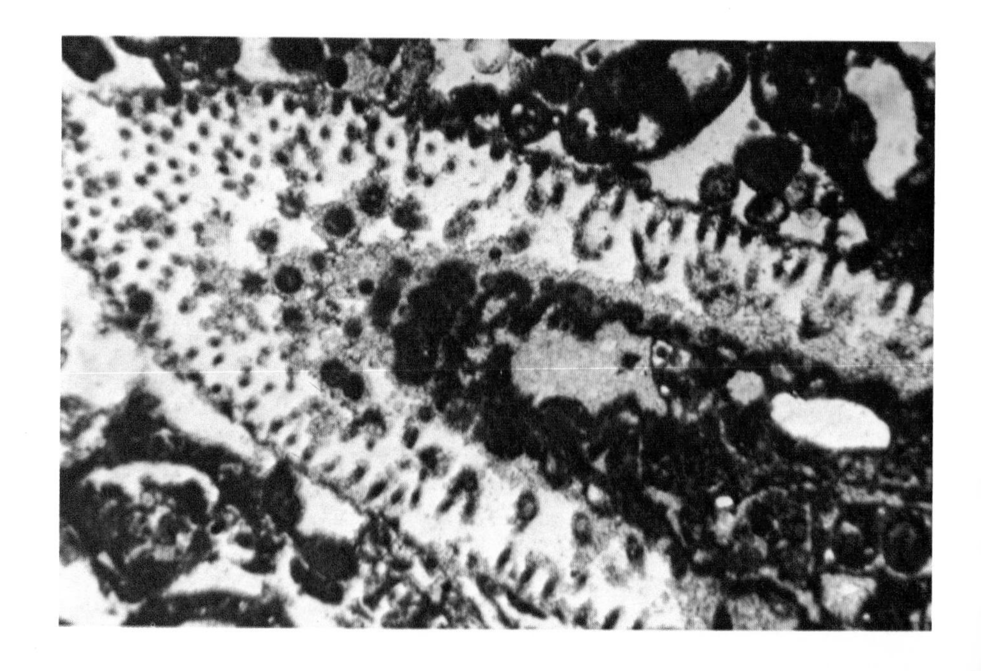

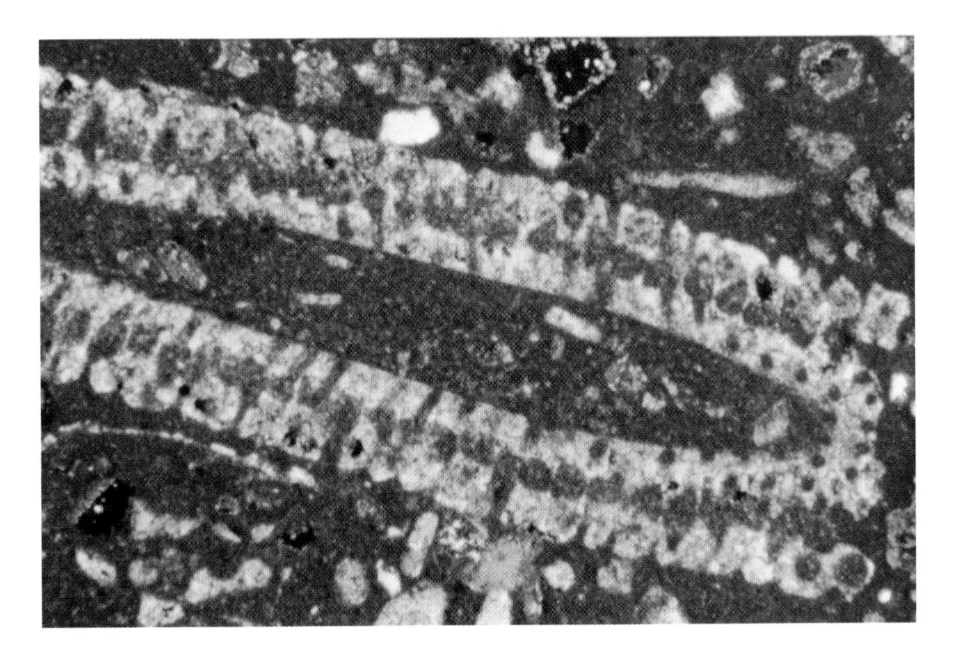

Up. Cretaceous limestone
Colombia

Large green-algal (Dasycladacean) grain with oblique cut through central cavity and radiating porous tubes. Presence of tubes is major feature allowing identification as a green algal fragment.

X.N. 0.25 mm

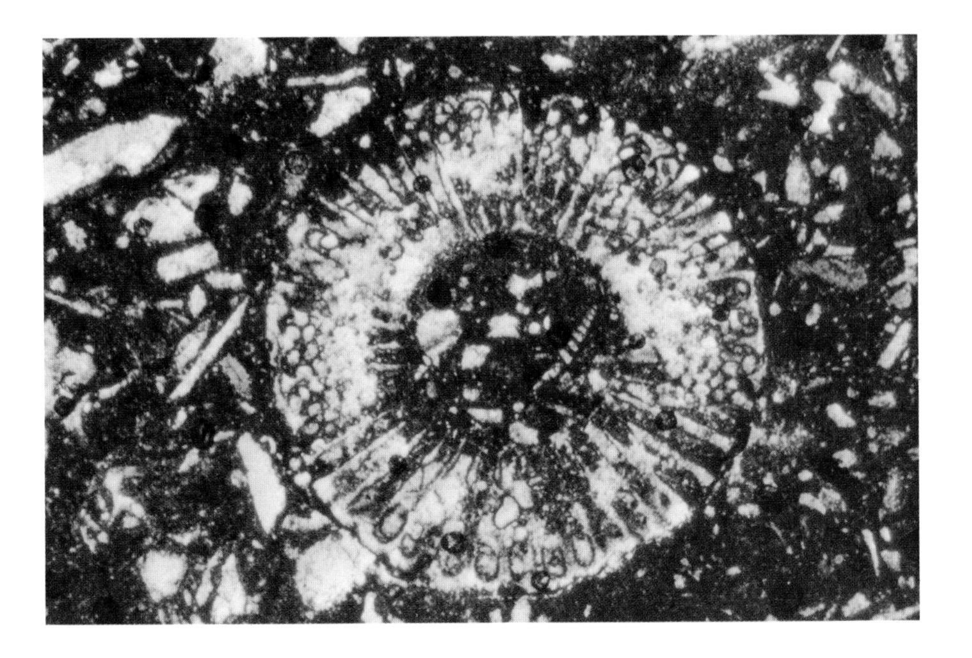

Cretaceous Tamabra Ls.
Mexico

Cross section directly across a Dasycladacean green alga. Shows radial symmetry of elements about the central cavity. The characteristic features which allow identification are the presence of radiating tubes and a central cavity coupled with poor preservation of wall structure.

0.64 mm

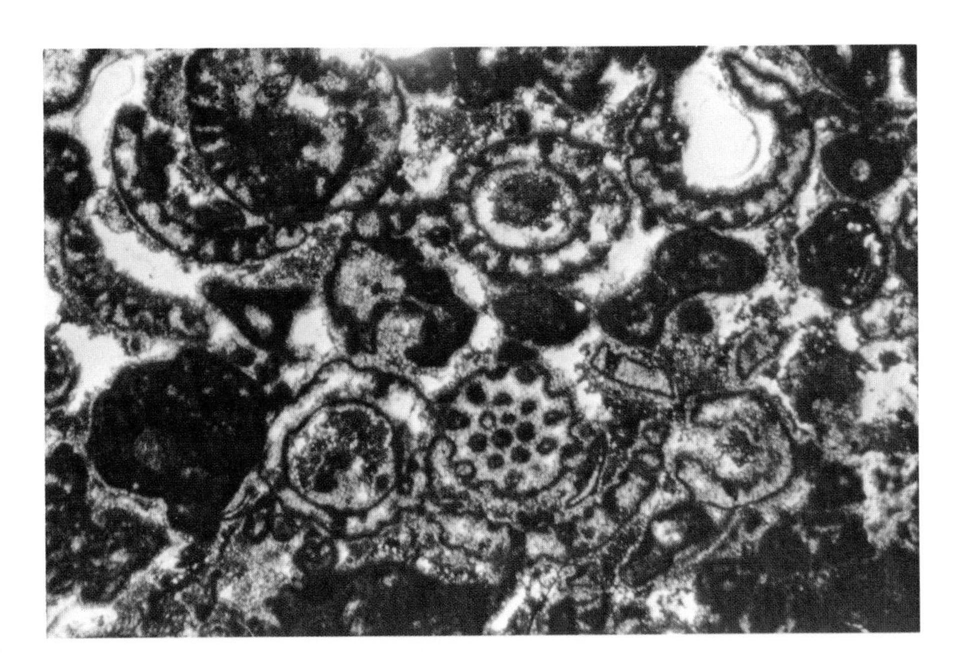

Up. Permian Capitan Ls.-Tansill Fm. transition
New Mexico

Abundant *Mizzia*, a type of Dasycladacean green algae very abundant in near-back-reef facies of the Permian. Characteristic radial symmetry, tubular structures, poor wall preservation, and central cavity are shown in a number of the grains. The green color is stained plastic in voids.

0.64 mm

Mid-Pennsylvanian Paradox Fm.
Utah

Phylloid algal grain showing relatively good preservation. Note parallel, short tubules which line the outer margins of the grain. These are quite characteristic when well preserved but must be distinguished from micrite envelopes or grain coating micrite. These grains are very important in Pennsylvanian bioherms.

0.22 mm

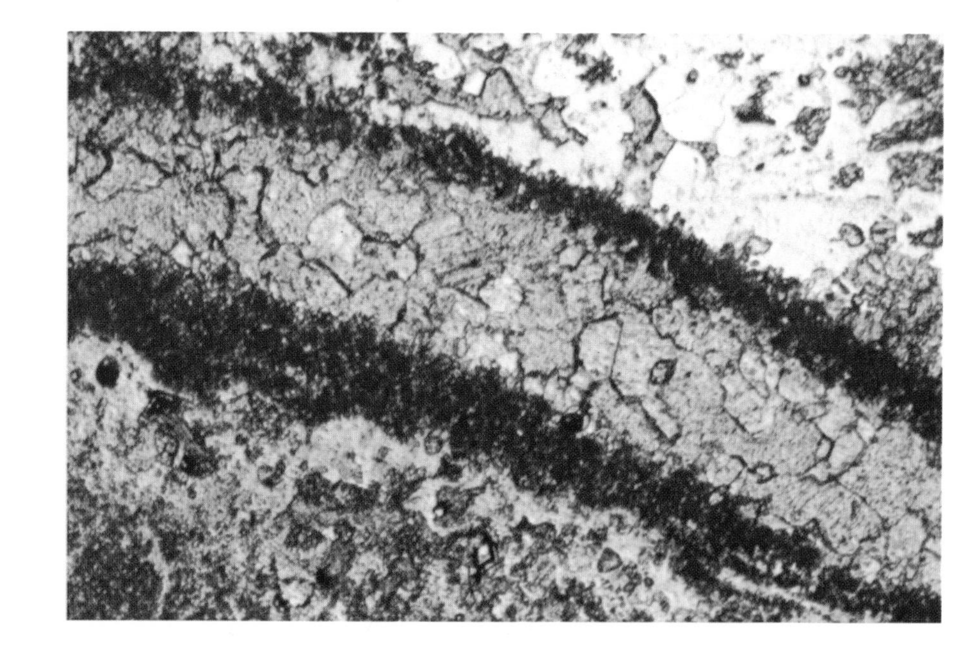

Pennsylvanian ls., Sacramento Mtns., New Mexico

Another relatively well preserved phylloid algal grain showing marginal tubules, some of which have collapsed into the grain interior during dissolution. Very poor preservation of all portions except the mud-filled tubes is quite typical.

0.25 mm

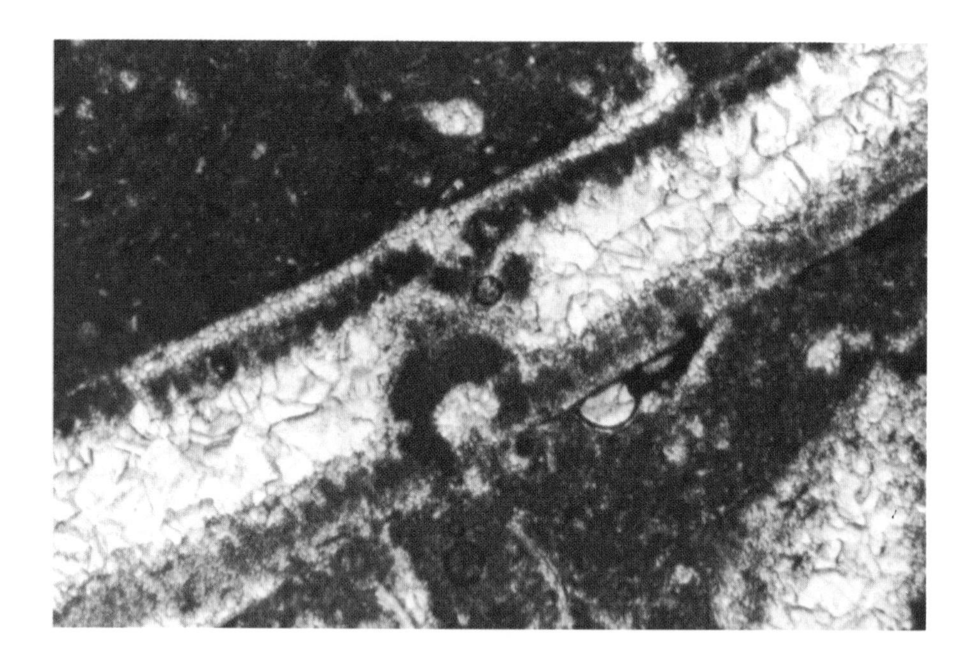

Mid-Pennsylvanian Minturn Fm.
Colorado

Poorly preserved phylloid algal plates (possibly of green algae). All internal structure has been obliterated and only structural outlines remain.

0.38 mm

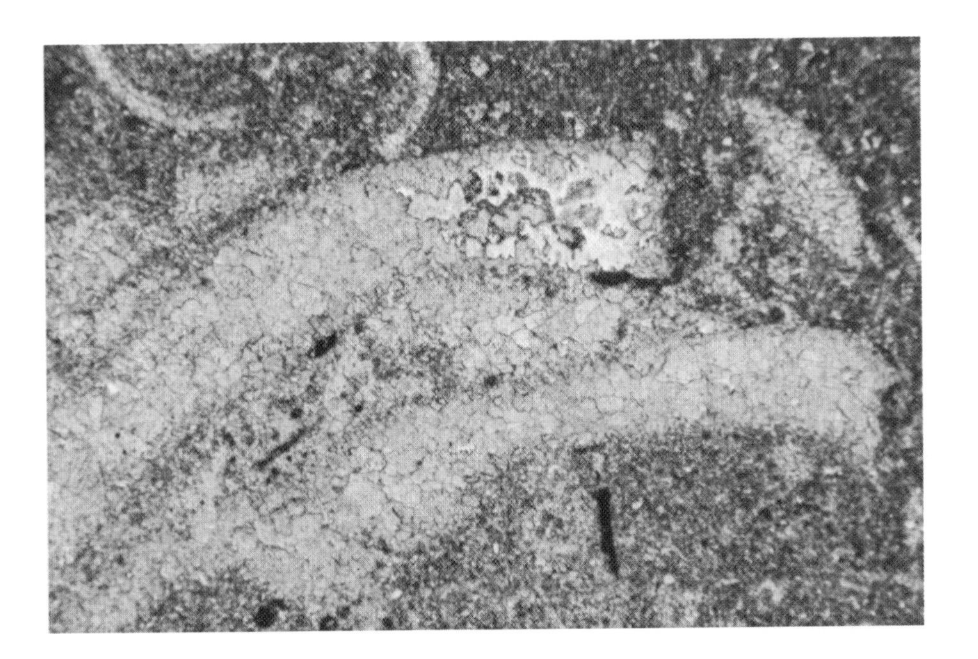

Mid-Pennsylvanian Minturn Fm.
Colorado

Typically poor preservation of probable green (phylloid) algae. Note complete replacement by sparry calcite with no preservation of internal structure.

0.38 mm

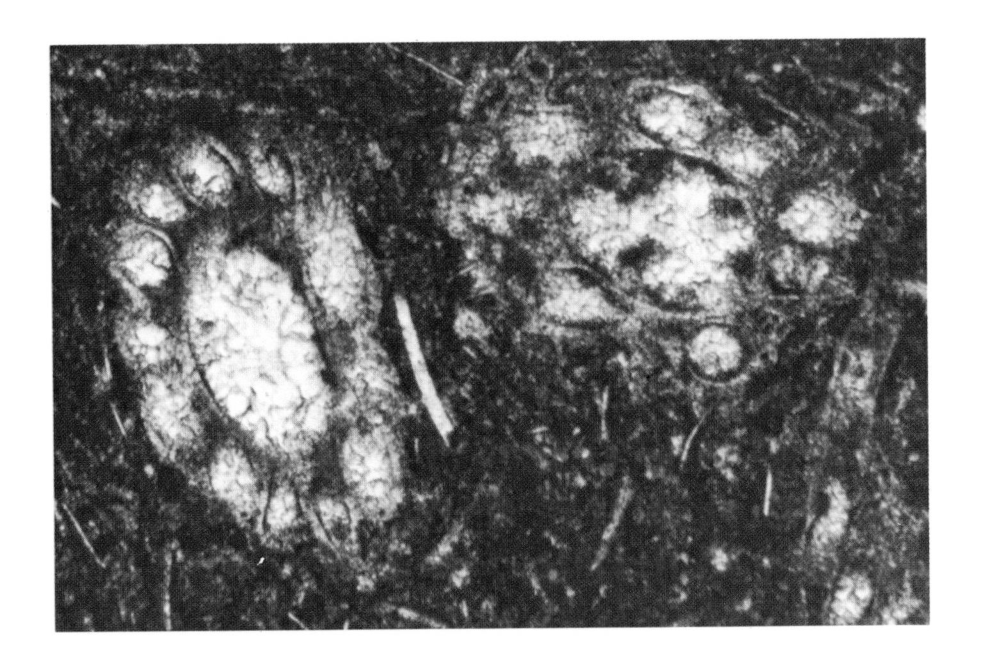

Lo. Cretaceous Newark Canyon Fm.
Nevada

Oblique sections through the vegetative parts of two charophytes. These plants are classed as green algae; modern as well as ancient forms appear to be restricted to nonmarine (especially lacustrine) environments. Charophytes can be major rock-forming elements as well as important biostratigraphic markers in nonmarine sediments.

0.30 mm

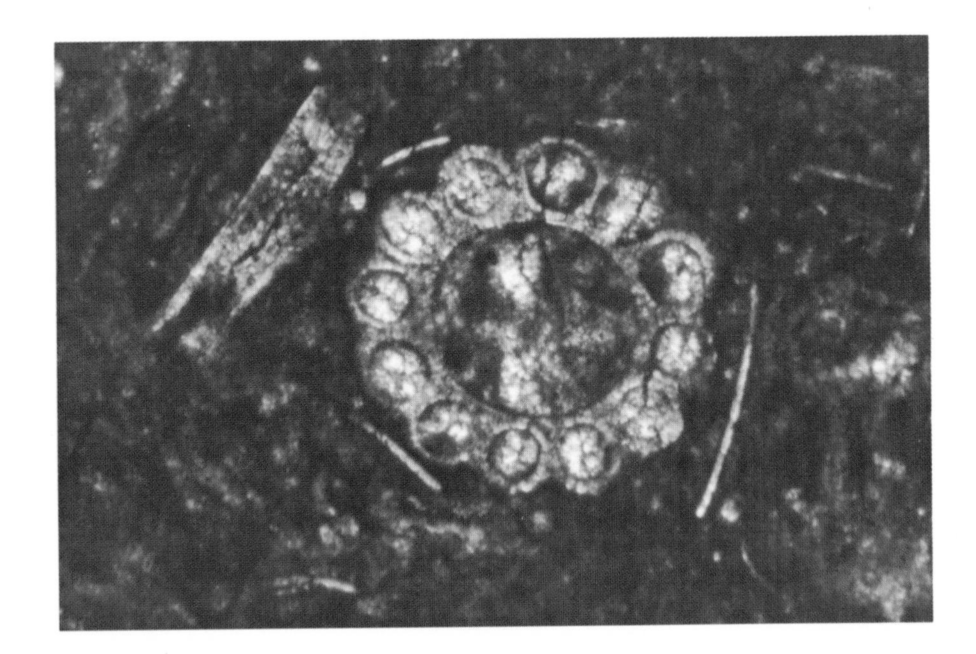

Lo. Cretaceous Newark Canyon Fm.
Nevada

Transverse section through the vegetative part of a charophyte. Shows characteristic central tube with surrounding cortical tubes. Calcified gyrogonites (oogonia) often occur in association with the vegetative remains.

0.30 mm

Holocene bottom sediment (Rodriguez Key)
Florida

Fragment of *Goniolithon* red-algal grain. Note characteristic fine cellular structure. Although red-algal grains invariably are converted from original high-Mg to eventual low-Mg calcite during diagenesis, the fine-scale cellular structure is normally preserved.

0.30 mm

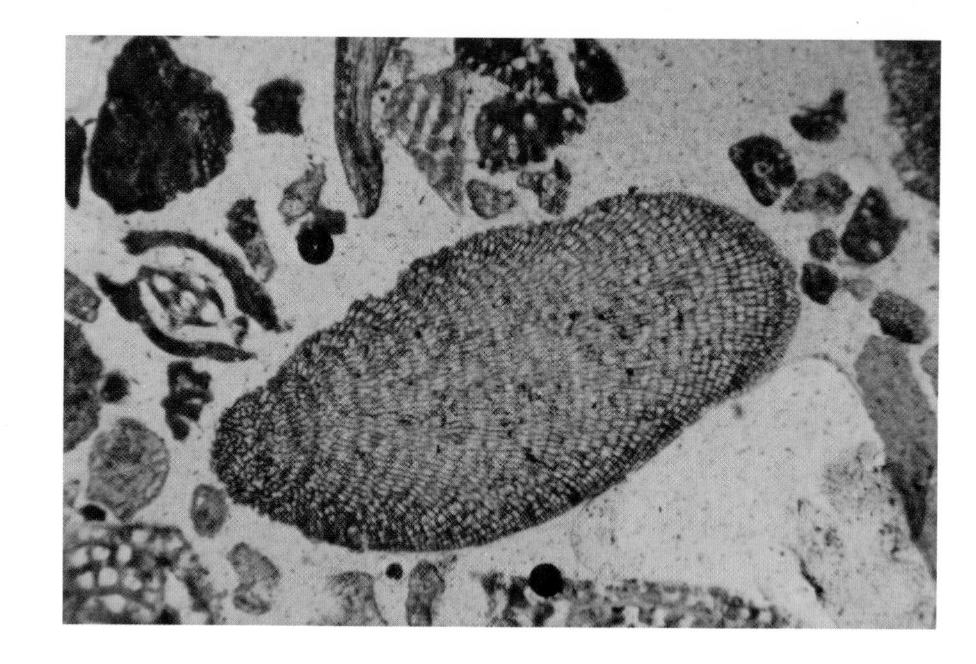

Mid. Eocene Coamo Springs Ls.
Puerto Rico

Lithophyllum red-algal grain showing cellular and laminated structure as well as spore cases (small round white spots).

0.24 mm

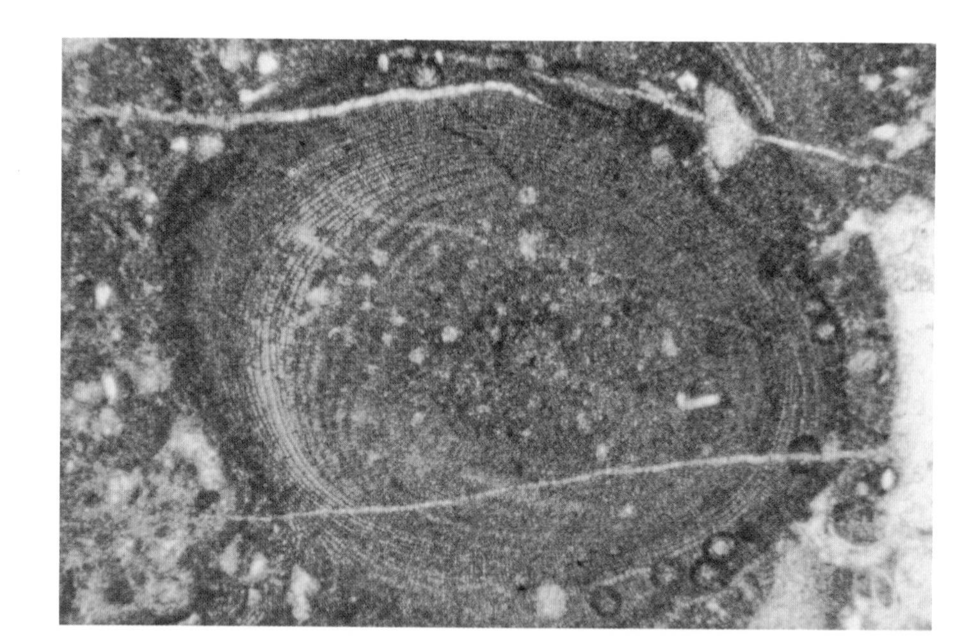

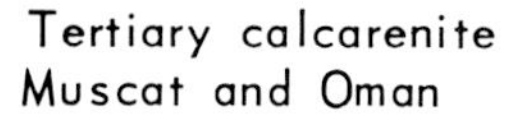

Tertiary calcarenite
Muscat and Oman

Very well preserved red alga, possibly *Lithophyllum*. This example shows differentiation of central hypothallus and the denser, external perithallus. The cellular structure, branching form, and good preservation distinguish these coralline algae from other algal types.

0.38 mm

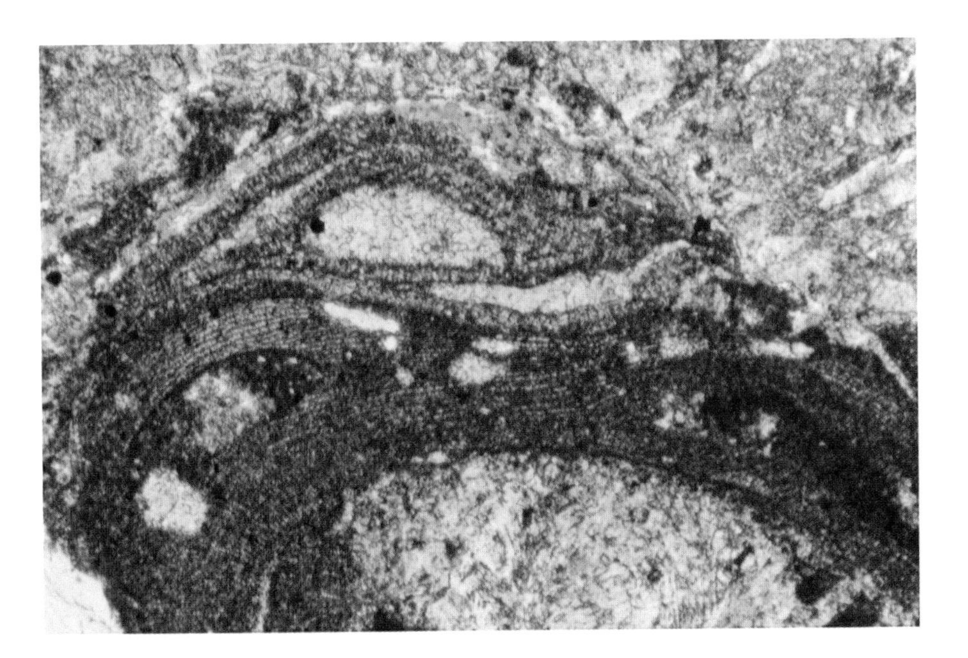

Up. Eocene limestone
Germany

An encrusting red alga, possibly *Lithoporella melobesioides*. Shows regular cellular structure with irregular encrusting strips and conceptacle chambers. Red-algal encrustations are very important in modern reef development as they provide much of the binding of grains into a wave-resistant framework.

0.22 mm

Mid. Eocene Coamo Springs Ls.
Puerto Rico

Large red-algal fragment with typical laminated and cellular structures. Cut by numerous dolomite rhombs (often selectively replacing red algae because of original high Mg content).

0.38 mm

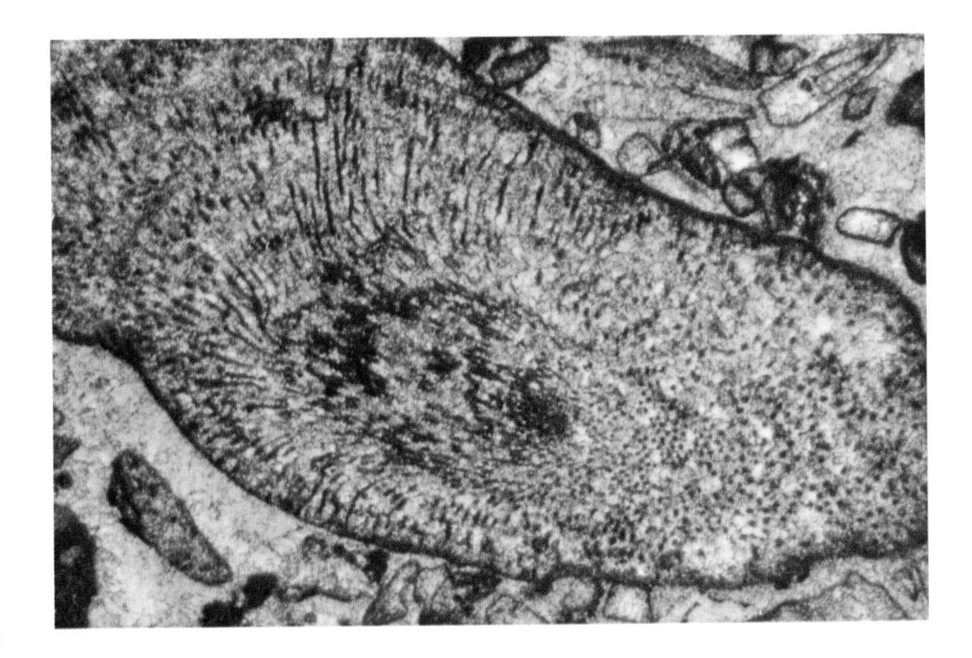

Cretaceous Tamabra Ls.
Mexico

A large algal grain from a rudistid bioherm. This could be a Codiacean green alga or possibly a red alga. Note abundant mud-filled tubes. Where mud has not infilled tubes recrystallization would have effectively obliterated this grain.

0.64 mm

Cretaceous Tamabra Ls.
Mexico

A probable Solenoporoid grain within a rudistid reef. Note radiating tubes with cellular structure. In many areas, this structure is largely obliterated by recrystallization. In this group, the cross partitions separating cells are weakly calcified and often do not preserve well, yielding a tubular structure rather than a cellular one.

0.64 mm

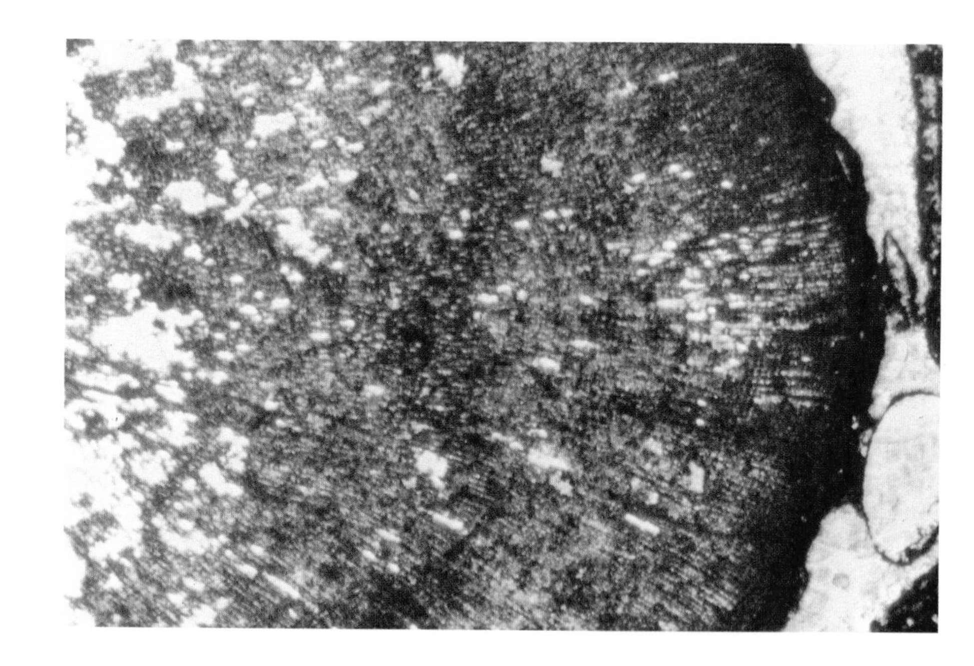

Cretaceous Tamabra Ls.
Mexico

Several types of encrusting algae. The large tubular form may be a red alga, whereas the dark, micritic, lumpy forms are probably blue-green algae. This illustrates the complex intergrowth of algal types and their importance in holding together reef material (such as these rudistid fragments).

0.64 mm

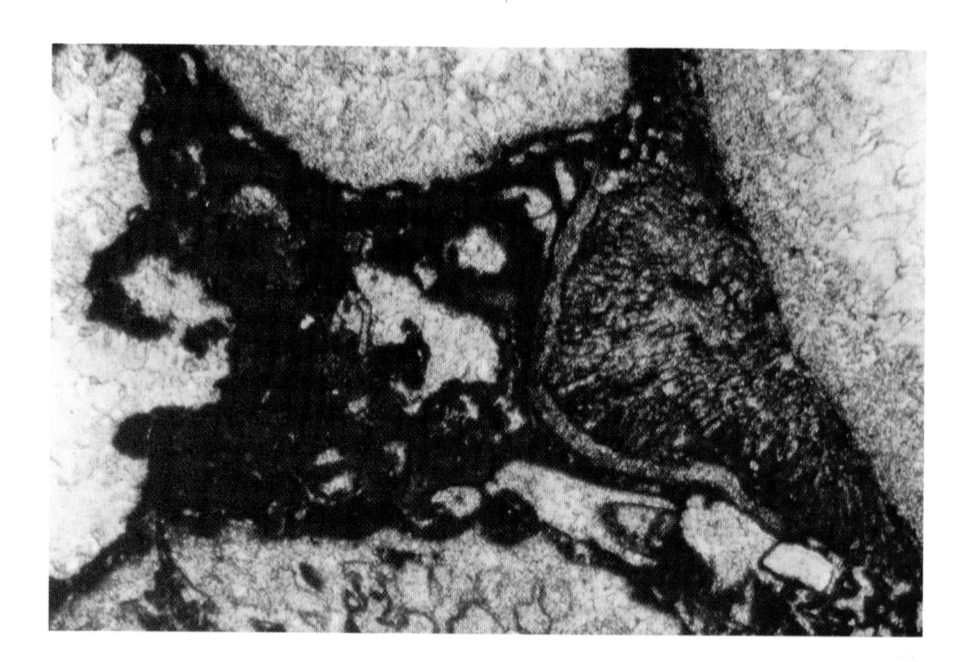

Miscellaneous Microfossils

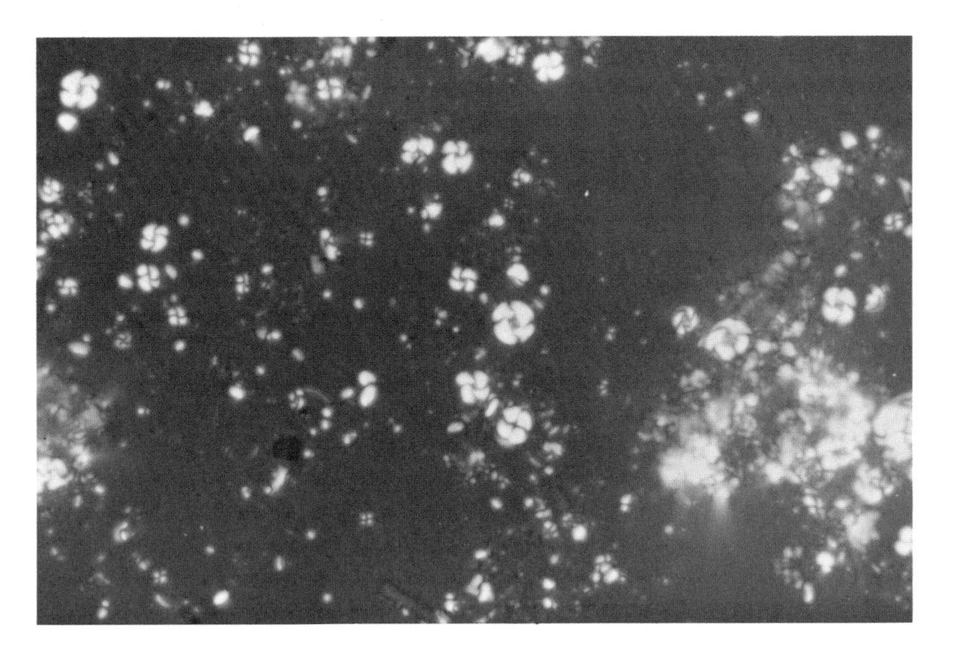

Eocene deep water ooze, DSDP leg 1
Bermuda Rise, N. Atlantic Ocean

Abundant coccoliths (smear mount) showing characteristic curved-cross extinction pattern. Recognition of coccoliths in normal-thickness thin sections is much more difficult and can only be accomplished along unusually thin edges of the slide. Coccoliths, members of the golden-brown algae, range from Jurassic to Holocene.

X.N. 0.016 mm

Mid. Miocene, DSDP Leg 12 Site 116
Hatton-Rockall Basin, N. Atlantic Ocean

SEM view of a coccolith ooze. Sediment is composed almost entirely of coccolith plates and fragments with subordinate discoasters and fragments of planktonic Foraminifera. Virtually no cement; sediment still has about 60 percent porosity. Some corrosion of coccoliths is evident, a common syn- and post-depositional feature in deep sea sediments.

SEM Mag. 5000X 2.5 μm

Up. Cretaceous Atco Fm. of Austin Group
Texas

An example of an exposed, onshore chalk. Some well-preserved coccoliths are visible along with numerous coccolith fragments, clay minerals, spines, and other skeletal fragments. This chalk was probably deposited in no more than 200 meters of water and still retains about 25-30 percent porosity.

SEM Mag. 3000X 4.3 μm

Up. Cretaceous Atco Fm. of Austin Group
Texas

SEM view of a single coccolith from the chalk in the previous photo. It is well preserved with virtually no dissolution or cementation affecting its outline.

SEM Mag. 14.500X 0.9 μm

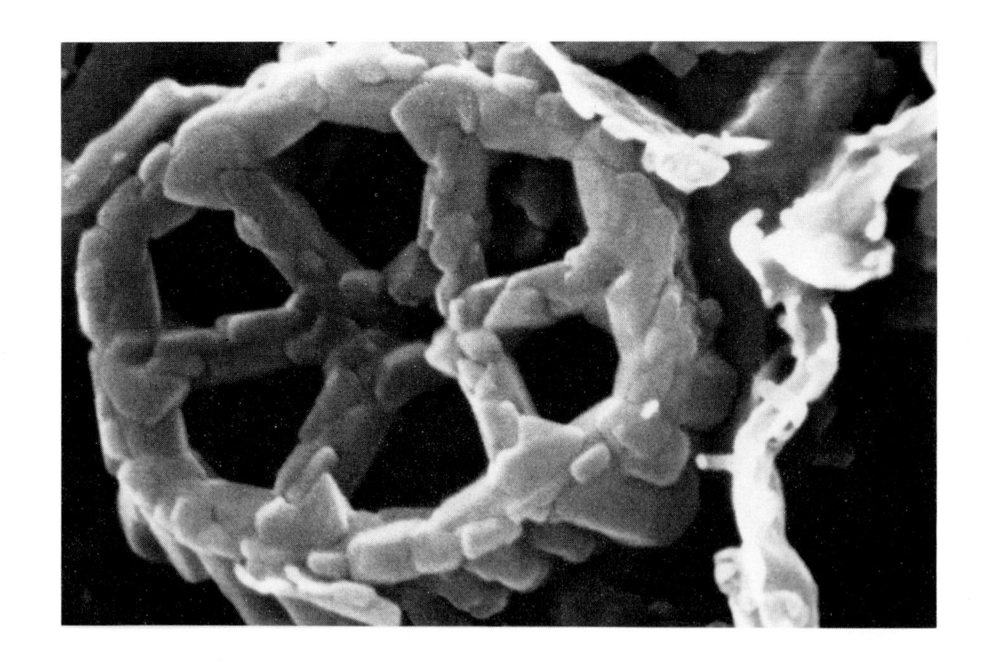

Holocene plankton
Belize (British Honduras)

A coccosphere of *Emiliania huxleyi*. Coccospheres are the original, undisaggregated hollow spheres formed by interlocking coccolith plates. They are found occasionally in undisturbed chalk sediments, although most tend to fall apart into constituent plates.

SEM Mag. 10,000X 1.3 μm

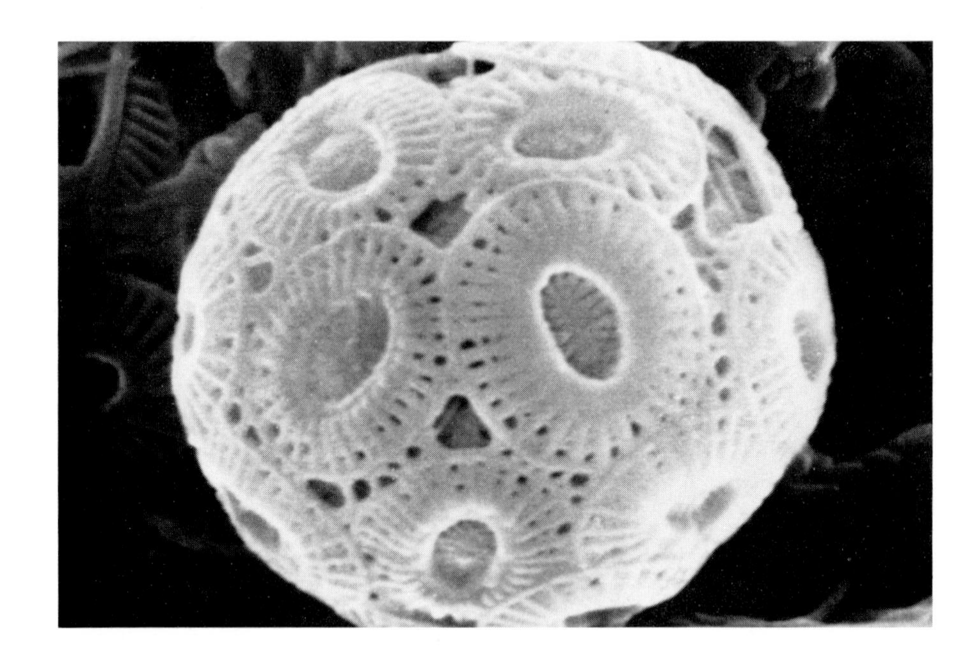

Holocene plankton
Belize (British Honduras)

A partly broken coccosphere of *Emiliania huxleyi* showing the fact that each coccolith is really a double plate and that adjacent coccoliths interlock in a tongue-and-groove manner to form a coccosphere.

SEM Mag. 10,000X 1.3 μm

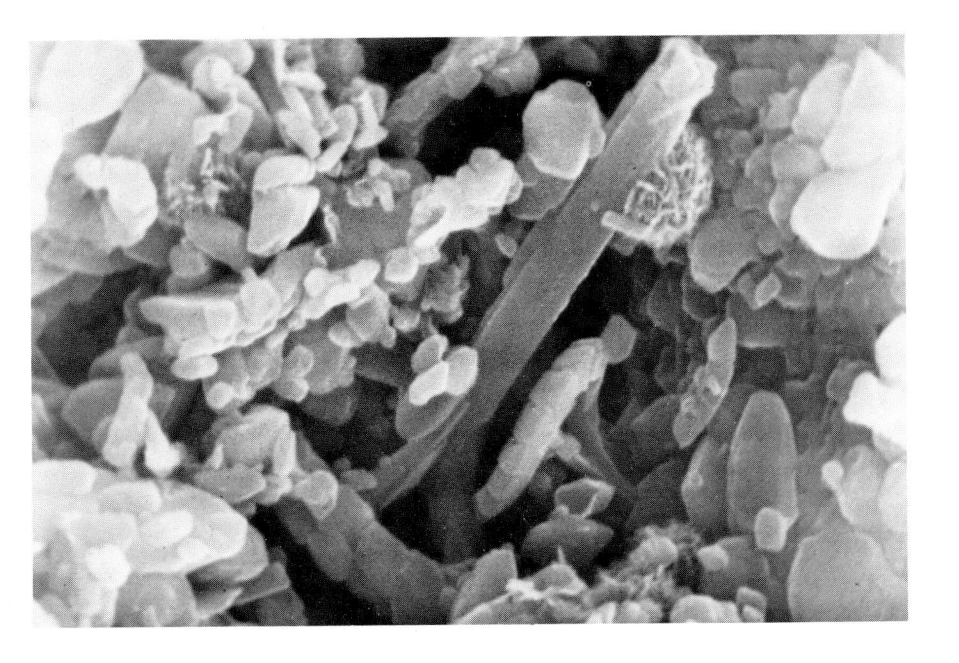

Maestrichtian chalk
Denmark

SEM view of a rhabdolith. This is a coccolith-type plate with a long spine attached to the outer surface. It is quite common to find the rhabdolith spines broken off; such spines can often compose a significant portion of chalks.

SEM Mag. 6000X 2.2 μm

Pliocene? marine sediment
Pacific Ocean

A discoaster in SEM view. These star-shaped nannofossils are especially important in Cenozoic stratigraphic zonation.

SEM Mag. ca. 10,000X 1.3 μm

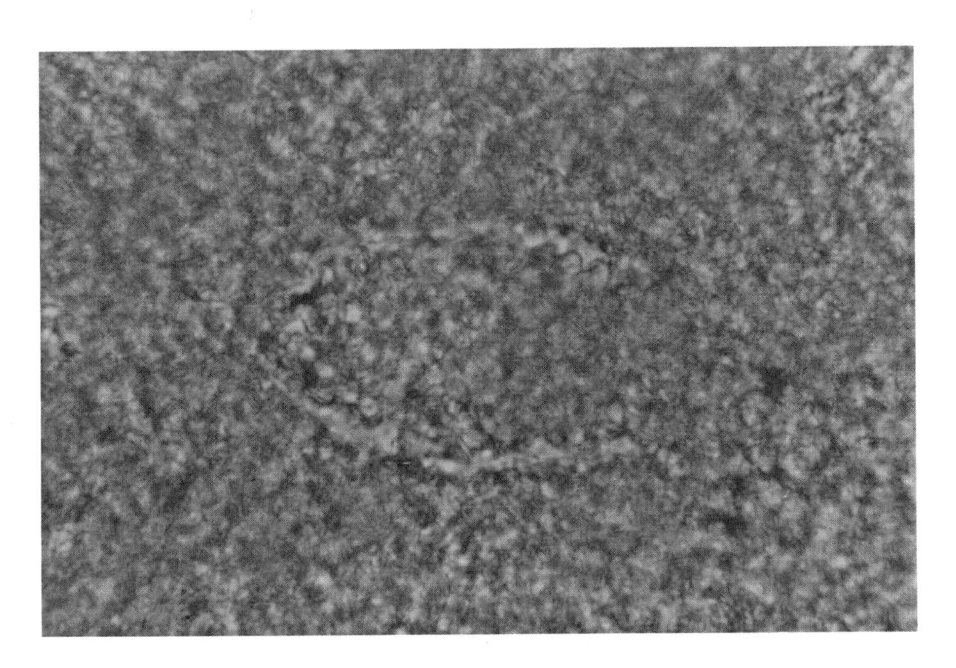

Up. Jurassic Calpionellid Ls.
Italy

A single U-shaped calpionellid. Visible only at relatively high magnification, these microfossils are quite important in Jurassic and Cretaceous deep-water sediment and are useful in the stratigraphic zonation of those deposits.

0.014 mm

Up. Jurassic Calpionellid Ls.
Italy

Section through a single calpionellid showing classic vase or urn shape. A large variety of shapes are known but many incorporate the feature of a straight or curved lip around the aperture, as seen here.

0.014 mm

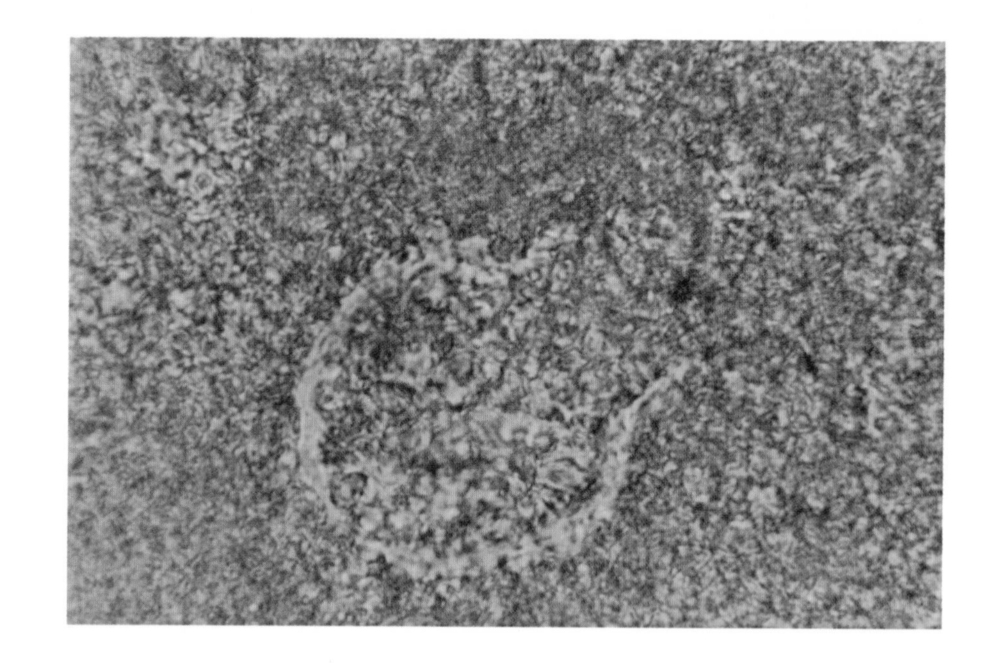

Up. Pennsylvanian Gunsight Ls. Mbr., Graham Fm.
Texas

Walled calcispheres (of possible green algal origin) associated with micritized and spar filled skeletal debris. In Devonian to Permian carbonates calcispheres are especially important in restricted or back reef environments.

0.38 mm

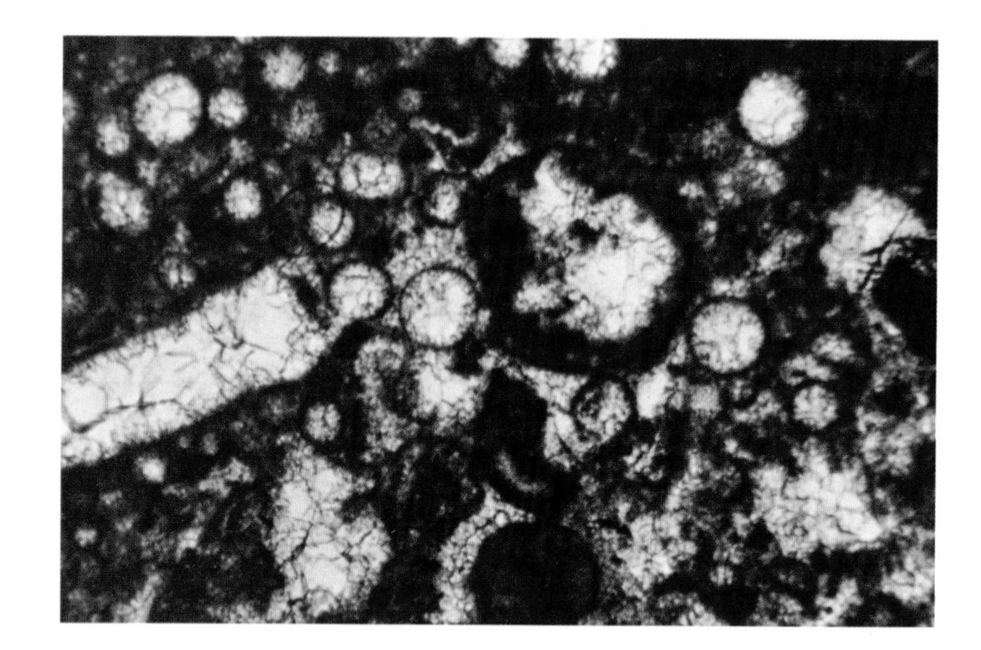

Up. Cretaceous (Cenomanian) Lower Chalk
England

A calcisphere limestone. The bulk of the microfossils in this section are calcispheres of the genera *Oligostegina* and *Pithonella*. They show a well-defined wall which is difficult to distinguish from single chambers of some planktonic Foraminifera. *Pithonella ovalis* can show a very elliptical outline with a large aperture. These grains are very common in Cretaceous pelagic deposits.

0.11 mm

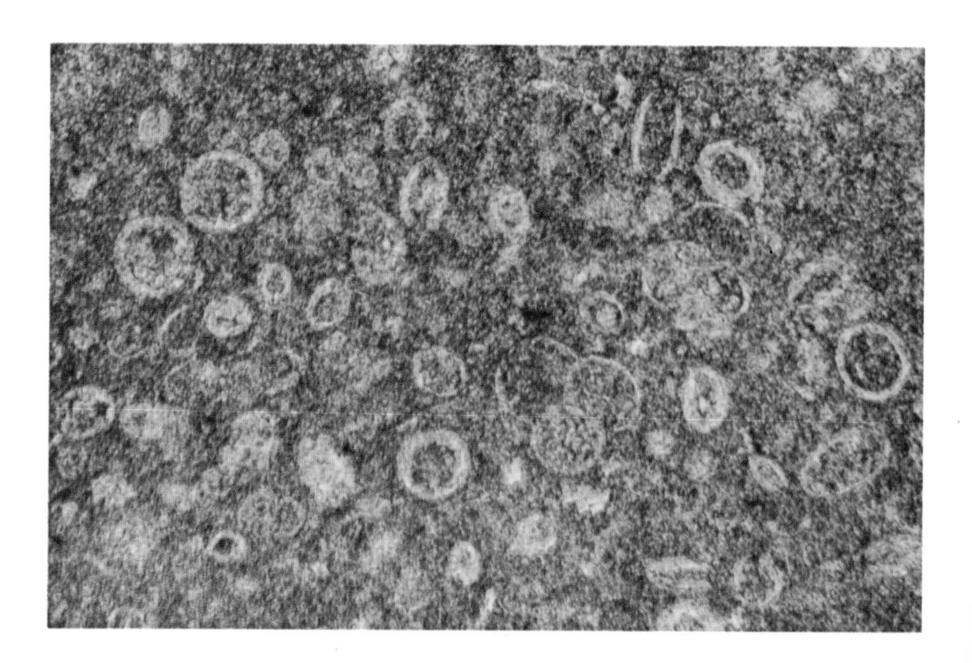

Up. Cretaceous Chalk
North Sea

A chalk with abundant calcispheres, possibly *Pithonella ovalis*. Note the single, spherical chambers and the distinctive wall texture of these grains as well as the small, associated coccoliths in the background.

SEM Mag. 1000X 12.5 μm

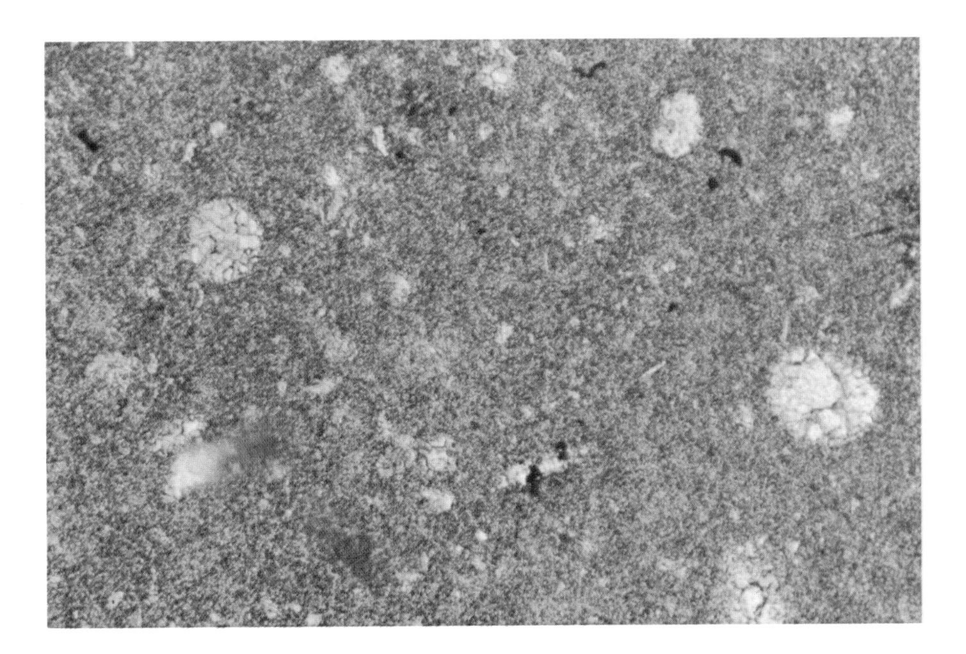

Up Cretaceous Buda Ls.
Texas

Unwalled calcispheres in a micrite matrix. These calcispheres may be of non-algal origin (e.g. alteration of radiolarians).

0.10 mm

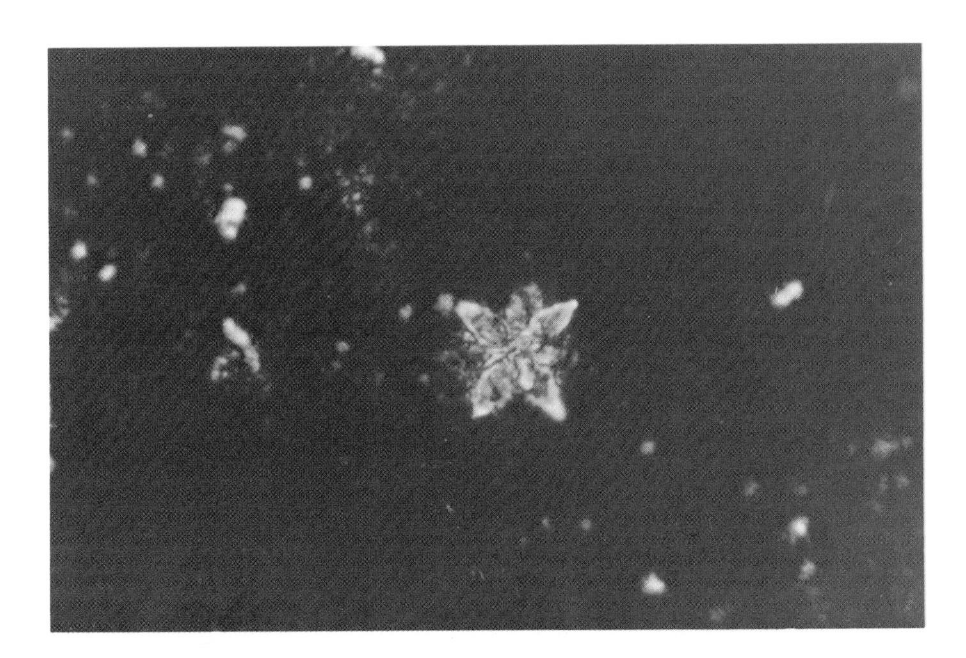

Holocene sediment
Belize (British Honduras)

Views of a single tunicate spicule. Tunicates are classed in the subphyllum Hemichordata, and some genera have calcitic spicules embedded in their fleshy tissue. They make up as much as 1 percent of the sediment in some modern settings in shallow water. They are characterized by their multi-rayed shape and very small size.

X.N. 0.016 mm

Up. Cretaceous Red Bank Sand
New Jersey

Stained preparation of a dinoflagellate *Deflandria diebeli*. Central portion (endocyst) and spinose appendages (horns) have stained differentially. Dinoflagellates can be useful in correlation of marine sediments of Triassic to Recent age.

0.027 mm

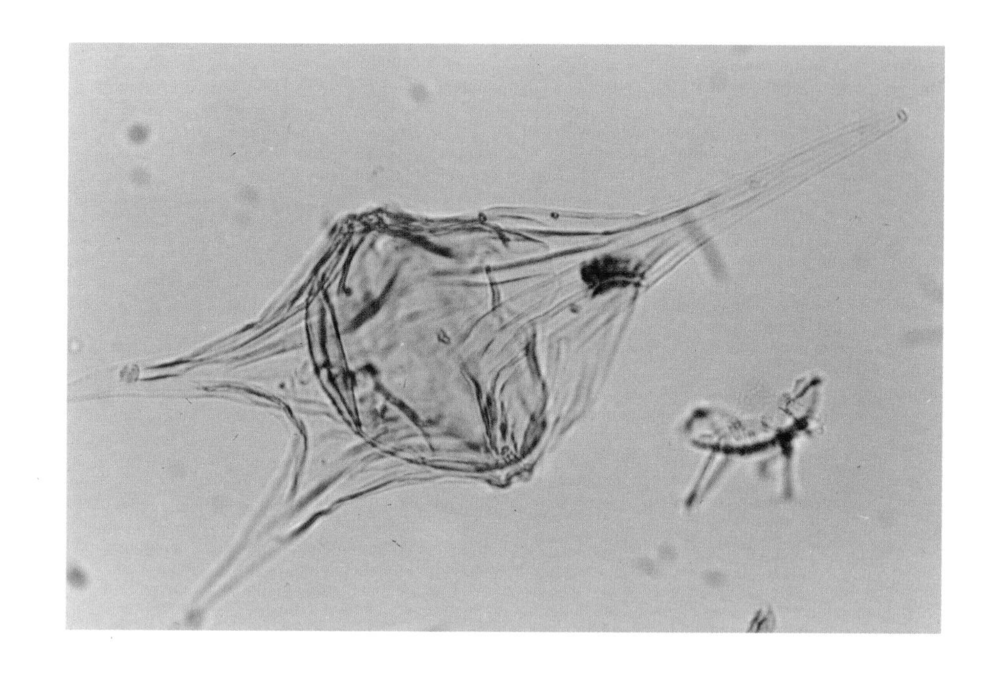

Up. Cretaceous Mount Laurel Sand
New Jersey

A stained preparation of the dinoflagellate *Spiniferites ramosus*. The central portion has taken a dark stain and the multiple, radiating, splay-tipped or trumpet-like appendages are clearly visible. These resting cysts are noncalcified.

0.022 mm

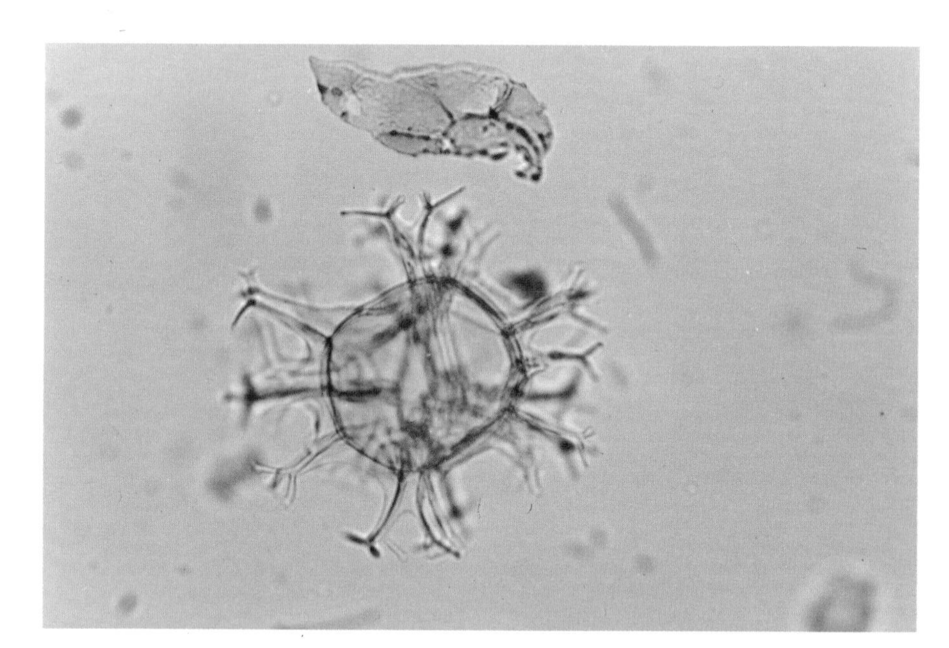

Cretaceous(?) sediment

SEM view of a single dinoflagellate, probably *Oligosphaeridium* sp. The central portion of the body and the radiatings appendages with their triangular terminations are shown.

SEM Mag. ca. 1000X 12.5 μm

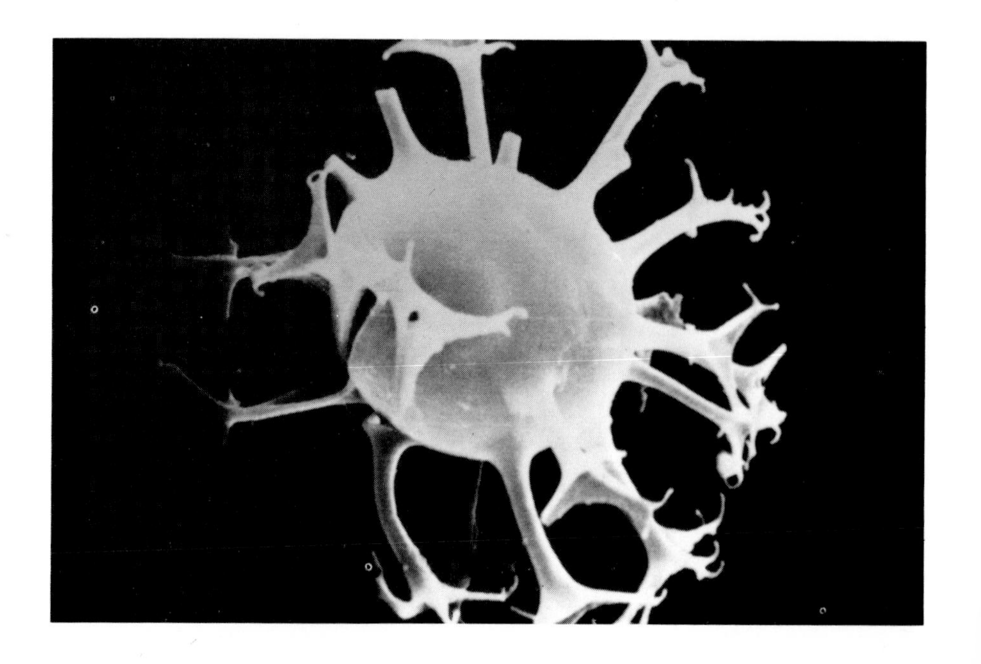

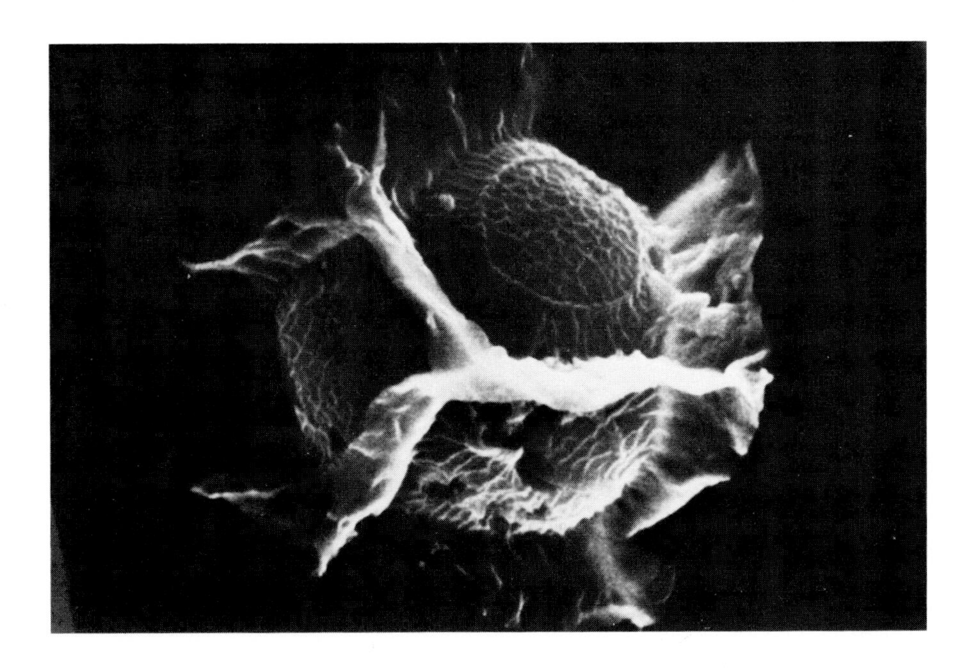

Paleozoic(?) sediment
United States

SEM view of a herkomorph acritarch. This is a fossil group of uncertain taxonomic affinity, likely produced by zooplankton as well as phytoplankton. They range from Precambrian to Holocene and are commonly found in palynological preparations (HCl-HF residue).

SEM Mag. ca. 800X   17 μm

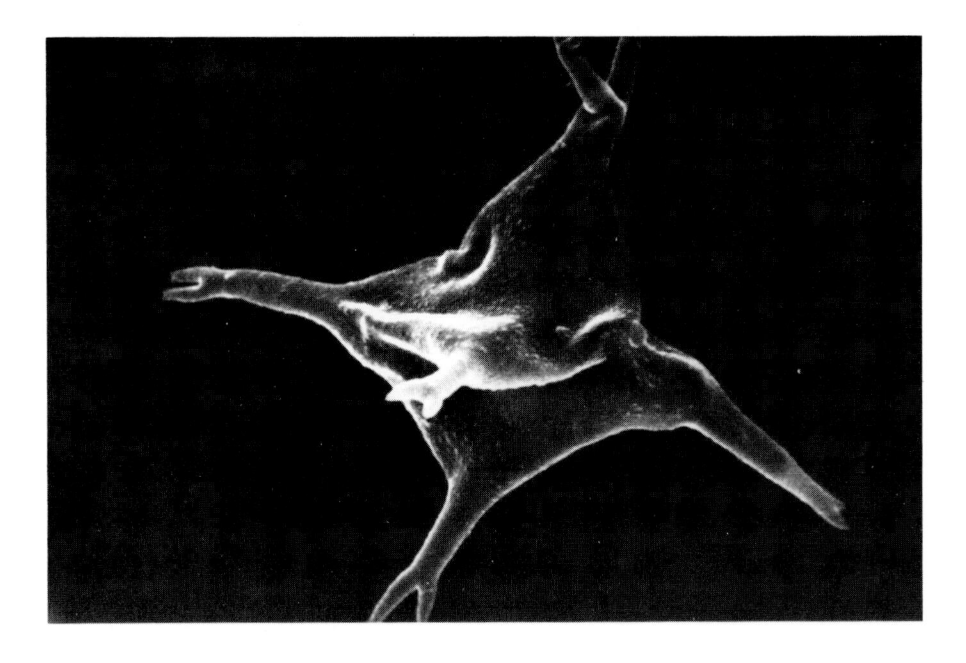

Paleozoic(?) sediment
United States

SEM view of another acritarch of totally different form of the group Polygonomorphitae. An algal affinity has been suggested for this group. Acritarchs, like dinoflagellates, are non-calcified but may frequently be found in calcareous rocks, especially in palynological preparations.

SEM Mag. ca. 800X   17 μm

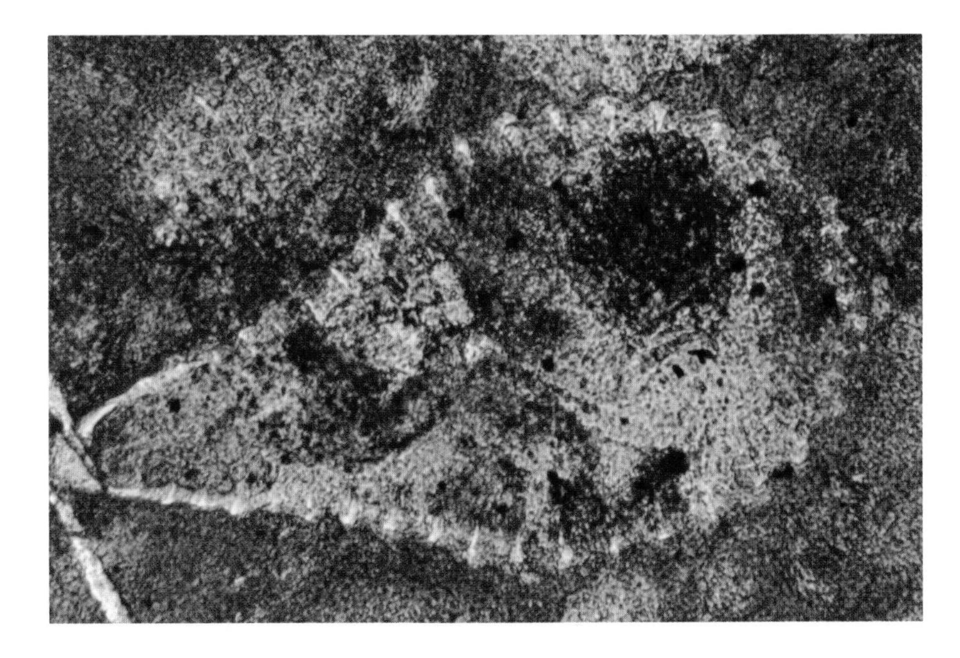

Jurassic (Tithonian) Point Sal ophiolite group
California

Longitudinal cross section of a single radiolarian text. Originally opal, this example has gone to quartz chert and still shows the coarse pore structure in the outer skeleton. This group is wide ranging and extremely important in formation of deep-water deposits especially in the Devonian, Jurassic, and Tertiary.

0.07 mm

Jurassic (Tithonian) Point Sal ophiolite group
California

Cross sections through two radiolaria showing good preservation of tests despite alteration from opal to chert. Coarse pore structure, size, and shape of tests are main criteria for identification.

0.07 mm

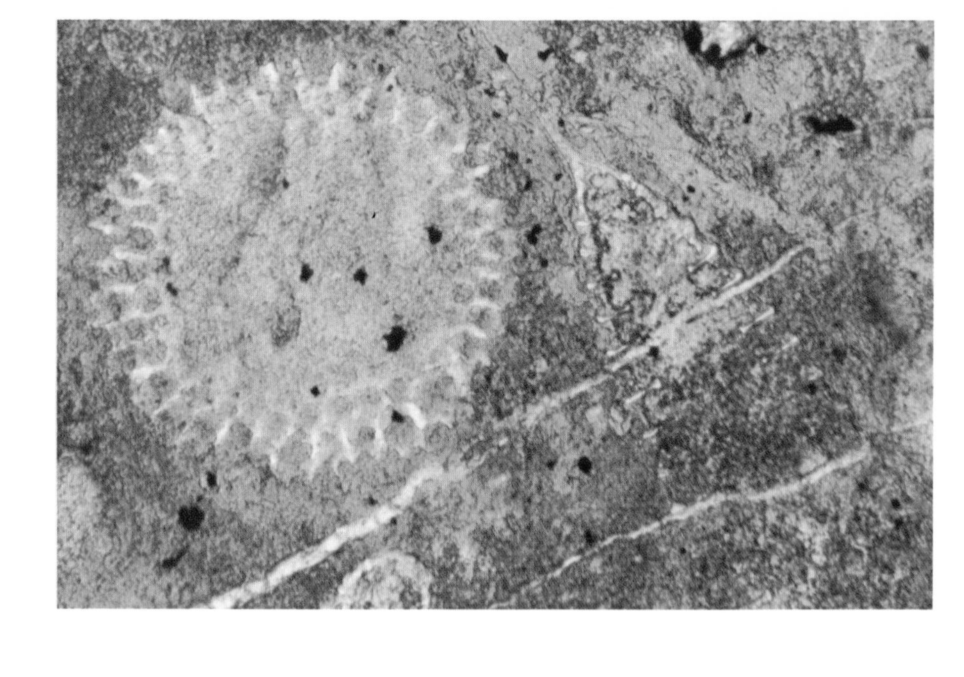

Up. Jurassic(?) radiolarite
Muscat and Oman

Tangential section through the poorly preserved outer wall of a radiolarian. In the alteration of opal tests to cristobalite and, finally, to quartz, much of the structural detail may be lost. But coarse pore structure is still visible.

0.03 mm

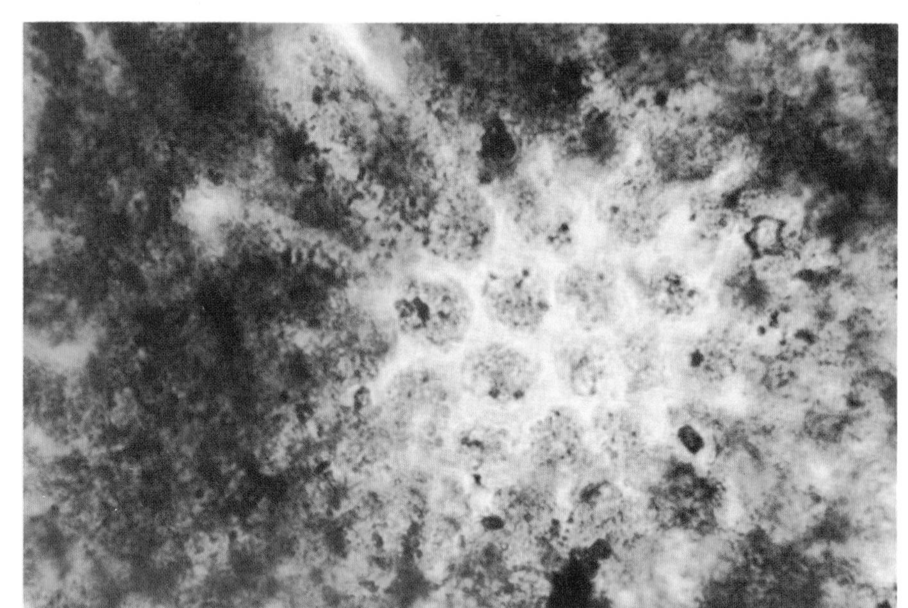

Up. Jurassic(?) radiolarite
Muscat and Oman

Section through a poorly preserved radiolarian test. Some traces of coarse, radial pore structure or outer wall is still visible.

0.03 mm

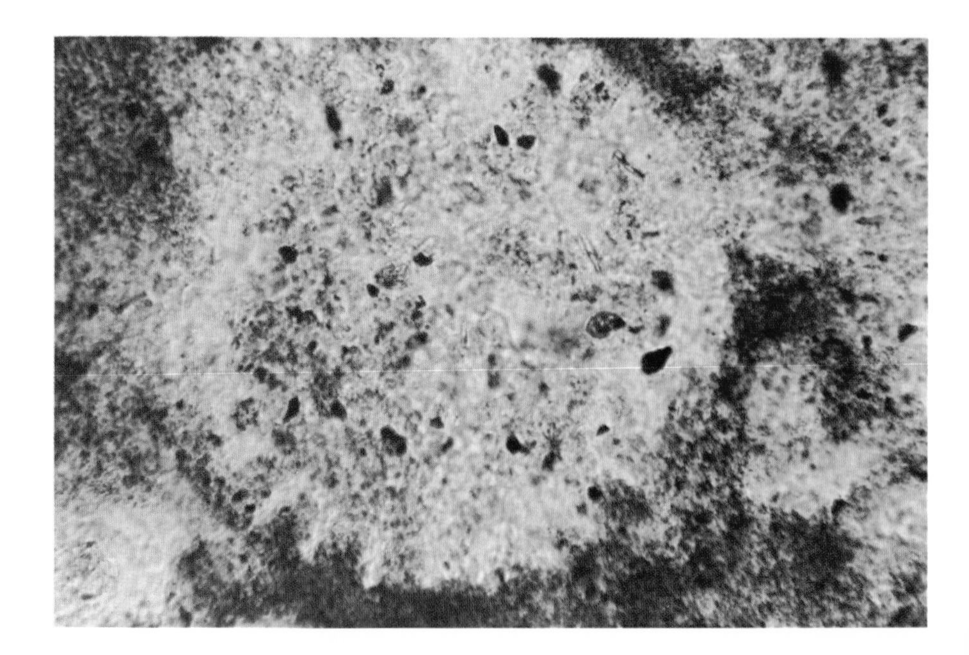

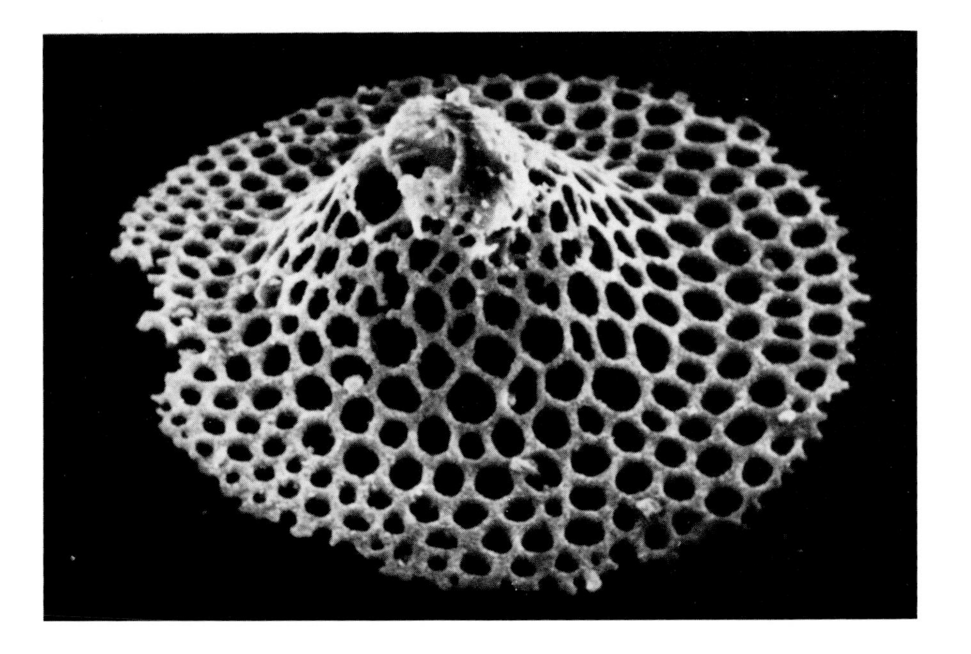

Up. Jurassic part of Franciscan Fm.
California

SEM view of a single radiolarian which has been etched out of a chert with HF acid. The radiolarian is also composed of chert but etches out selectively. Note coarse pore structure and shape.

SEM Mag. ca. 300X 41 μm

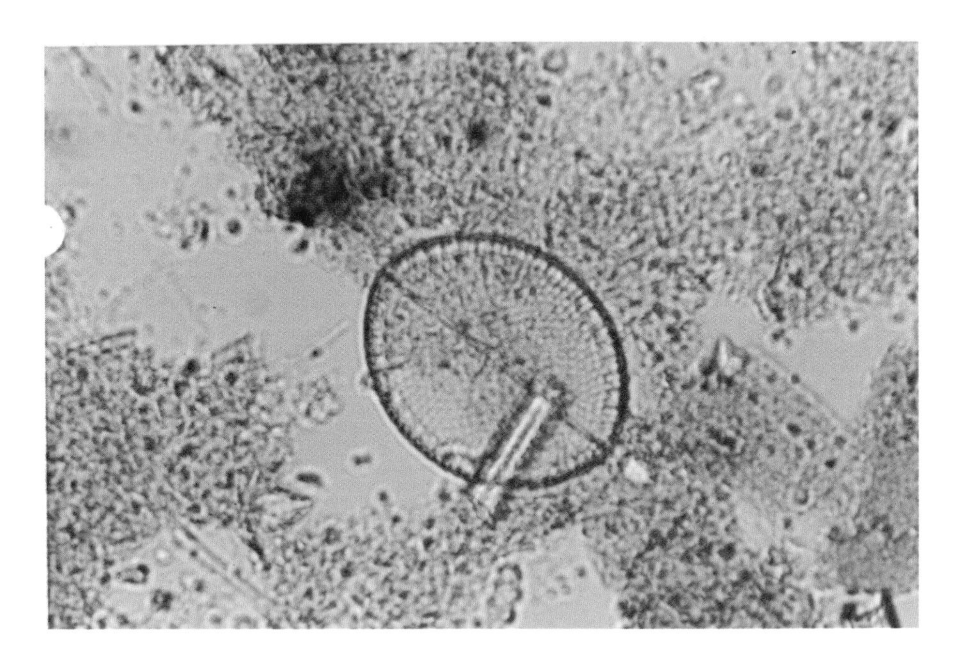

Holocene sediment
Mexico

An opaline diatom test. Diatoms are important in Cretaceous and Tertiary sediments from a number of environments. Originally composed of opal, they alter to other forms of silica, often with very considerable loss of textural detail. Note small bit of a broken sponge spicule which overlaps diatom.

0.018 mm

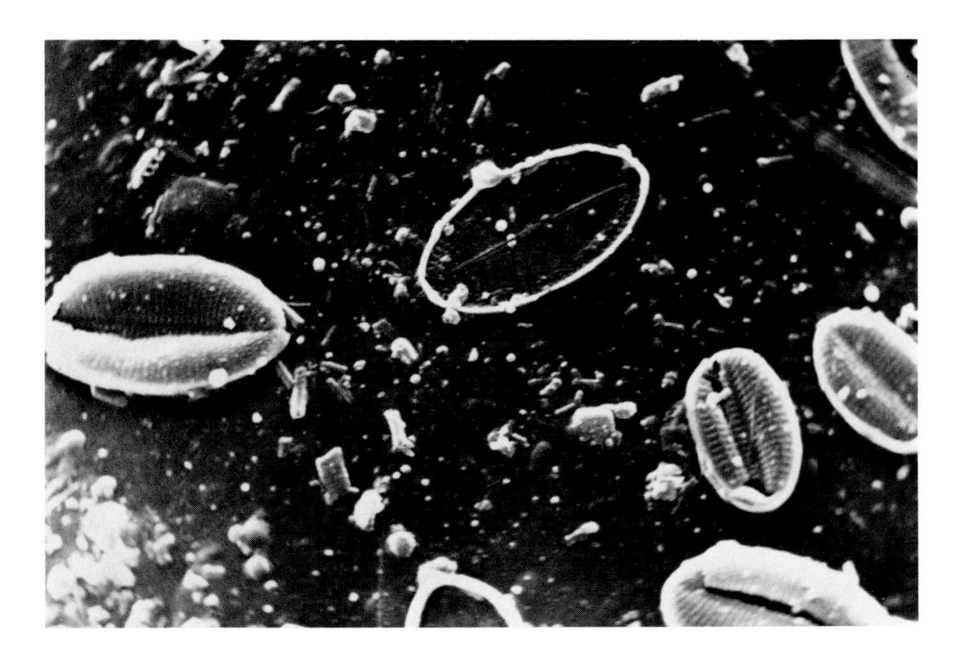

Holocene sediment
Belize (British Honduras)

SEM view of numerous marine diatoms on a sponge spicule. Although most are oval-shaped, one elongate diatom can be seen in the lower left. Diatoms (and their deposits---diatomites) can be found in both marine and fresh-water settings, often in close association with carbonate sediments, particularly in the Miocene of the circum-Pacific area.

SEM Mag. 3000X 4.6 μm

Miocene Monterey Shale
California

A siliceous shale which can be seen to consist almost exclusively of fragmented marine diatoms. These are still opal; conversion to cristobalite or quartz-chert normally involves the loss of much of the textural detail seen here.

SEM Mag. 6000X   2 μm

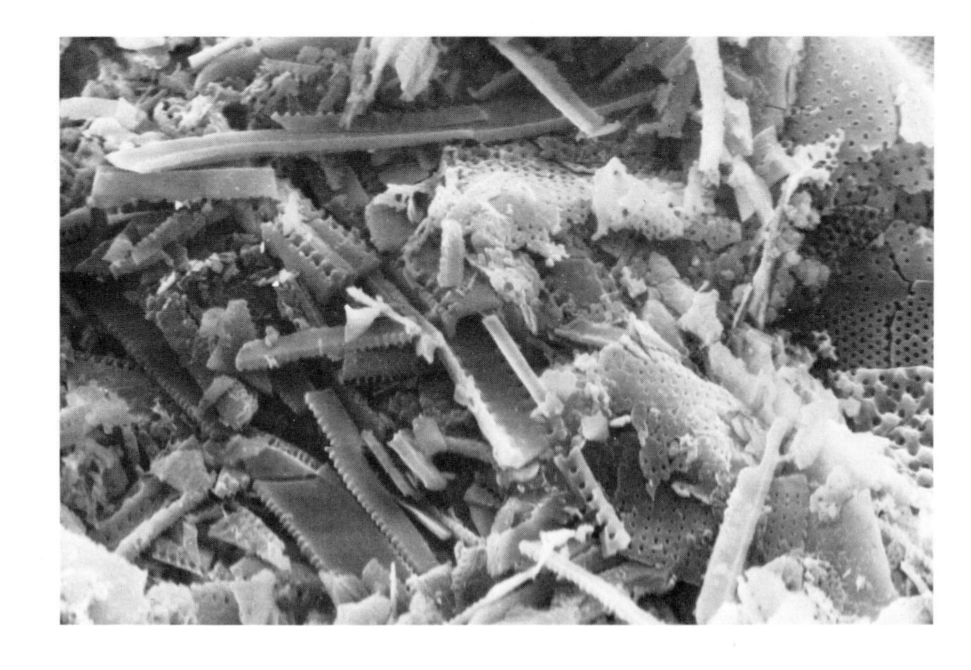

Holocene sediment, Shagawa Lake
Minnesota

Four specimens of *Stephanodiscus* sp., a nonmarine diatom from a eutrophic freshwater lake. Diatoms, a group of siliceous algae, are known from sediments at least as old as Jurassic and can be useful in paleoecologic as well as biostratigraphic interpretations.

SEM Mag. 4750X   2.5 μm

Lo. Mississippian Cottonwood Canyon Mbr. of Lodgepole Ls.
Montana

Abundant conodonts, mainly from the genus *Siphonodella*. These minute tooth-like or plate-like fossils are of uncertain biological affinity but are extremely important in Paleozoic biostratigraphic zonation. Although rarely found in this abundance, they are common in carbonate rocks. The color of conodonts in transmitted light has been used as an index of the degree of thermal alteration of the sediments.

0.38 mm

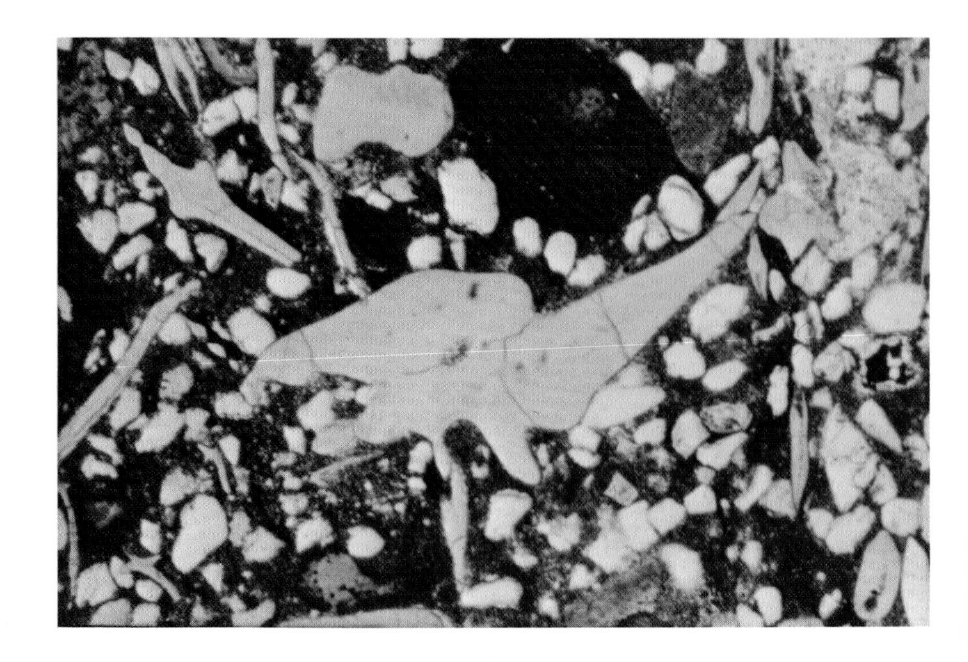

Lo. Mississippian Cottonwood Canyon Mbr. of Lodgepole Ls.
Montana

Several longitudinal and transverse sections of conodonts (mainly *Siphonodella* sp.) are visible. Note characteristic plates with pronounced denticles. Section also contains abundant quartz sand and some dark-brown fish debris.

0.38 mm

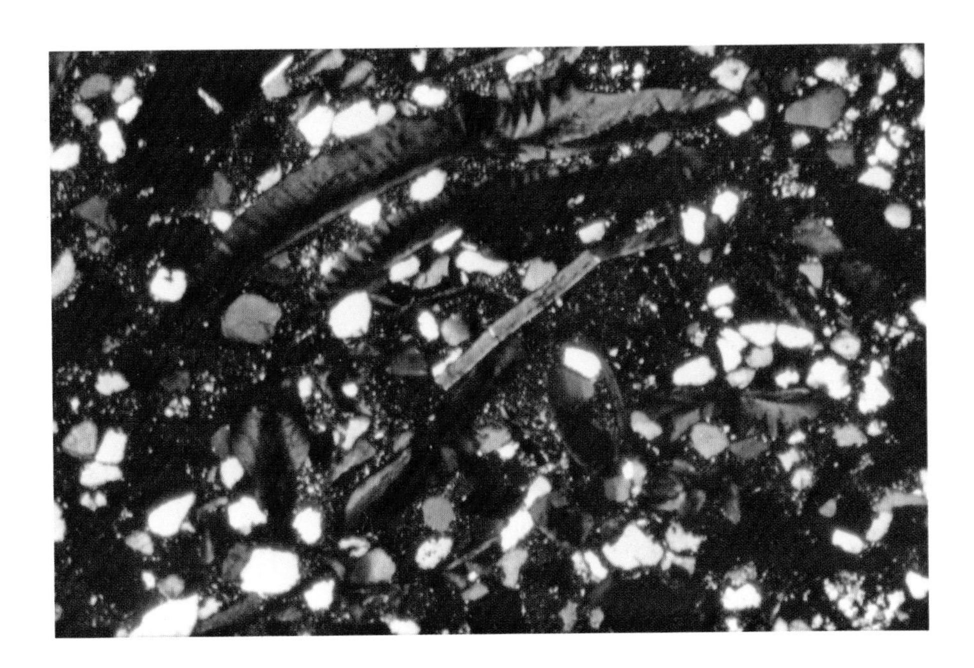

Lo. Mississippian Cottonwood Canyon Mbr. of Lodgepole Ls.
Montana

Same view as above but under cross-polarized light. Illustrates typical extinction behavior of conodonts with pronounced appearance of "white matter" within saw-toothed extinction pattern.

X.N. 0.38 mm

Foraminifera

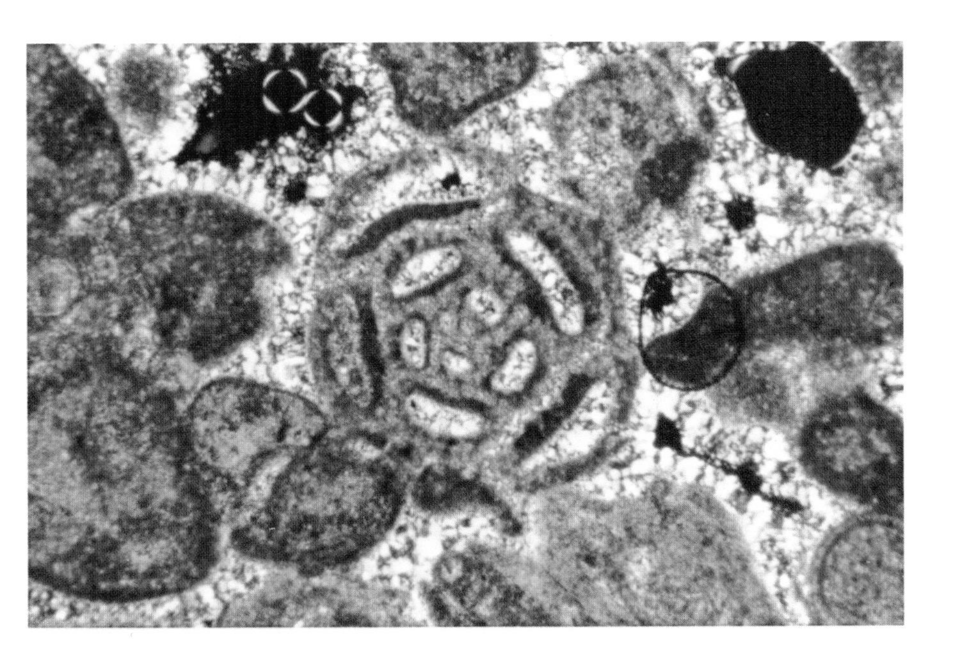

Up. Eocene Ocala Ls.
Florida

Miliolid Foraminifera with walls largely micritized and pores filled with chalcedonic chert.

X.N. 0.24 mm

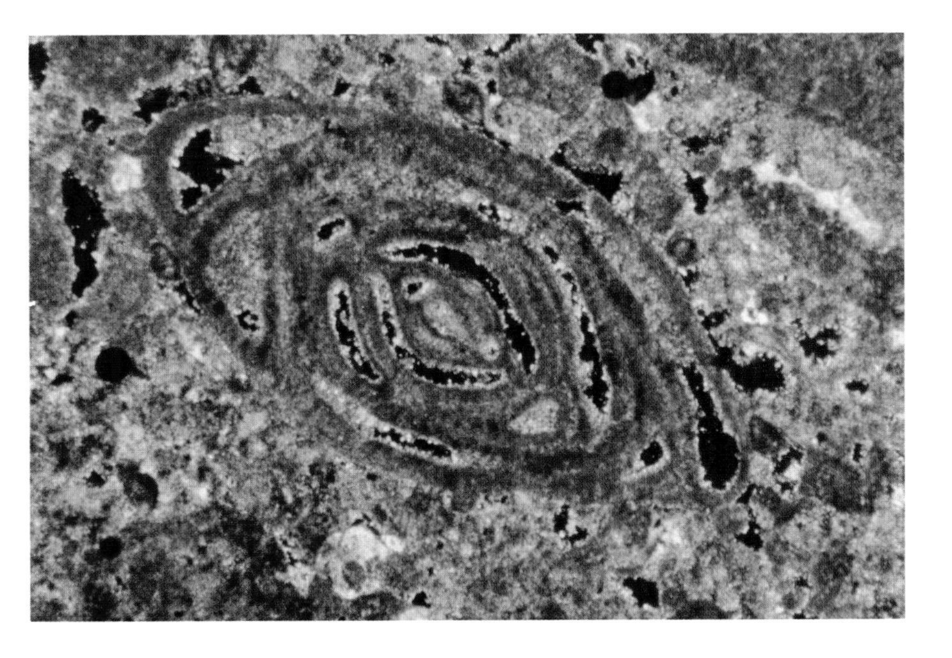

Up. Eocene Ocala Ls. (Inglis)
Florida

Miliolid Foraminifera with largely obliterated wall structure and partially cemented porosity.

X.N. 0.24 mm

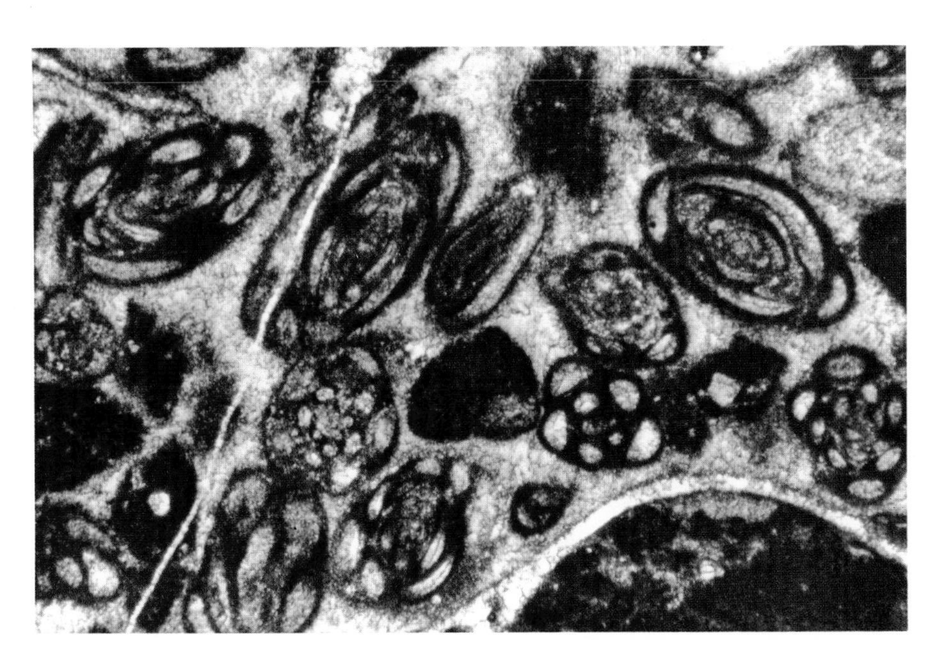

Lo. Cretaceous Cupido Fm.
Mexico

A limestone composed largely of miliolid Foraminifera. These are especially common in slightly restricted back-reef or bank facies. This example shows some of the variety of shapes which can result from random slices through miliolids.

0.22 mm

Holocene sediment
Belize (British Honduras)

Peneroplid Foraminifera showing characteristic shape, structure and shell color (for Holocene specimens only).

0.24 mm

Holocene sediment
Belize (British Honduras)

Peneroplid Foraminifera showing typical shapes and structures, and illustrating early diagenetic changes within depositional environment-- some alteration of wall texture has taken place and much of the intragranular porosity has been lost.

X.N. 0.24 mm

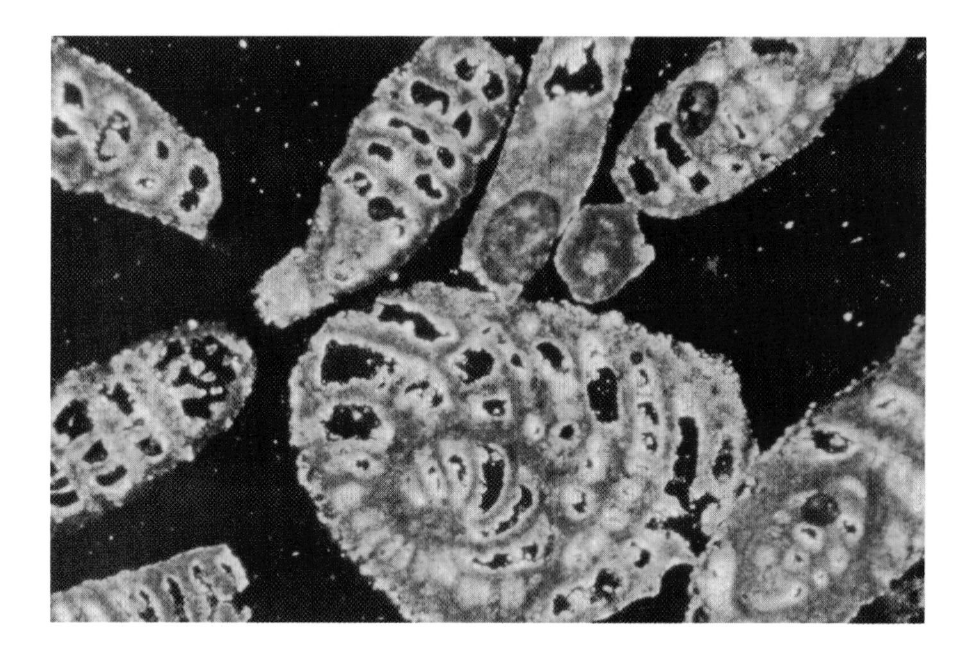

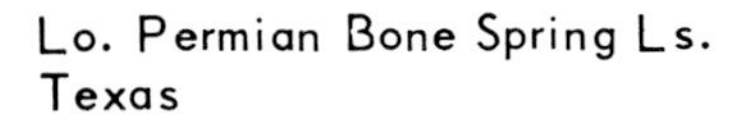

Lo. Permian Bone Spring Ls.
Texas

Fusulinid Foraminifera in section across long axis. Note chamber shapes and radial wall structure. Very common in Pennsylvanian and Permian sediments.

0.38 mm

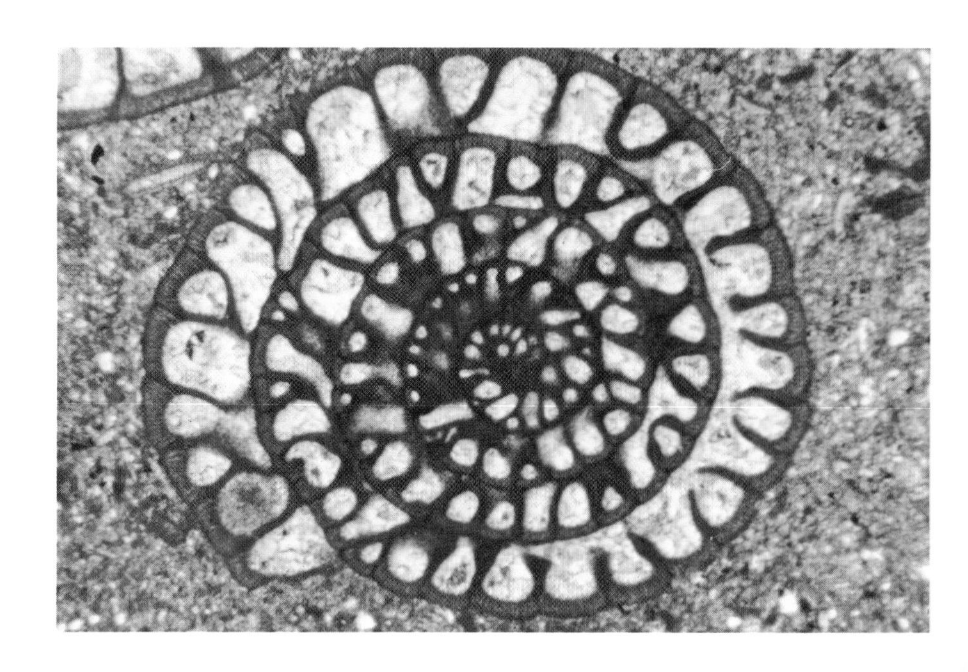

Eocene Nummulite ls.
Jugoslavia

Long-axial section of a single Foraminifera, *Nummulites* sp. These large Foraminifera can be major rock forming elements in Eocene sediments.

X.N. 0.38 mm

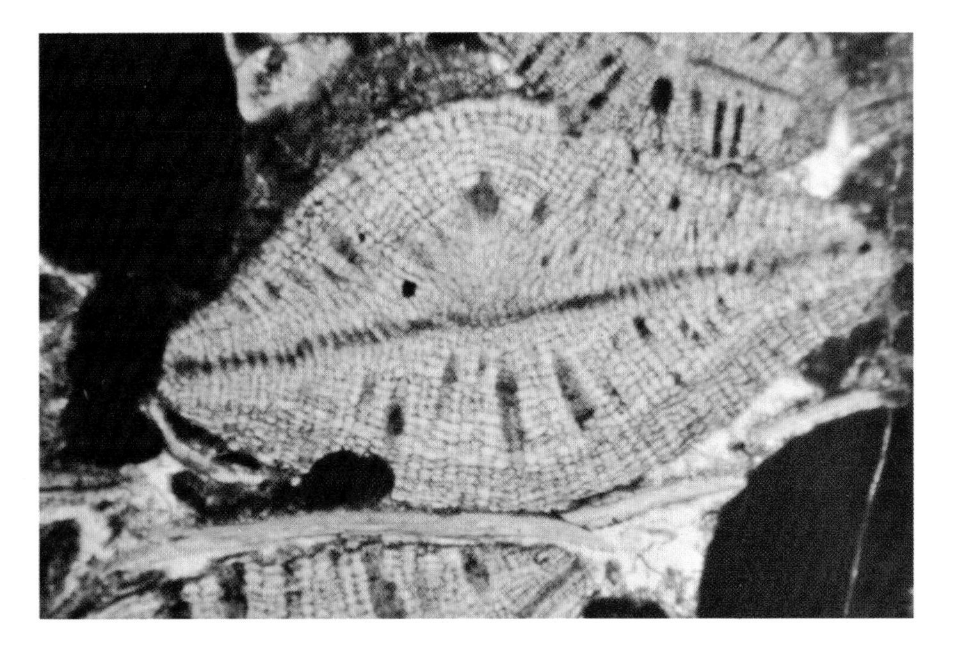

Mid. Eocene Coamo Springs Ls.
Puerto Rico

Long-axial section of a single Foraminifera, *Discocyclina* sp.

0.38 mm

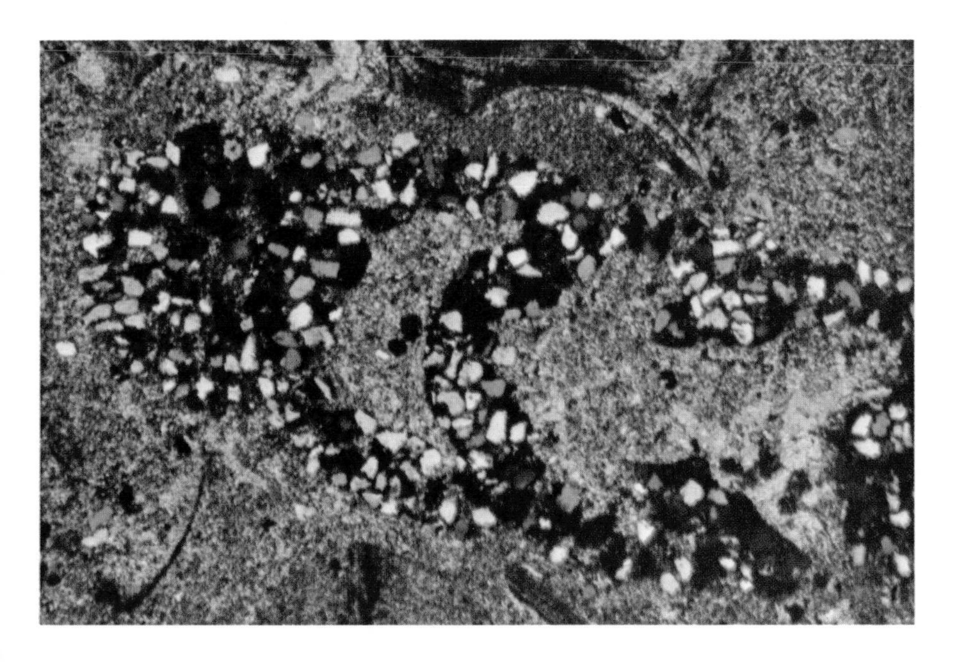

Up. Cretaceous Del Rio Clay
Texas

Arenaceous Foraminifera showing chambers composed of agglutinated individual quartz-silt grains.

X.N. 0.38 mm

Up. Cretaceous Lower Chalk
England

Section through a single agglutinating Foraminifera--in this case one which selects both carbonate and non-carbonate grains. These Foraminifera are recognizable by the chamber-shaped grain arrangements rather than the otherwise random arrangement of grains in the rest of the rock.

0.10 mm

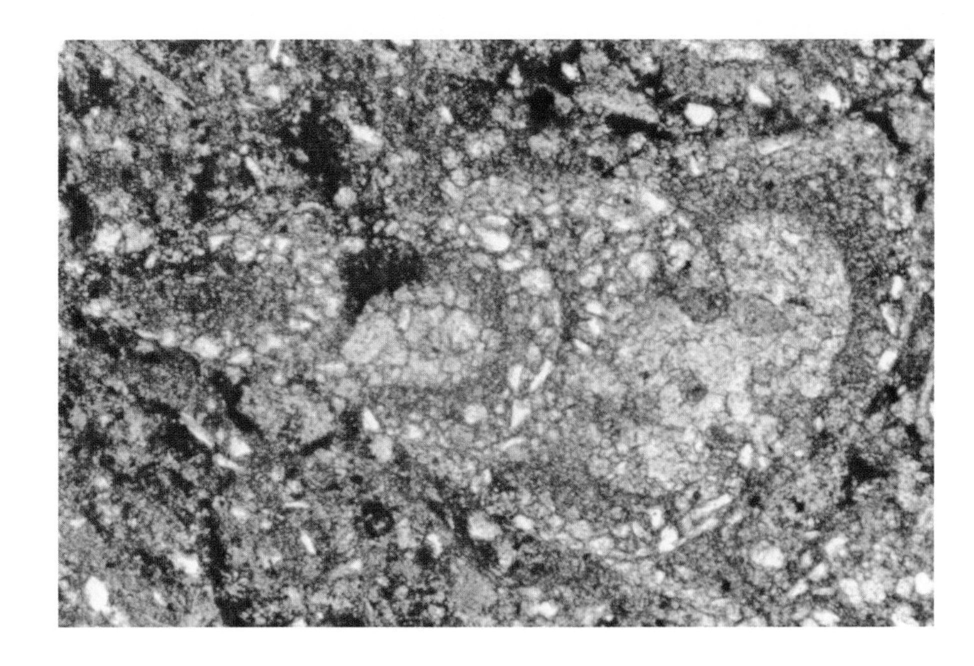

Mid. Pennsylvanian Paradox Fm.
Utah

Sections through uniserial and biserial Foraminifera. Note micritized wall structures and typical chamber shapes.

X.N. 0.25 mm

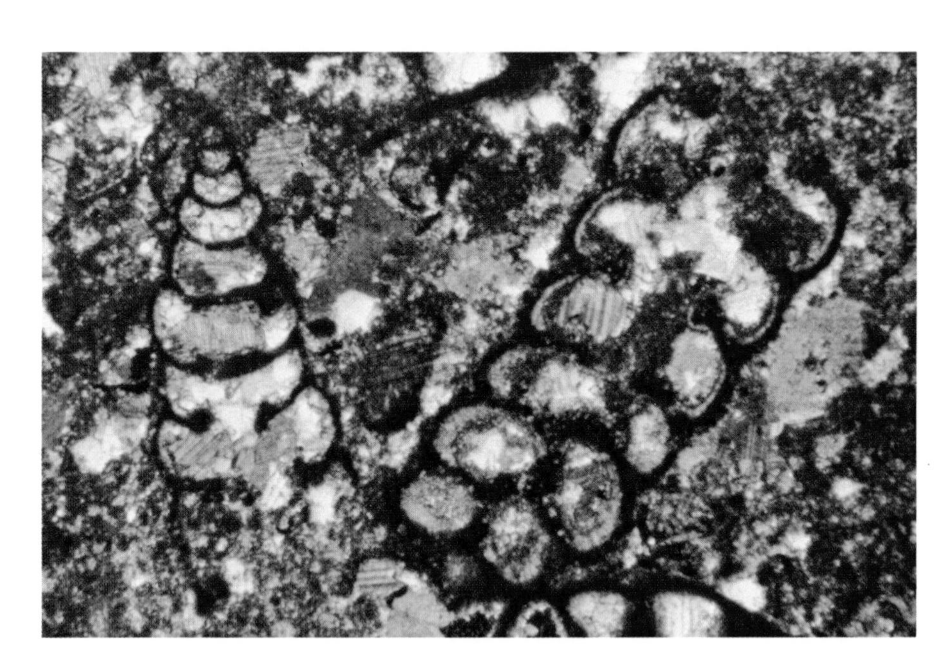

Cretaceous (Cenomanian) ls.
Austria

Section through a single, large Orbitolinid Foraminifera. These Foraminifera can be important rock formers in Cretaceous sediments.

0.37 mm

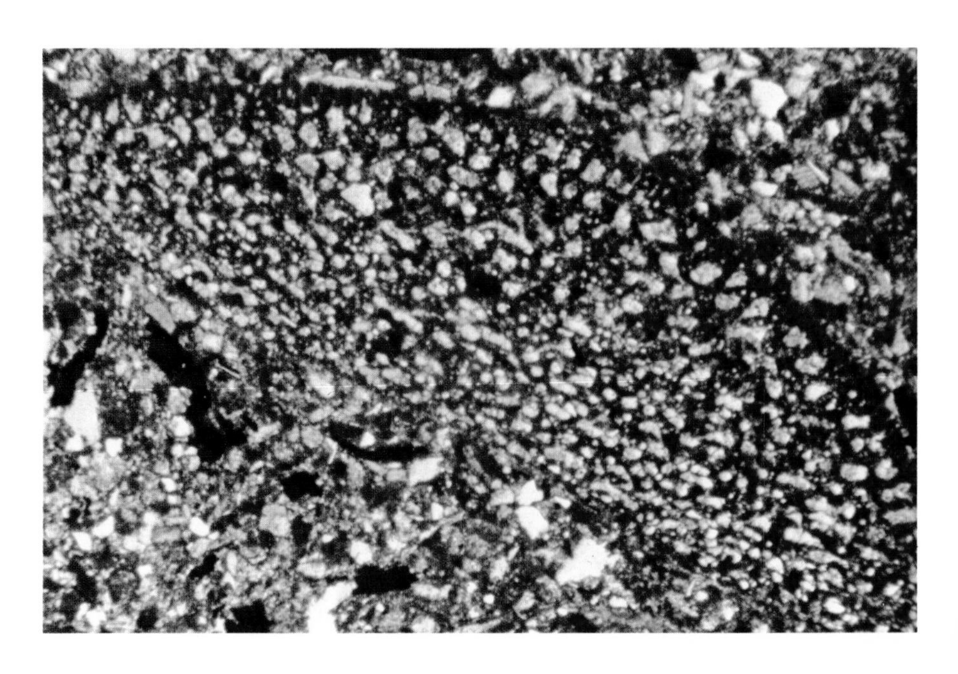

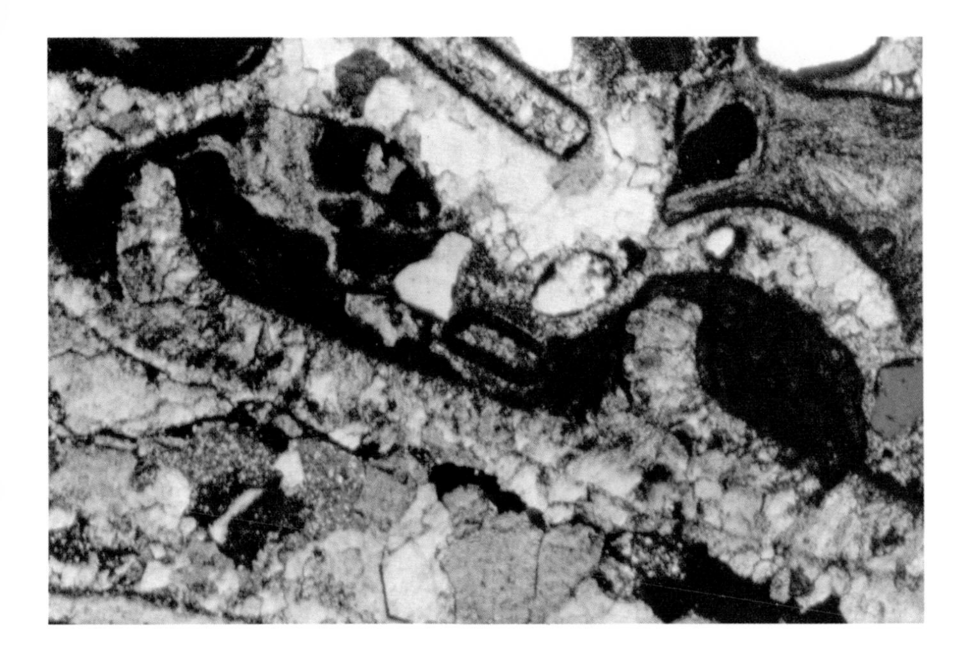

Lo. and Mid. Pennsylvanian Bloyd Fm.
Oklahoma

Encrusting Foraminifera on a neomorphosed pelecypod shell. Globular, chambered foraminiferal structure is vaguely visible within the dense, micritic areas. Encrusting foraminifera are commonly misinterpreted as encrusting algae.

X.N. 0.22 mm

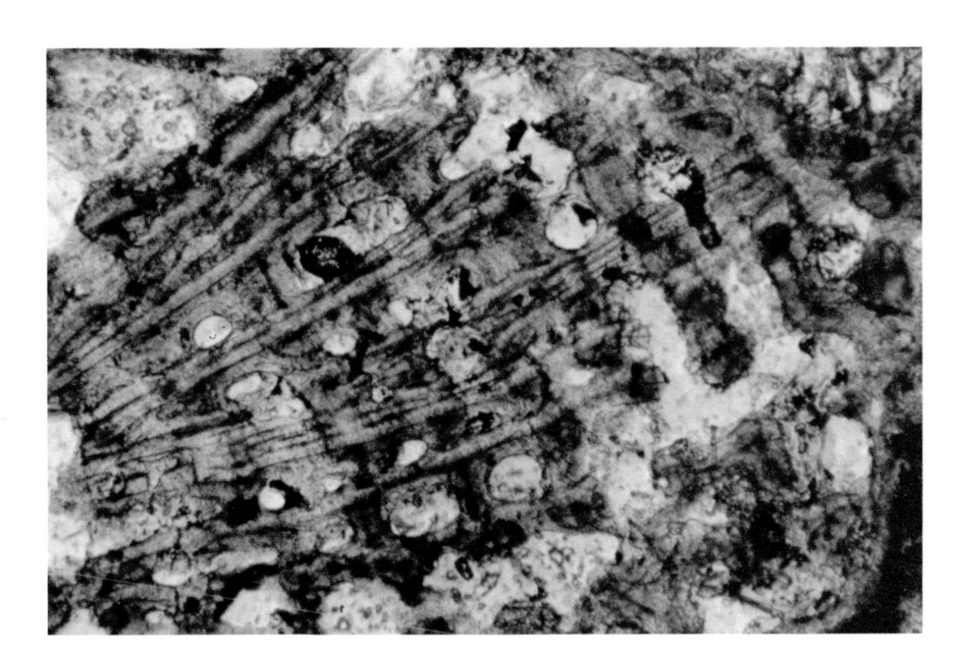

Holocene coral-rubble beach ridge
Belize (British Honduras

Cross section of a large encrusting Foraminifera--*Homotrema rubrum* with very regular wall and cell structure. Reddish tint is characteristic of this species.

0.38 mm

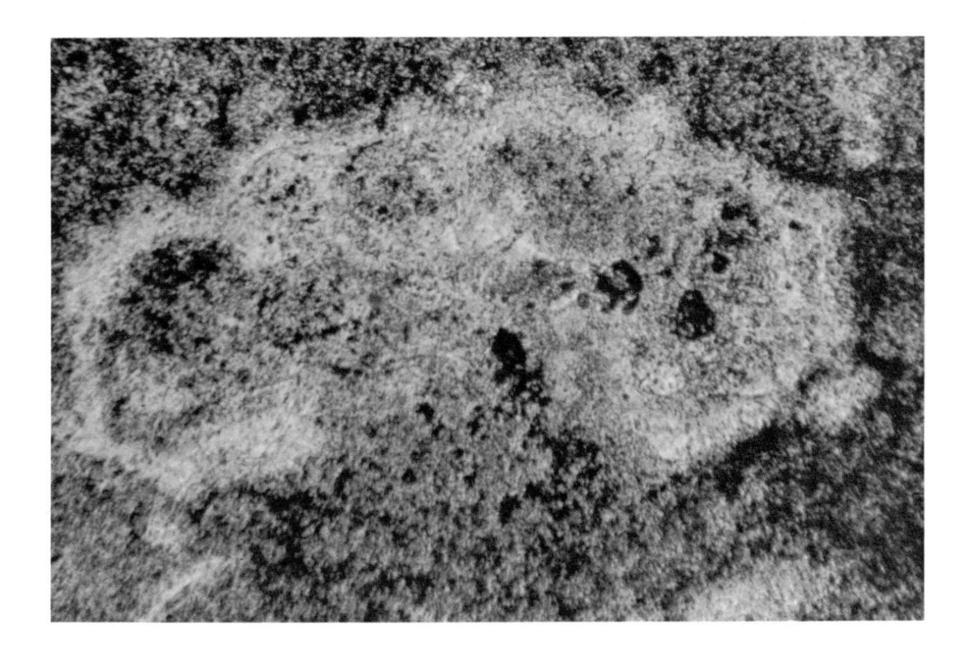

Up. Cretaceous Upper Chalk
England

A *Globotruncana* sp. planktonic Foraminifera. Quite common in Cretaceous sediments, this example shows the very poor wall preservation that is frequently found. Note the keeled chambers.

0.04 mm

Holocene ooze, 1000 meter depth,
Coral Sea, Pacific Ocean

A modern pelagic ooze filled with *Globorotalia*. Globorotalids and globigerinids have spherical chambers, radially porous wall structures and commonly very thick walls. They are an extremely important component of the planktonic assemblage in Late Cretaceous and Tertiary deeper water sediments.

0.17 mm

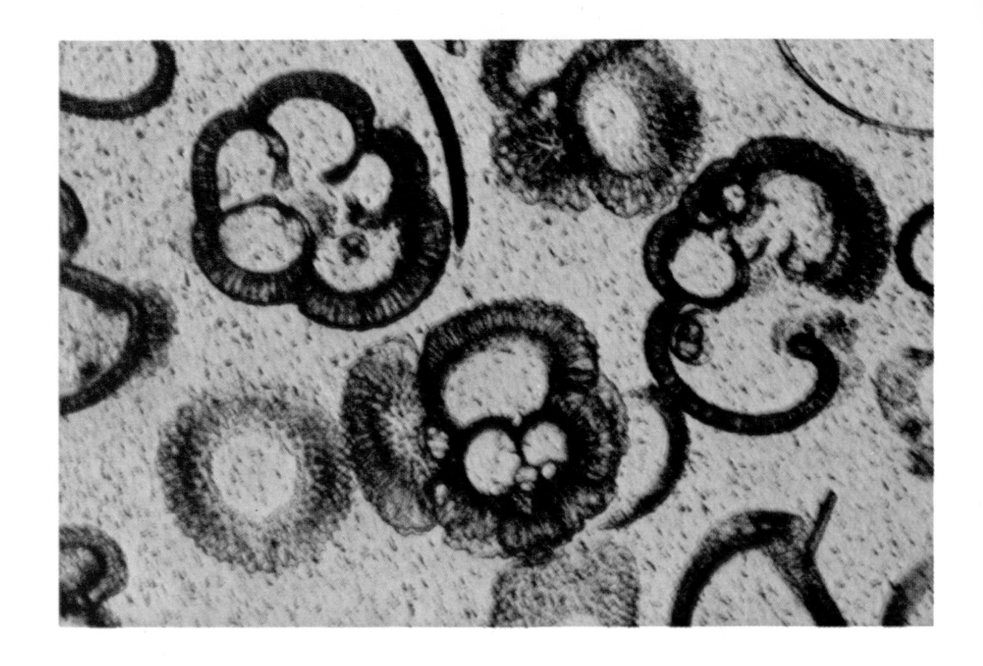

Pleistocene(?) ooze
Offshore Florida (Miami Terrace)

Globorotalid Foraminifera showing well-preserved porous radial wall structure and micrite filling of chambers. These organisms are planktonic and normally are most abundant in outer shelf to oceanic sediments.

X.N. 0.10 mm

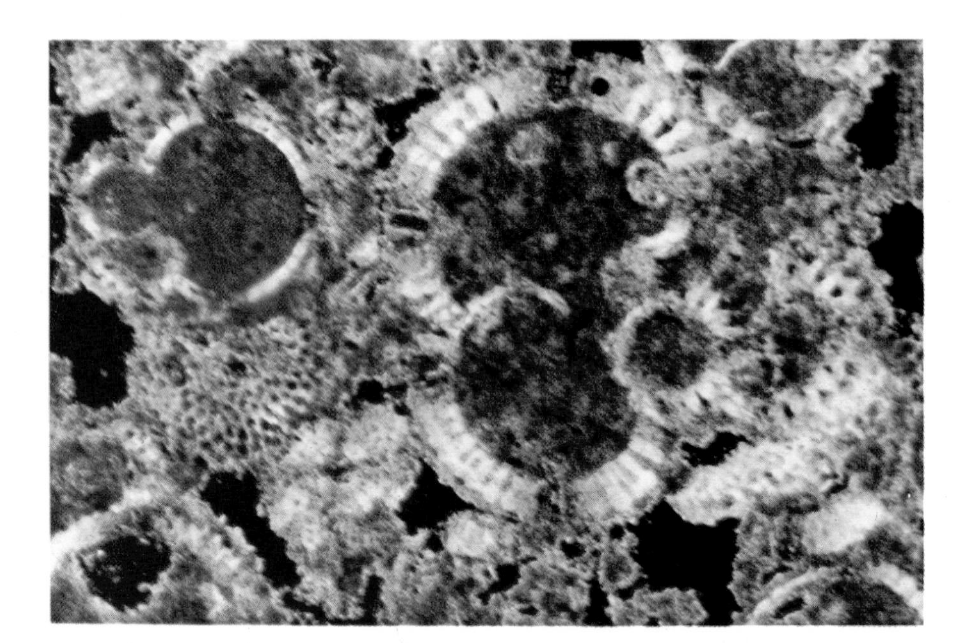

Up. Cretaceous Greenhorn Ls.
Colorado

Thin-walled globigerinid Foraminifera with spar-filled chambers.

X.N. 0.06 mm

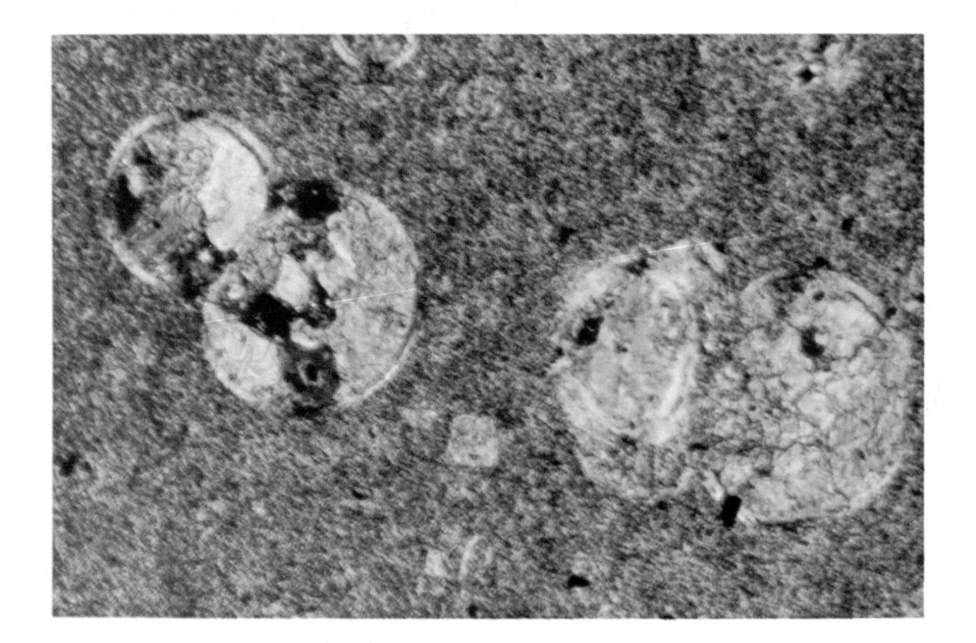

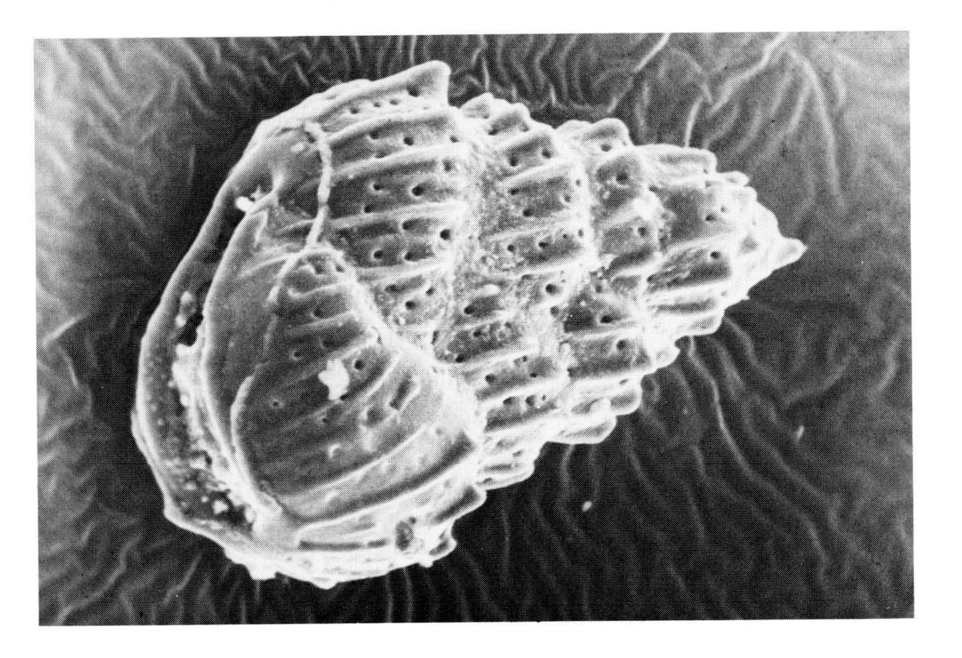

Holocene sediment
Belize (British Honduras)

SEM view of a single specimen of *Sagrina (Bolivina) pulchella.* It shows biserial chambers, pores in walls and external ornamentation.

SEM Mag. 300X 43 μm

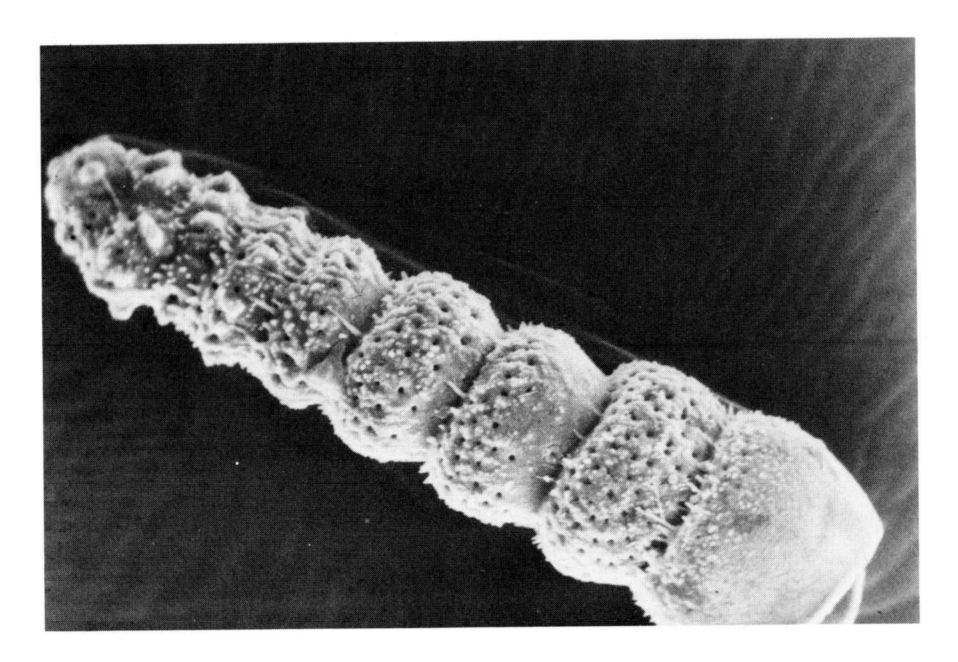

Holocene sediment
Belize (British Honduras)

SEM view of a single specimen of *Rectobolivina adrena.* This uniserial Foraminifera shows porous wall and attached spines.

SEM Mag. 300X 43 μm

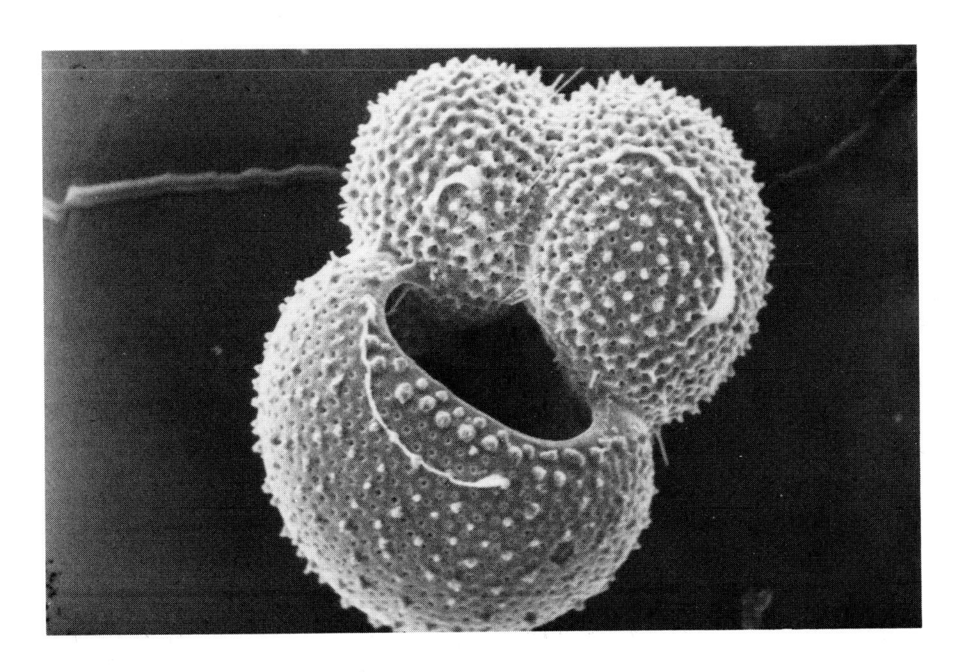

Holocene sediment
Belize (British Honduras)

SEM view of a single specimen of *Globigerinoides rubra,* a planktonic Foraminifera. Globular chambers, small spines and pores in wall, as well as large aperature, are clearly visible.

SEM Mag. 150X 86 μm

Holocene sediment
Belize (British Honduras)

SEM view of a single specimen of *Pyrgo* sp. Shows smooth, non-perforate wall and globular chambers. Next photo shows detail of test surface texture.

SEM Mag. 150X 86 μm

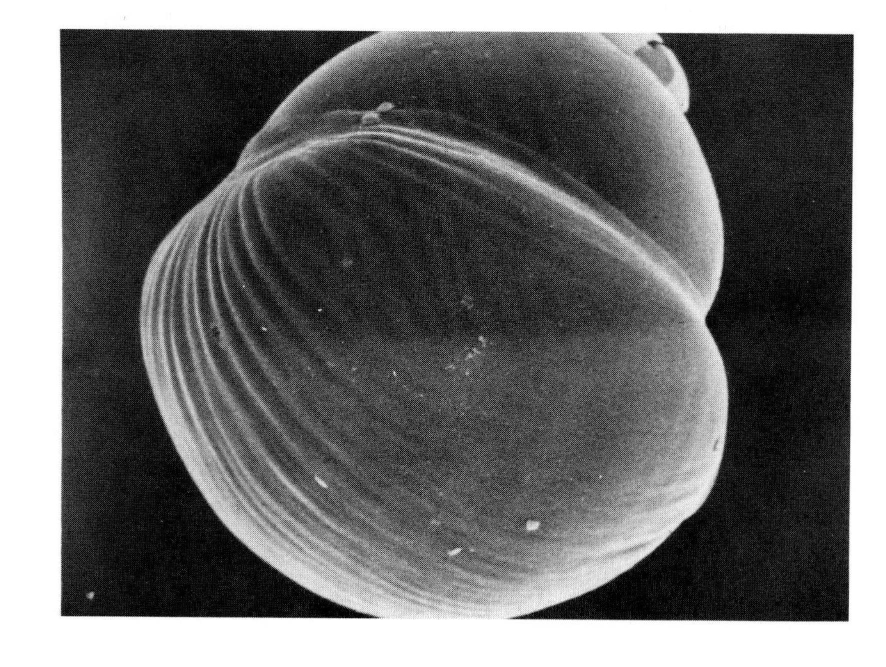

Holocene sediment
Belize (British Honduras)

SEM closeup of *Pyrgo* sp. Note tightly packed, oriented laths of calcite which make up wall. With degradation of the organic binding material, or grinding of the shell by natural processes, these laths could contribute to the production of carbonate mud.

SEM Mag. 10,000X 1.3 μm

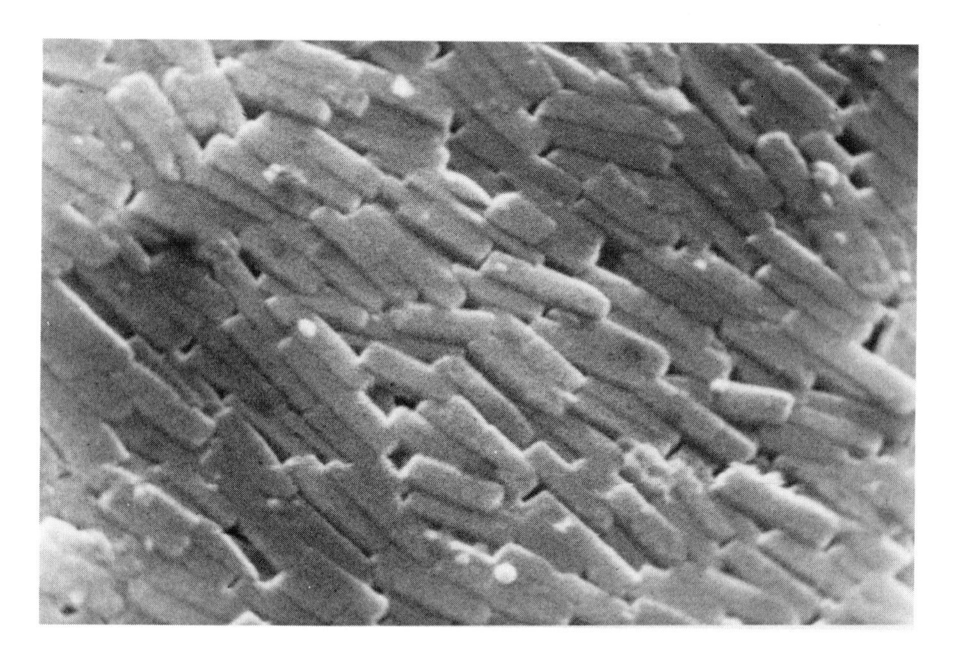

Holocene sediment
Belize (British Honduras)

SEM closeup of *Spiroloculina* sp. View is of a fractured wall showing inner and outer shell surfaces, as well as a cross section of the wall itself. Note the bladed crystals which make up the wall and their changing orientation toward the shell surface.

SEM Mag. 10,000X 1.3 μm

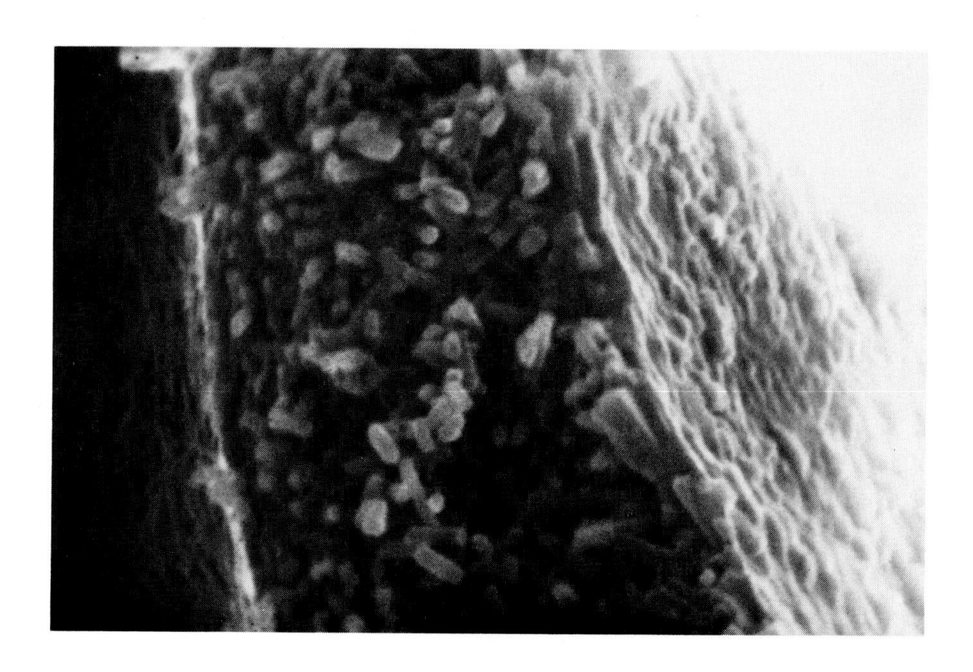

Annelids and Related Groups

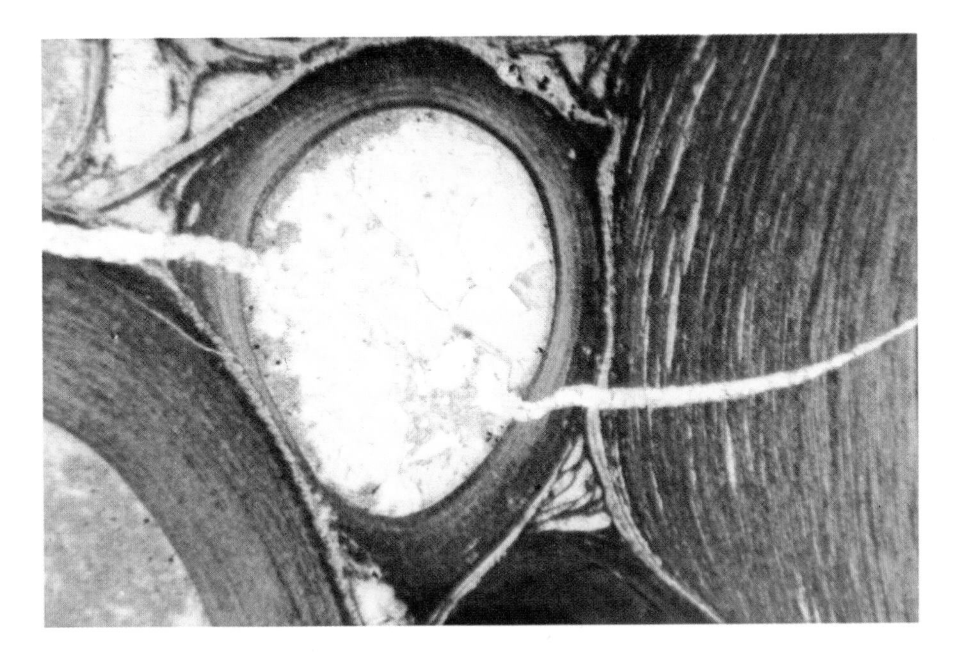

Lo. Cretaceous Glen Rose Ls.
Texas

Large serpulid worm tubes built of concentric laminae of calcite, and often adjoining and overlapping. Elongate, spar-filled voids parallel to laminar wall structure are characteristic of serpulids.

0.38 mm

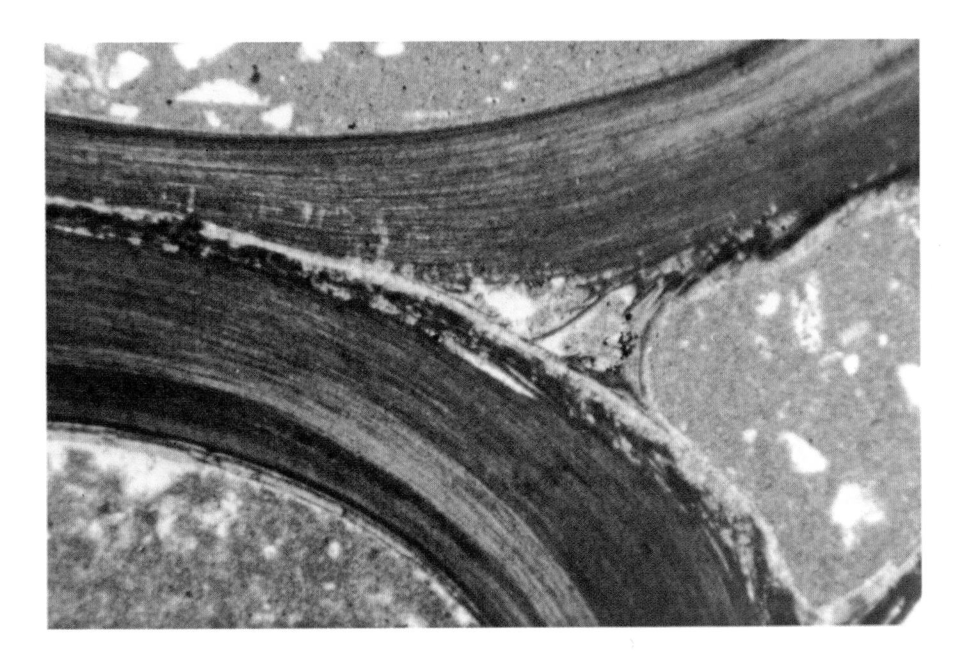

Cretaceous Glen Rose Ls.
Texas

Detail of large serpulid worm tube; shows junction of two tubes and micrite filling of tubes. Note coarse, slightly irregular, concentric lamination.

0.38 mm

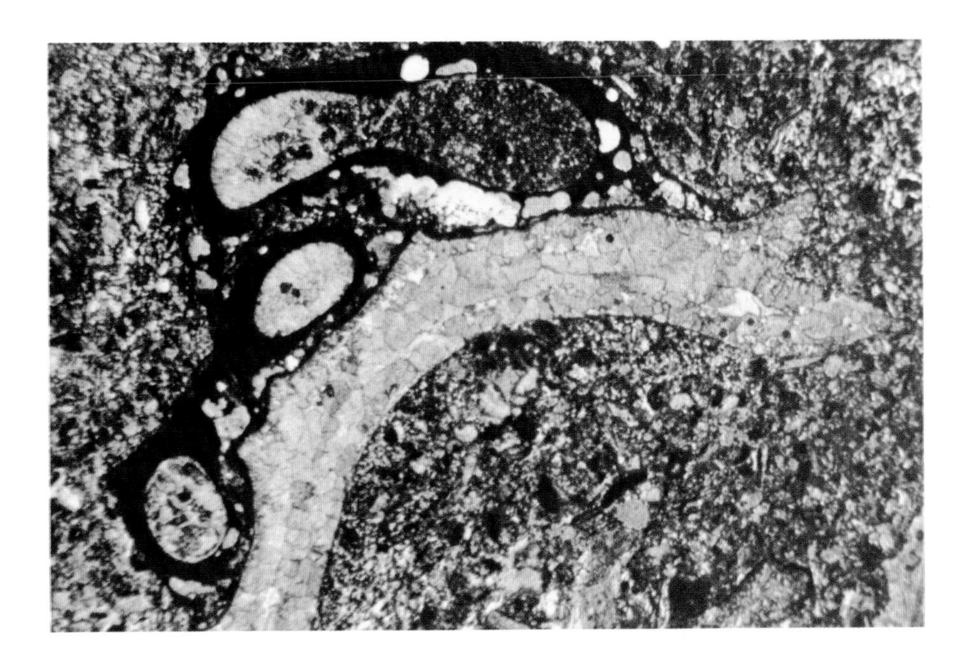

Jurassic Corallian Series, *Trigonia* beds
England

Serpulid worm tubes can be seen encrusting on an altered *Trigonia* (pelecypod) shell. The small, yellow to white patches within the serpulid shell walls are areas of silicification.

1.18 mm

Holocene sediment
Florida

Sabelariid worm tubes encrusting mangrove roots (not seen). These tubes are agglutinated and are visible only as oriented, imbricated grains which surround circular pore space. Green tint is stained impregnating medium.

1.10 mm

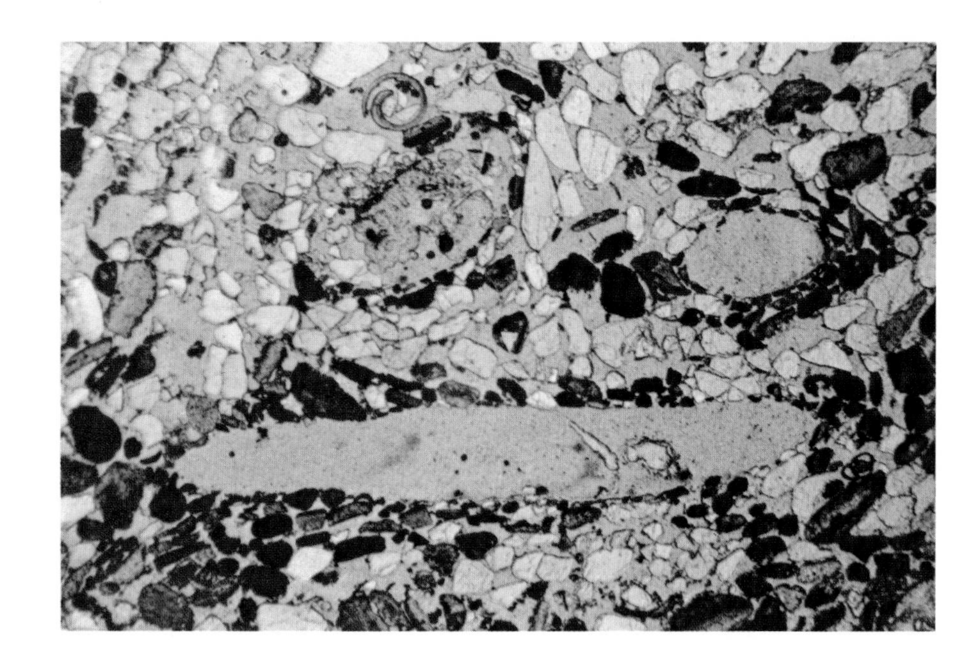

Holocene sediment
Florida

Closeup of a single Sabelariid tube showing the nonselective use of several types of grains, their orientation with long axes tangential to tube outline, and the very sparse binding material. Green-stained areas are filled with stained impregnating medium.

0.34 mm

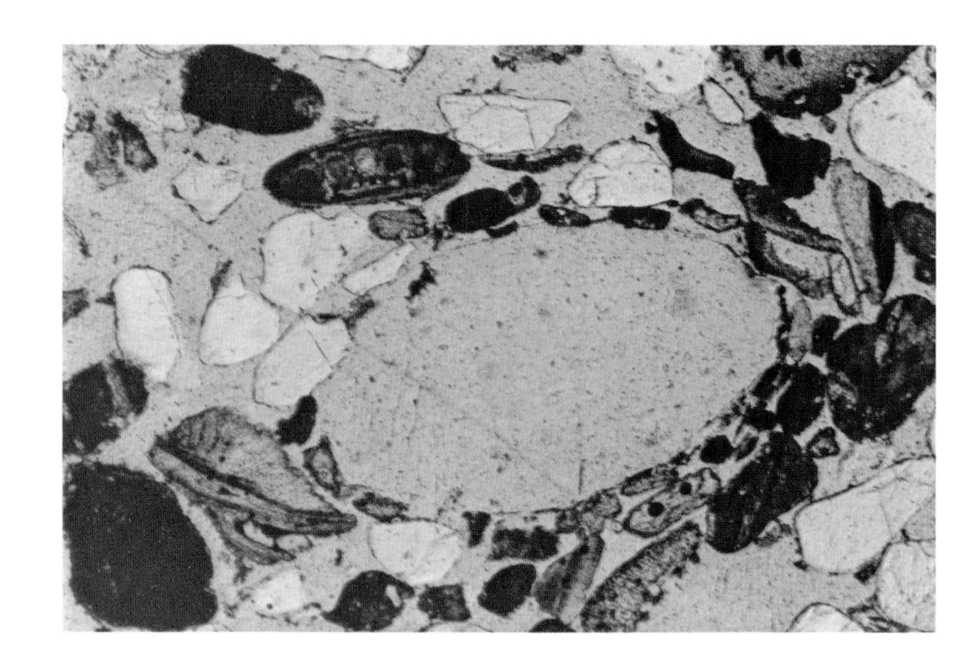

Up. Devonian Genundewa Ls.
New York

Abundant examples of *Stylolina fissurella*. These are small, conical fossils which are classed as conulariids and placed with the Annelids and other worms by some paleontologists. Stylolina and other similar forms are important rock-formers and index fossils at some levels in the Paleozoic.

0.11 mm

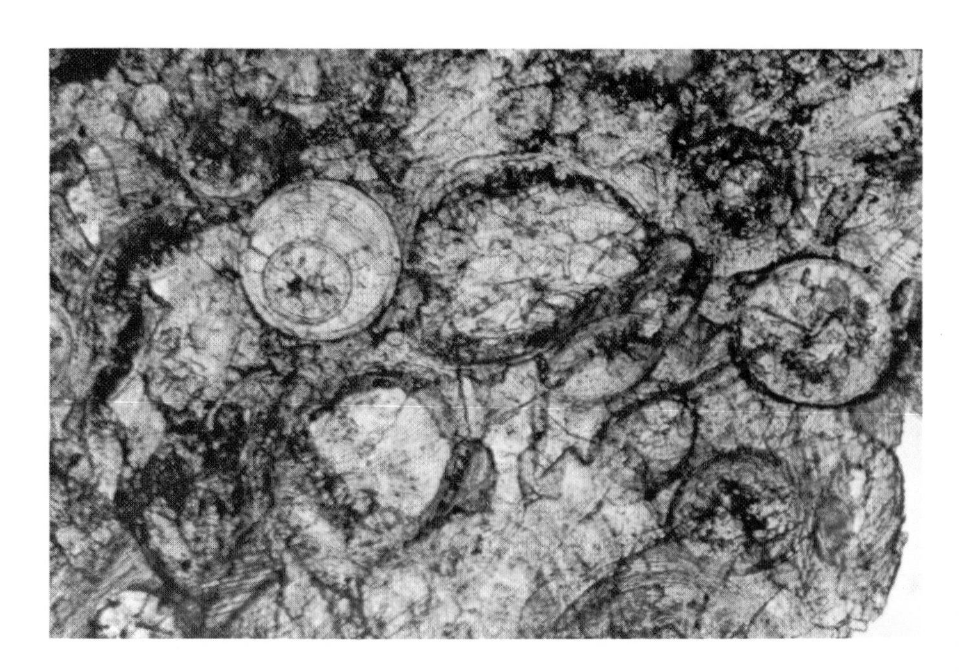

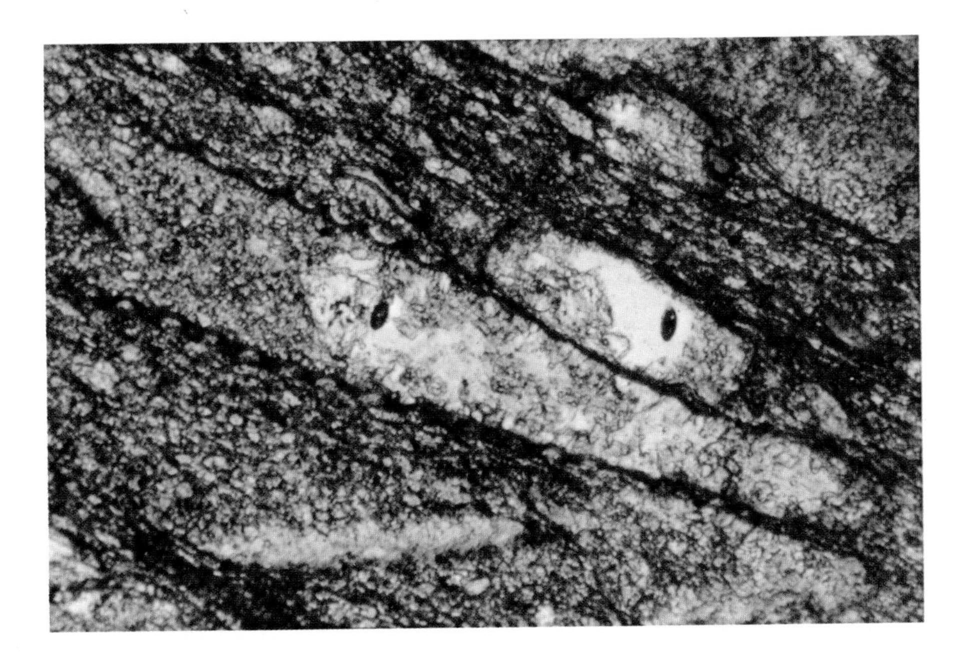

Devonian Tentaculiten Knottenkalk
Germany

A cut through a tentaculite parallel to the long axis. Shows conical shape and crenulate or corrugate exterior. These are similar to *Stylolina* except for the external ornamentation, and are classed by some as belonging to the worms; others place them with the mollusks.

0.11 mm

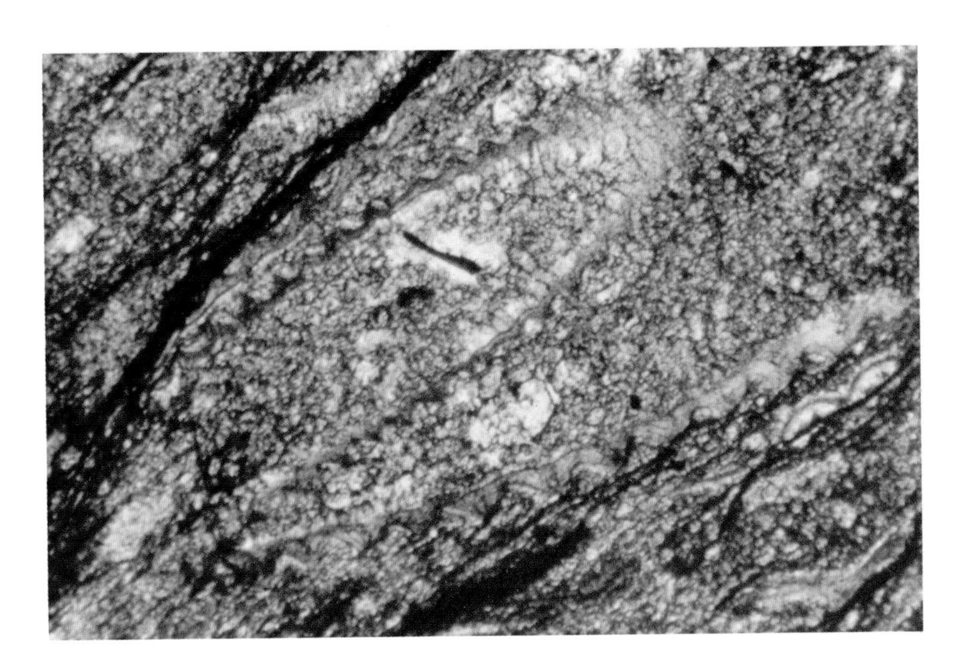

Devonian Tentaculiten Knottenkalk
Germany

Several oblique cuts through *Tentaculites* sp. showing the pronounced corrugation of the shell. Also note wavy, laminated shell structure.

0.11 mm

Archaeocyathids and Sponges

Cambrian Ajax Ls.
Australia

Cross section of an Archaeocyathid wall. Central cavity is at the top with double wall in center and external sediment at bottom. The inner and outer walls confine a series of radial compartments. These organisms may be related to sponges (although they are commonly placed in a separate phyllum), were restricted to the Cambrian, and built reefs or bioherms in many areas.

0.64 mm

Cambrian Ajax Ls.
Australia

Cross section of an Archaeocyathid wall with central cavity in upper right. Shows radial compartments formed by double wall structure with septate partitions.

0.25 mm

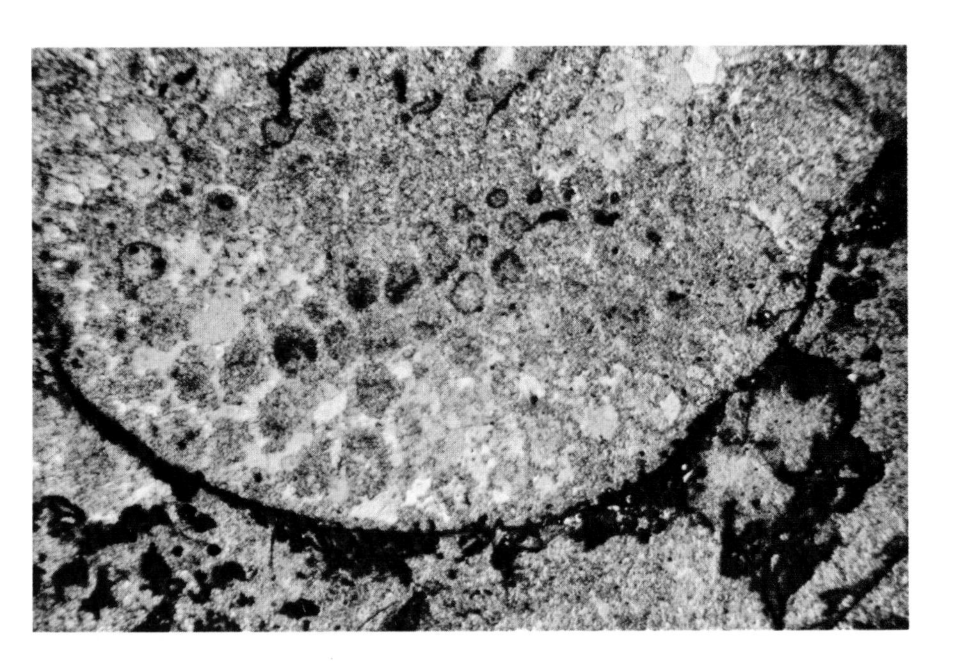

Up. Permian Capitan Ls.
NewMexico

A large calcareous sponge with encrusting blue-green algae. As with most sponges, skeletal preservation is poor but vague outlines of chambers can be seen. These sponges were major contributors to the Permian reef complexes of New Mexico and Texas.

1.24 mm

Lo. Permian Getaway Ls. Mbr. of Cherry Canyon Fm.
Texas

Large, originally calcareous sponge which shows large cavities preserved as voids or with sediment infill. Sponge wall structure has been obliterated by silicification. Irregular chamber shapes and large size aid in identification.

0.64 mm

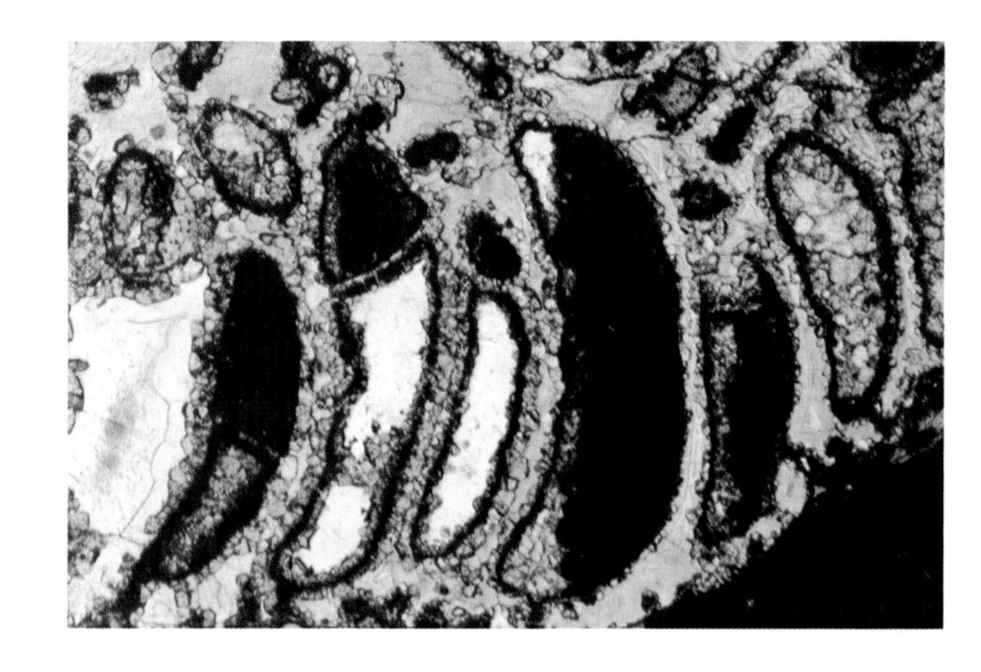

Up. Pennsylvanian Graford Fm.
Texas

A rather well preserved sponge--*Maeandrostia* sp. (a calcisponge). Note central cavity and irregular lateral chambers. Although little of the original wall structure is preserved, the original outlines of wall positions are clearly marked by changes in color due to inclusions in neomorphic calcite.

0.64 mm

Lo. Ordovician part of El Paso Group
Texas

A part of a large calcareous sponge showing irregular canals filled with micrite. Original wall structure has been completely obliterated through dissolution and reprecipitation of sparry calcite. The vague, meandering structural pattern is typical of many sponges.

0.64 mm

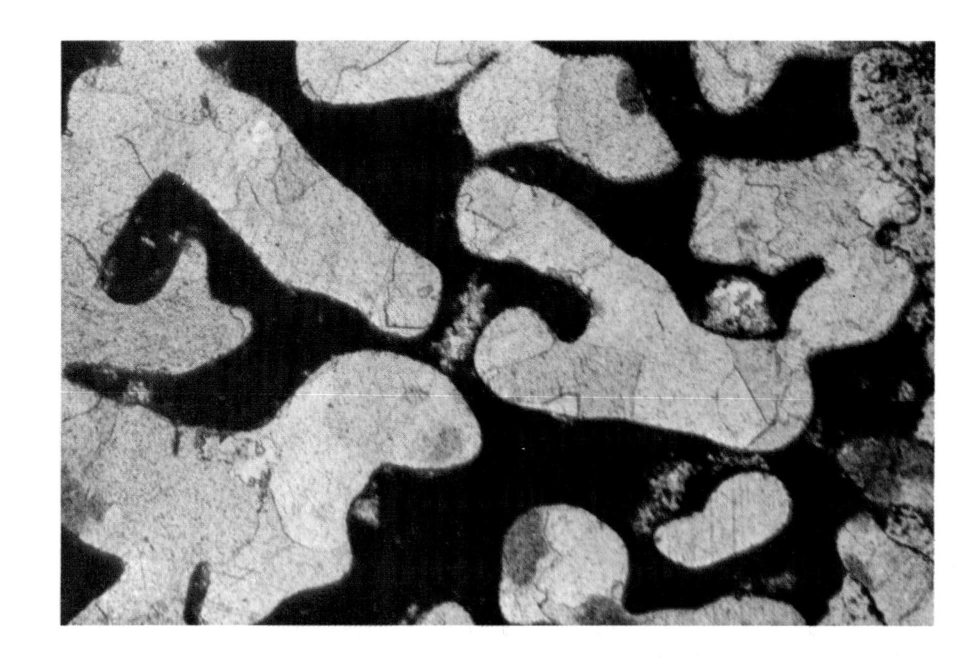

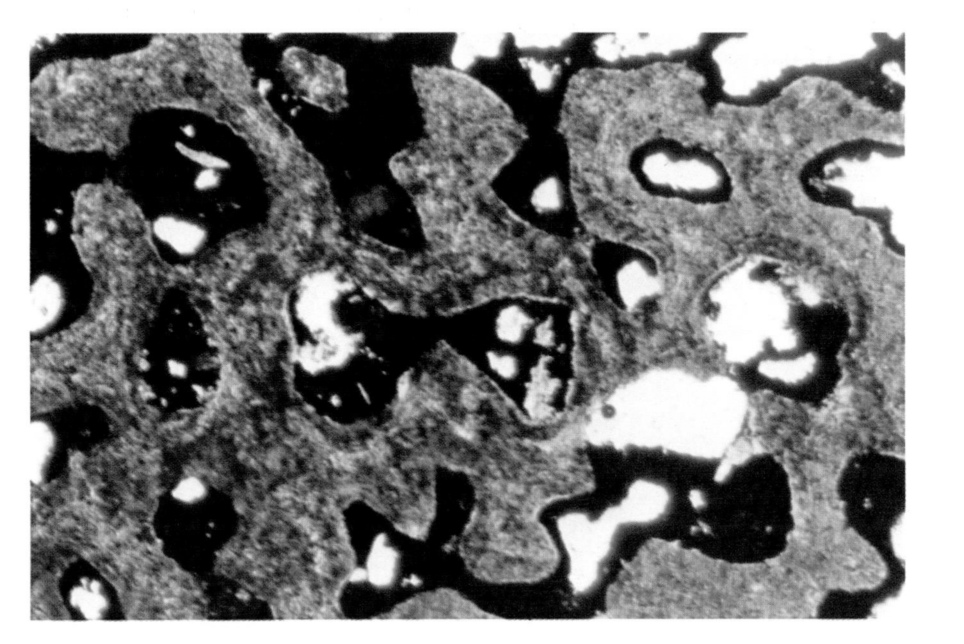

Lo. Cretaceous Lower Greensand
England

A cross section of a small portion of the calcareous sponge *Raphidonema farringdonense*. Note meandering canal structure and well-preserved wall structure. Canals are partly filled with hematite.

0.38 mm

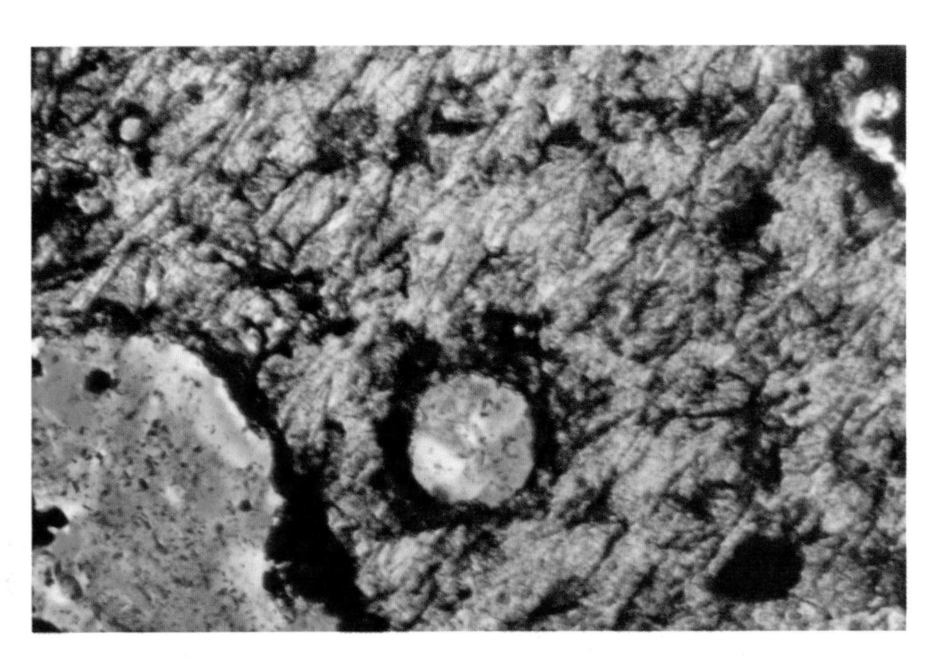

Holocene reef sediment
Jamaica

Longitudinal section of a modern sclerosponge (*Ceratoporella* sp.(?)). This organism contributes greatly to the deeper-water parts of modern Caribbean reefs. Skeleton is aragonite arranged in conical bundles. These organisms have been grouped with the sponges, coelenterates, or stromatoporoids.

G.P. 0.38 mm

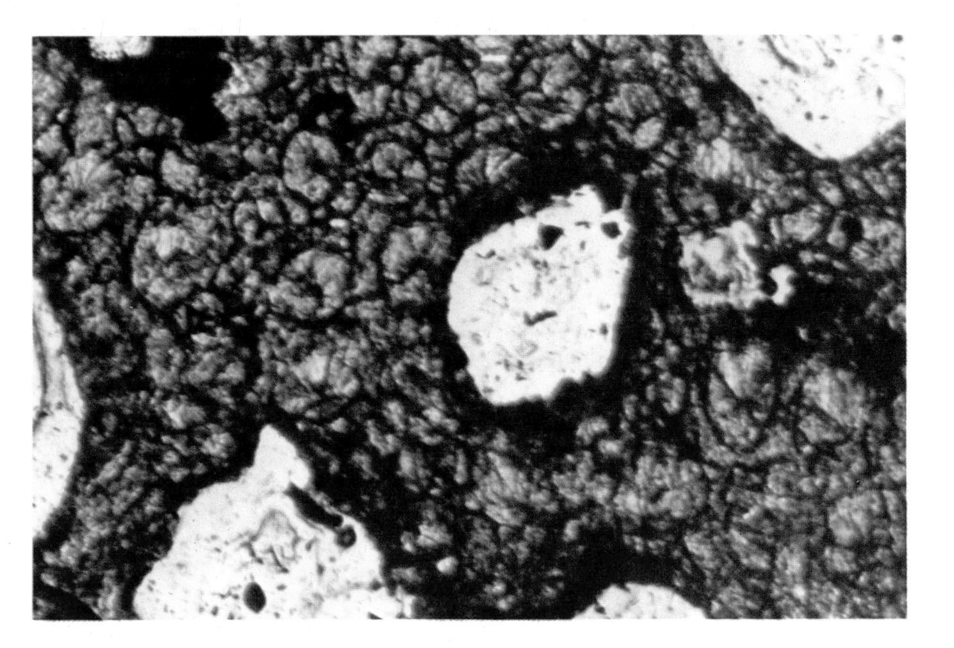

Holocene reef sediment
Jamaica

Transverse section of the same specimen as above. Note circular outlines of aragonitic fiber bundles in this section.

0.38 mm

Silurian Brownsport Gp.
Tennessee

Interlocking six-rayed (in this plane) spicules of the sponge *Astraeospongia meniscus.* Each spicule is a single calcite crystal with unit extinction; interlocking spokes make up the skeletal framework of the sponge.

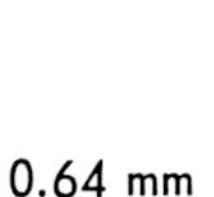

0.64 mm

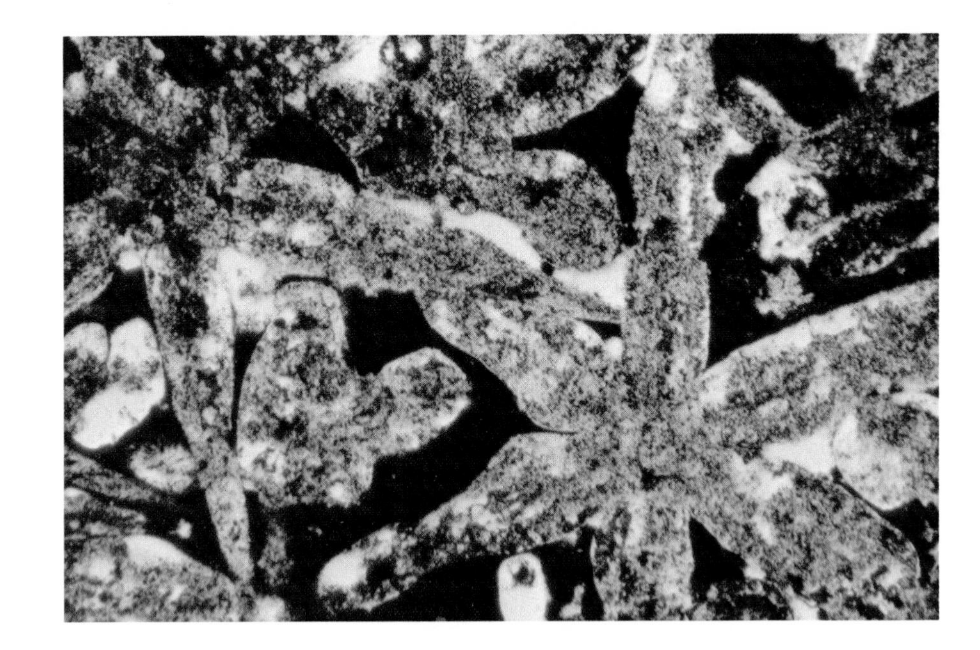

Lo. and Mid. Pennsylvanian Marble Falls Ls.
Texas

Sponge spicules in a clayey matrix. Most of these spicules are monaxons and show characteristic central canal and spicule shape. All were originally opaline silica; now some are chert, some calcite. Not all sponge spicules have a central canal, however.

0.24 mm

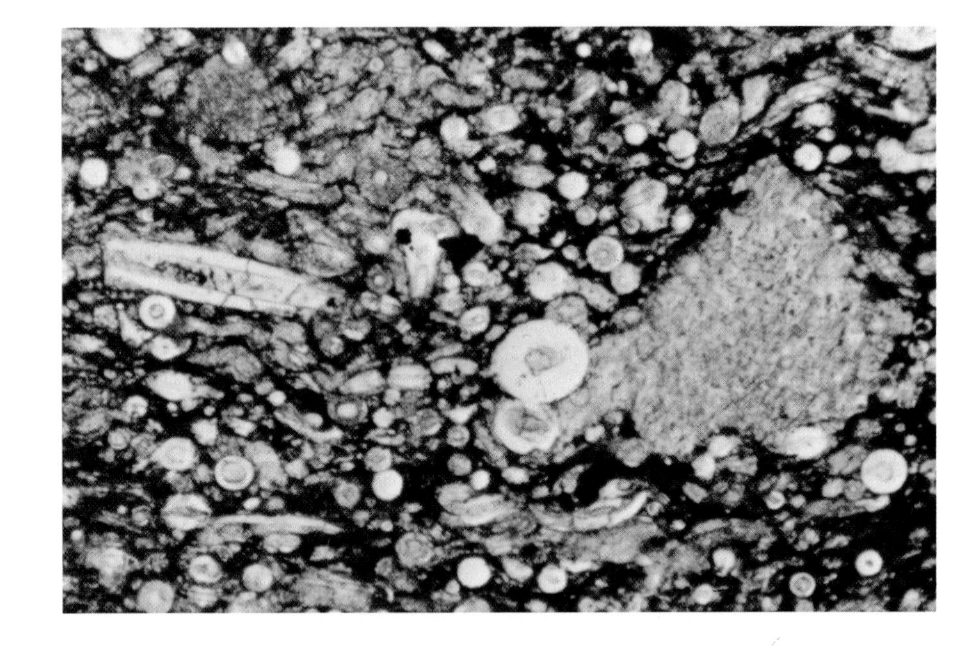

Lo. and Mid. Pennsylvanian Marble Falls Ls.
Texas

Sponge spicules. Shows same area as previous picture but under polarized light. Chert and calcite replacements of spicules are visible.

X.N. 0.24 mm

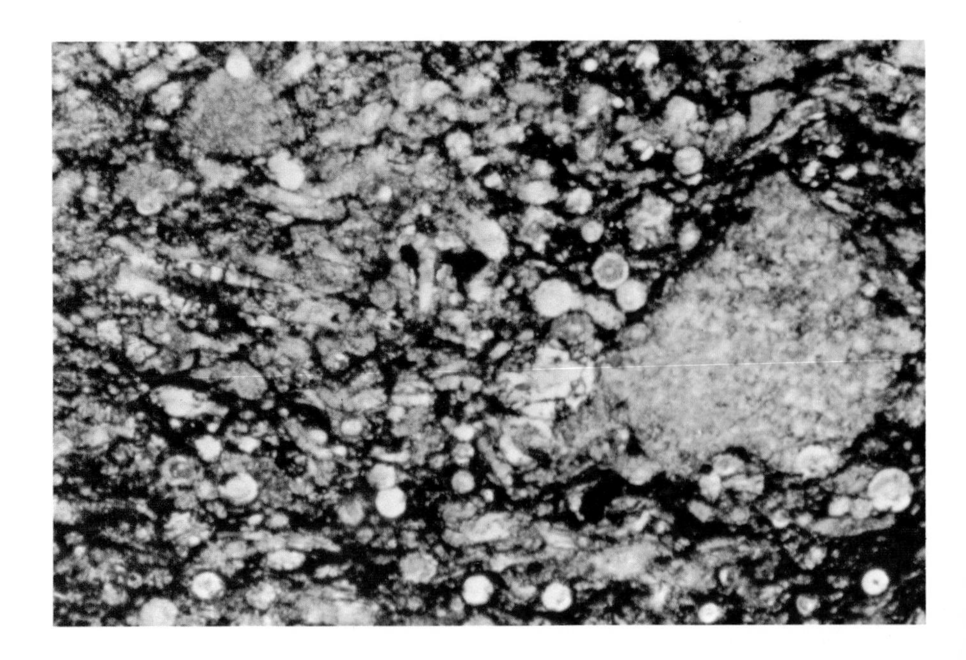

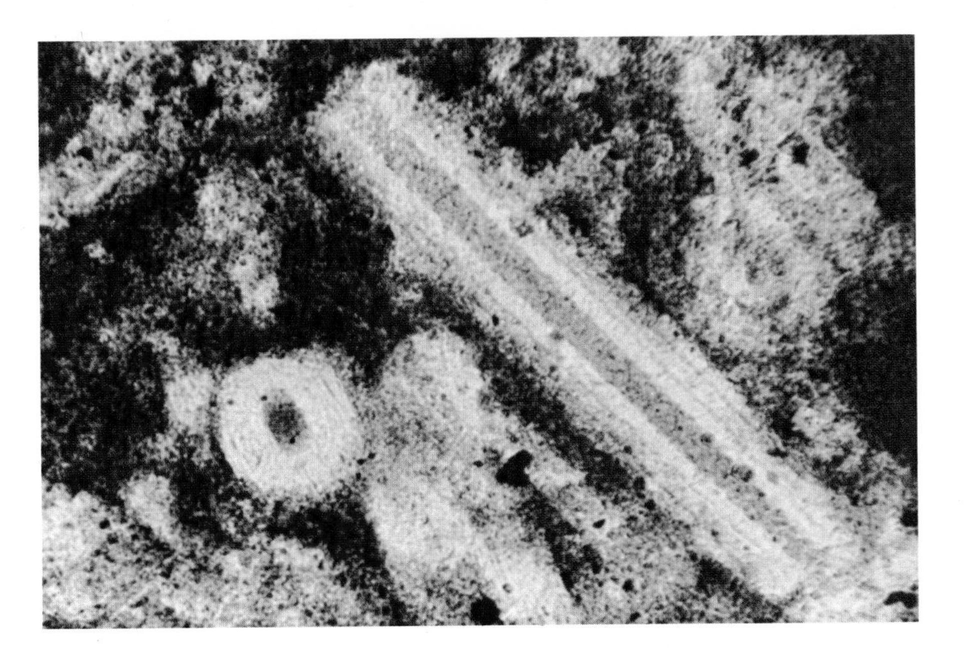

Up. Jurassic or Lo. Cretaceous Radiolarite
Muscat and Oman

Cross sections of Monaxon sponge spicules. One is a transverse section across a spicule, the other parallel to the long axis of a spicule. Note the thick walls and central canal characteristic of many originally opaline spicules.

0.04 mm

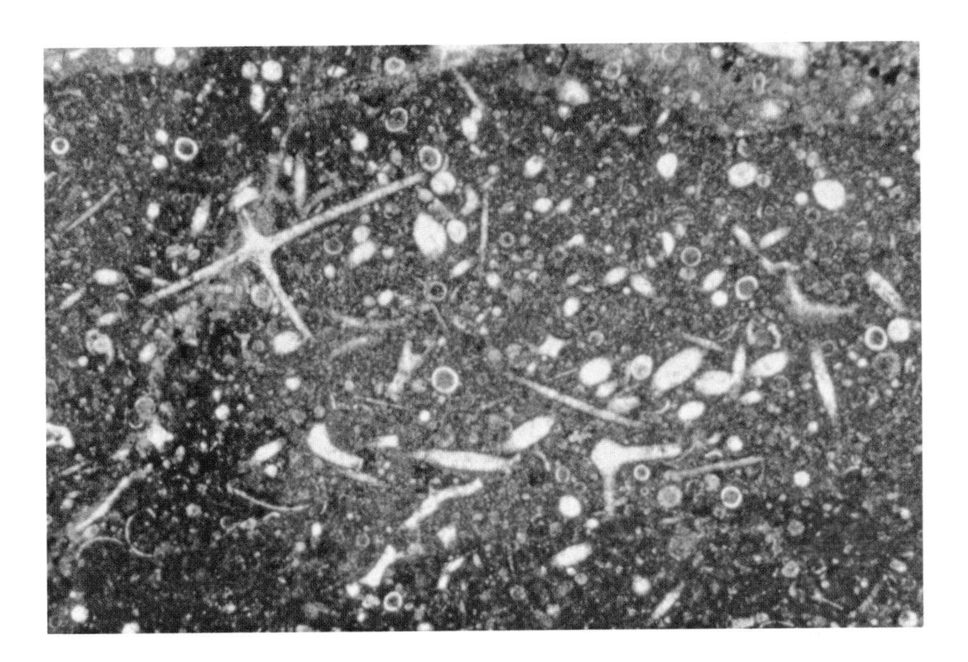

Up. Cretaceous White Limestone
Northern Ireland

A spicule-rich sediment showing outlines of sponge spicules (which are now found as voids). Note the mixture of monaxons and multiaxoned spicules. The grains with central fillings are calcispheres and planktonic Foraminifera.

0.25 mm

Holocene sediment
Belize (British Honduras)

SEM view of a modern siliceous sponge showing smooth, interlocked, diversely oriented opaline spicules. Upon death of the sponge, these spicules can be widely dispersed and loose spicules are commonly found in modern sediments from shallow-water and deep-water environments.

SEM Mag. 300X 43 μm

Stromatoporoids

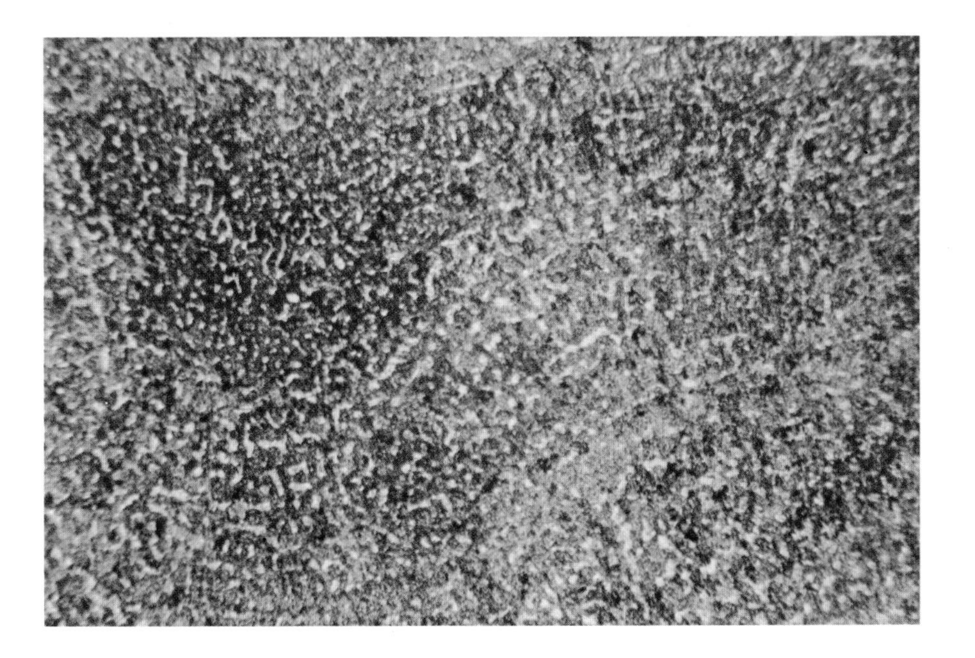

Up. Devonian ls.
Iowa

An *Actinostroma* sp.(?) stromatoporoid showing the general undulating, head-shaped colonial form with pore structures paralleling the exterior of the colony as well as pores which run perpendicular to the exterior. This yields a very crude and irregular box-work structure seen in more detail in the next photograph.

1.18 mm

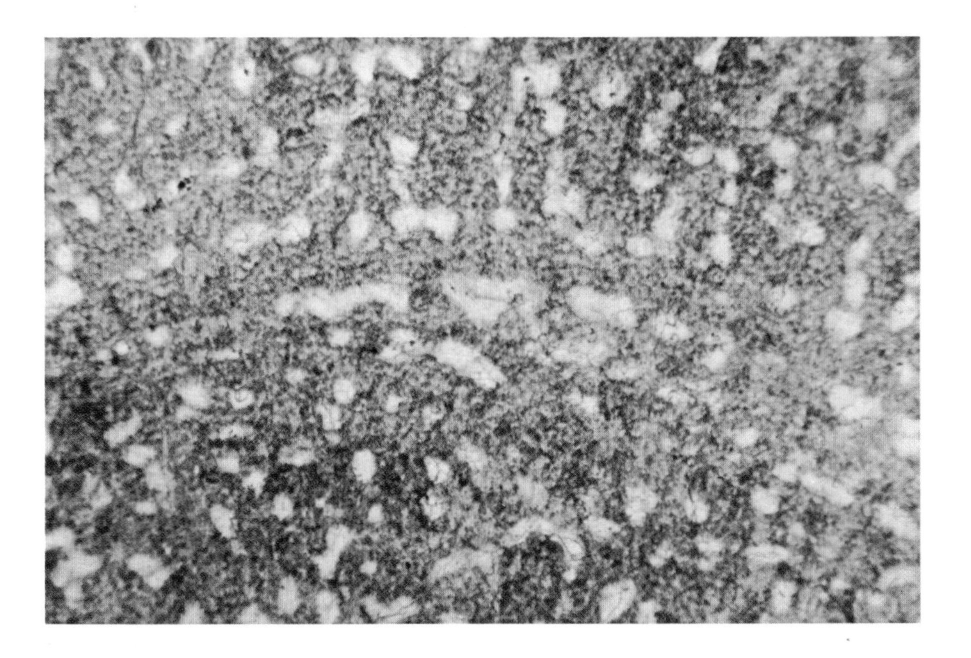

Up. Devonian ls.
Iowa

An *Actinostroma* sp.(?) stromatoporoid showing sheeted structure with regular and irregular open spaces. This colony has a head-shaped form. Stromatoporoids are major rock forming elements in many Upper Paleozoic limestones and contributed extensively to the framework formation of reefs and bioherms.

0.38 mm

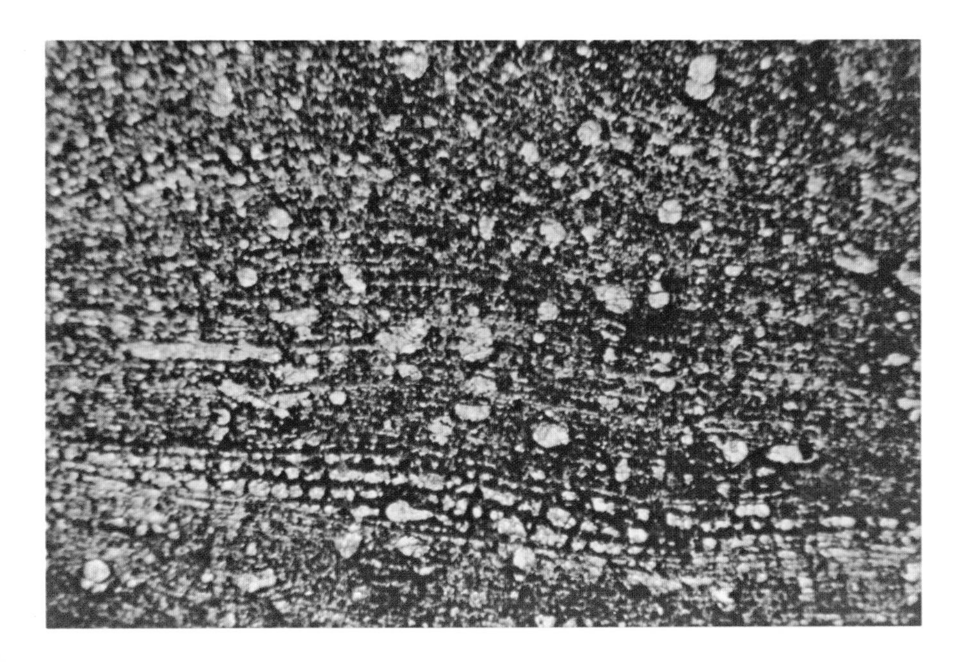

Up. Silurian Cobleskill Ls.
New York

A stromatoporoid (*Stromatopora constellata*) showing layered boxwork structure plus ovoid-to-round spar-filled holes. Preservation average to slightly better than average.

1.18 mm

Cretaceous El Abra Fm.
Mexico

A probable sphaeractinid (e.g. *Parkeria*) or other stromatoporoid-related hydrozoan. They occur as spherical, marble-to-golf-ball-sized nodules in the near-back-reef facies. A coarse pore structure and a tubular wall texture are visible.

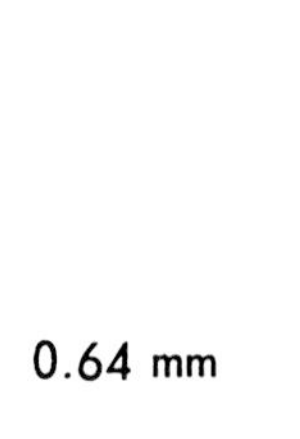

0.64 mm

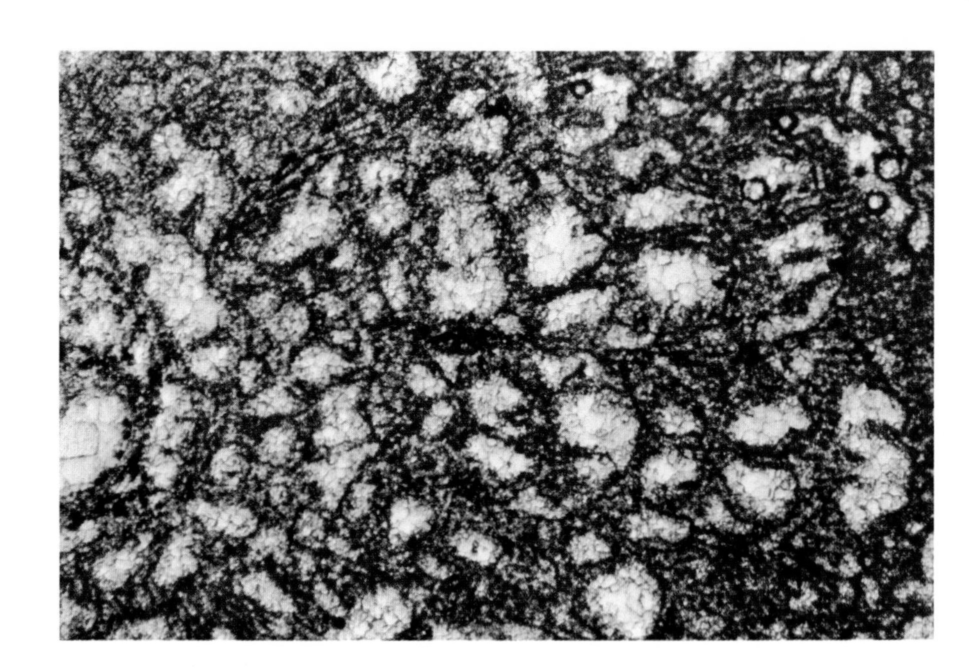

Corals

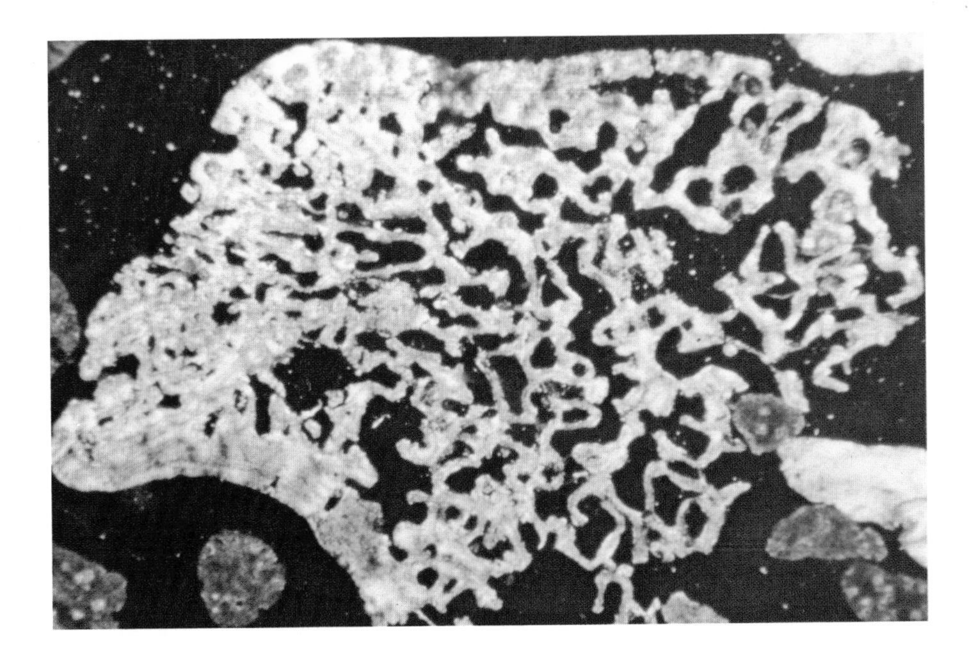

Holocene back-reef beach sand
Belize (British Honduras)

Relatively unaltered (but rounded) Scleractinian coral fragment illustrating high initial intragranular porosity. Note irregular corallite shapes in this section.

X.N. 0.38 mm

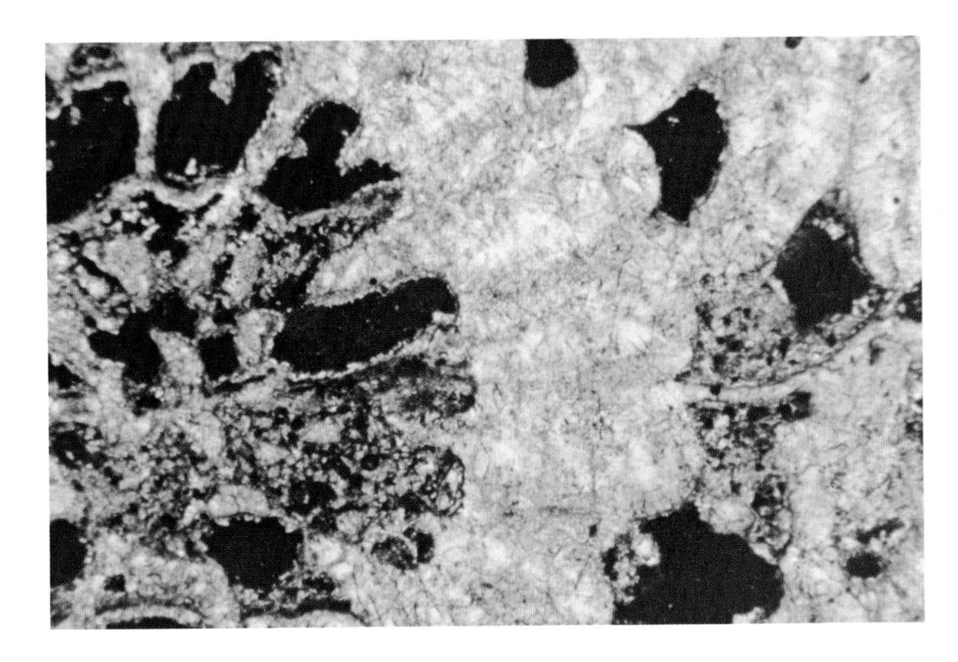

Pleistocene back-reef ls.
Belize (British Honduras)

Closeup of a segment of a large Scleractinian coral showing original septa (central dark line with surrounding trabecular structure) and partial filling of living chambers. Considerable recrystallization has occurred even in samples as young as Pleistocene.

X.N. 0.38 mm

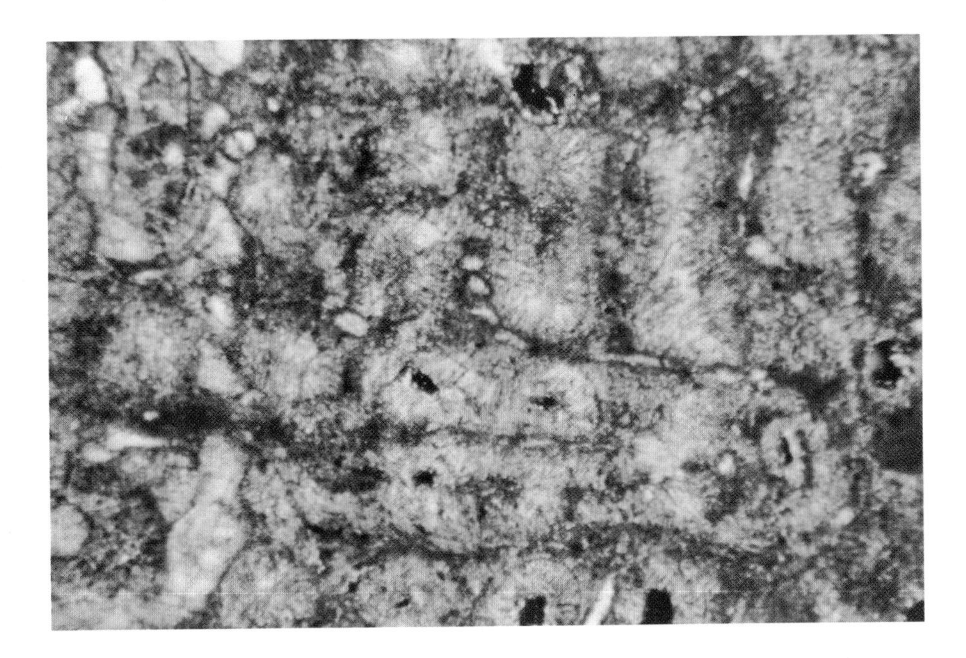

Pleistocene Key Largo Ls.
Florida

Segment of a large Scleractinian coral head showing growth lines with trabecular (fibrous bundle) structure and complete filling of intragranular porosity by carbonate cement.

X.N. 0.24 mm

Paleocene (Danian) Fakse Ls.
Denmark

Oblique section through the Scleractinian coral *Dendrophyllia candelabra* showing radial arrangement of septa. Wall structure entirely obliterated during inversion from aragonite to calcite. Recognizable on the basis of the shape of internal chambers and septa as well as their size.

0.64 mm

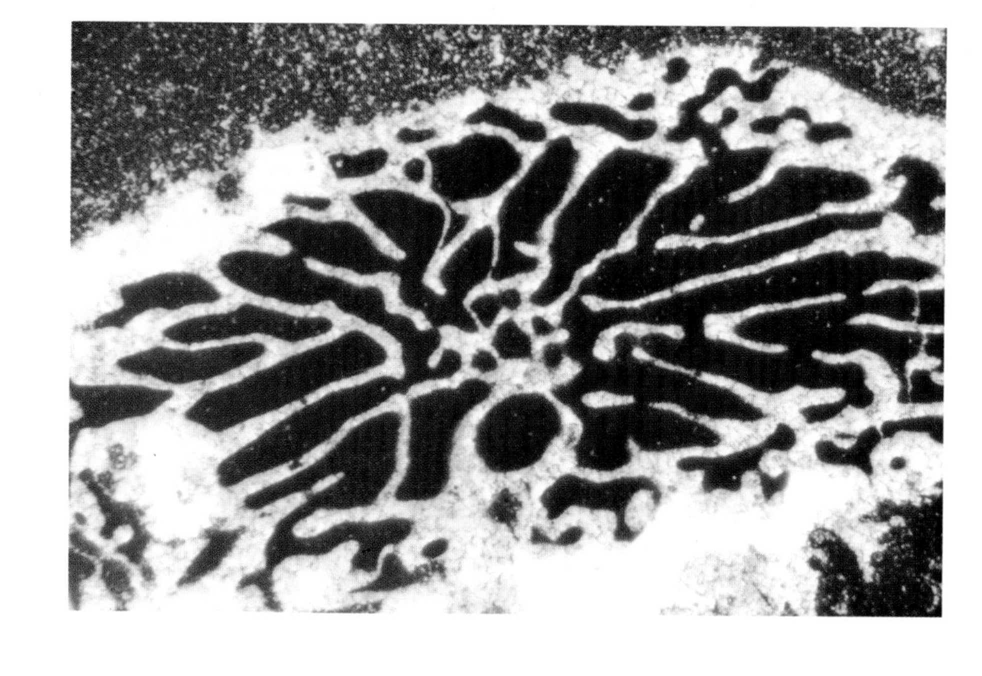

Holocene sediment
Belize (British Honduras)

SEM view of a broken section of *Montastrea annularis.* The trabecular structure with crystals radiating out from points can be seen. This grain is entirely aragonitic and has seen virtually no alteration.

SEM Mag. 800X 16 μm

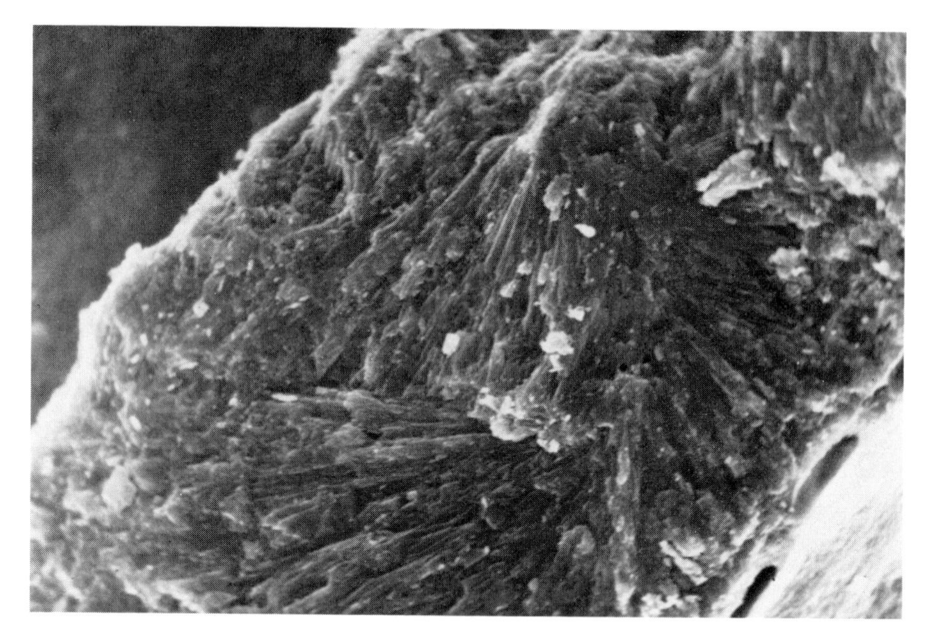

Holocene sediment
Belize (British Honduras)

SEM view of the margin of chamber of *Montastrea annularis* showing the early diagenetic cementation of the porosity of corals. Long needles of aragonite are seen growing into the chamber while algal(?) tubes add another small component. Such early cementation may allow preservation of some of the wall structure during later diagenesis.

SEM Mag. 1000X 13 μm

Holocene sediment
Belize (British Honduras)

SEM view of broken portion of the scleractinian coral *Agaricia agaricides*. One can see bundles of bladed crystals which make up the trabecular wall structure.

SEM Mag. 3000X 4.3 μm

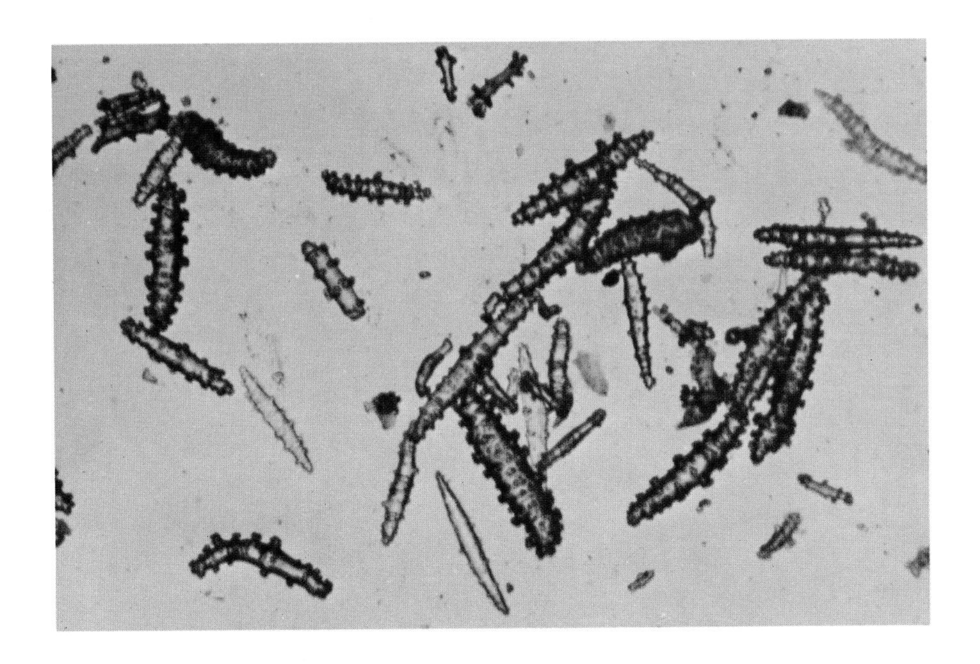

Holocene sediment
Florida

The clear and purple structures are spicules which normally are embedded in the tissue of Gorgonians (Alcyonaria). This is a clorox residue of a single coral colony and shows the types of spicules which may be released to the sediment upon decay of the organic matter of Gorgonians.

0.24 mm

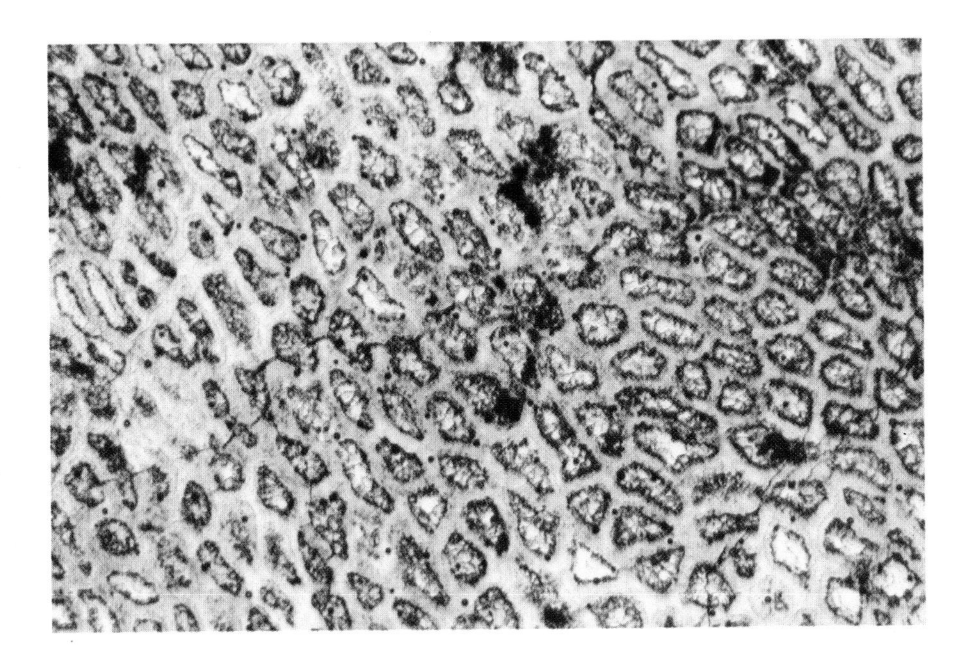

Mid. Pennsylvanian Paradox Fm.
Utah

A completely silicified tabulate coral. In this section one does not see the tabulae. Only the elliptical to circular tubes which are growing closely-spaced and parallel to one another are seen.

0.64 mm

Up. Pennsylvanian Graham Fm.
Texas

Cross section of a solitary rugose coral *Lophophyllidium proliferum*. The simple, radiating septa and central (or axial) complex are well preserved and show outlines of original septal texture with trabecular structure. All these features, as well as the large size of the corallites, help to differentiate coral and bryozoan fragments.

0.64 mm

Up. Devonian ls.
Iowa

A portion of a large colonial rugose coral--*Pachyphyllum woodmani*. This example shows moderate preservation of radiating septa and the axial complex. The Rugosa are especially important as reef and bioherm builders in Devonian to Pennsylvanian time.

1.24 mm

Up. Devonian ls.
Iowa

Section of the colonial coral *Pachyphyllum woodmani* showing differential preservation of main septal walls and crosswalls.

X.N. 0.38 mm

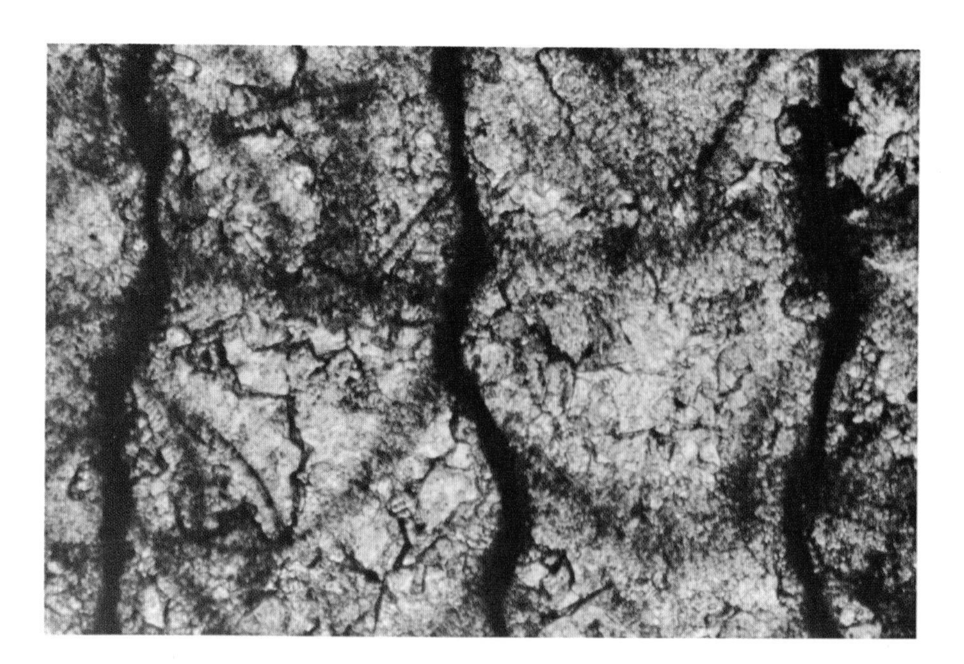

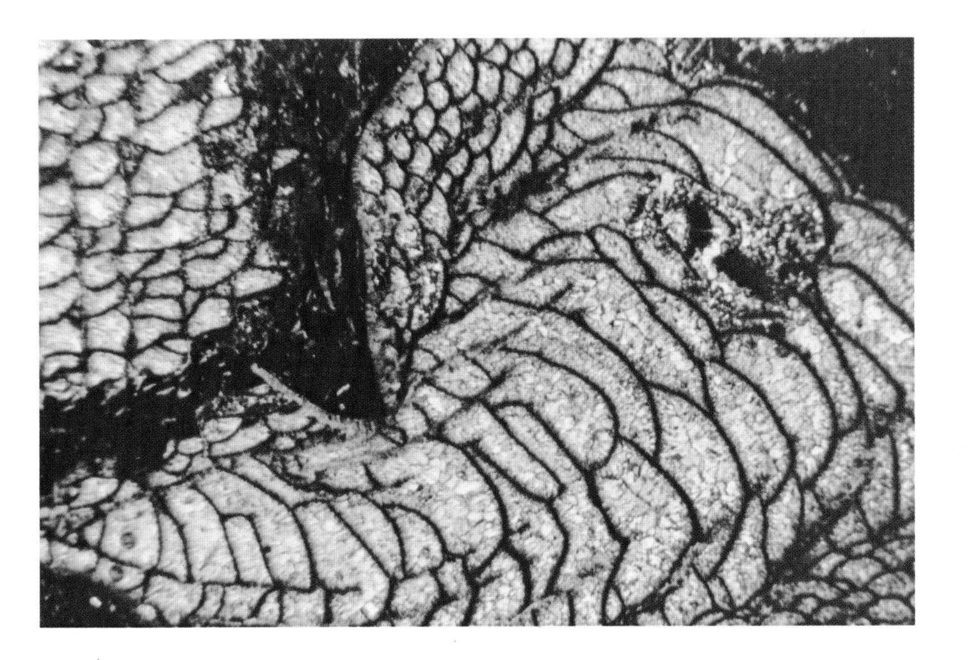

Lo. Carboniferous Coral Ls.
England

Large solitary rugose coral in oblique cut. See subsequent photos for more detailed views of wall structure. Large corallites, radiating septa and dissepiments, variations in corallite shape and differences in thickness of various types of walls are used to distinguish corals from superficially similar bryozoans.

1.24 mm

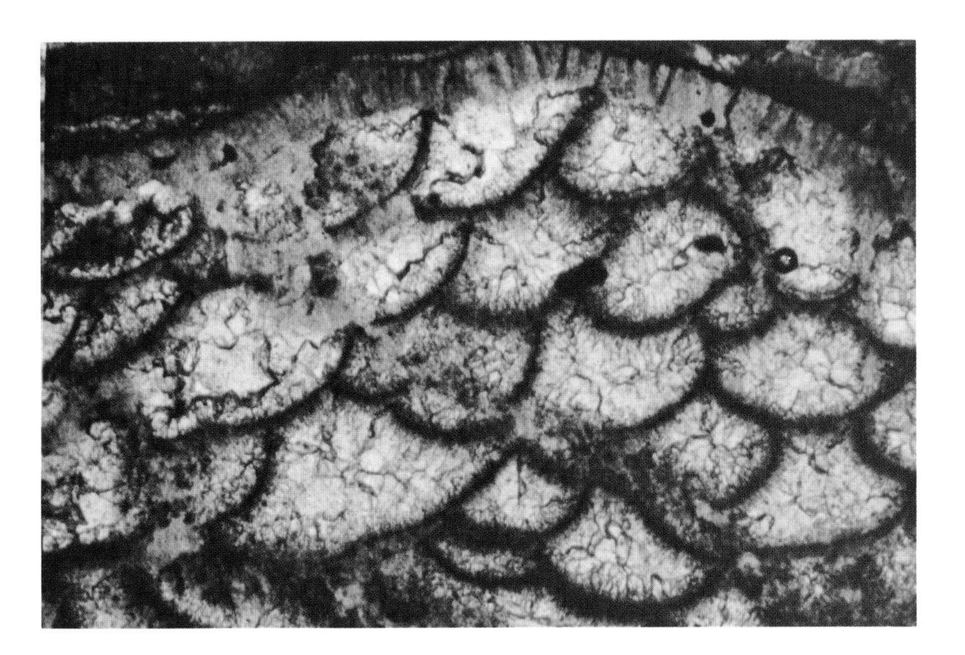

Lo. Carboniferous Coral Ls.
England

Closeup of large solitary coral. Has thin, dark growth lines of original dissepiments plus spar fillings of chambers. Note thicker wall in upper right-hand corner; this is characteristic of coral skeletons.

X.N. 0.38 mm

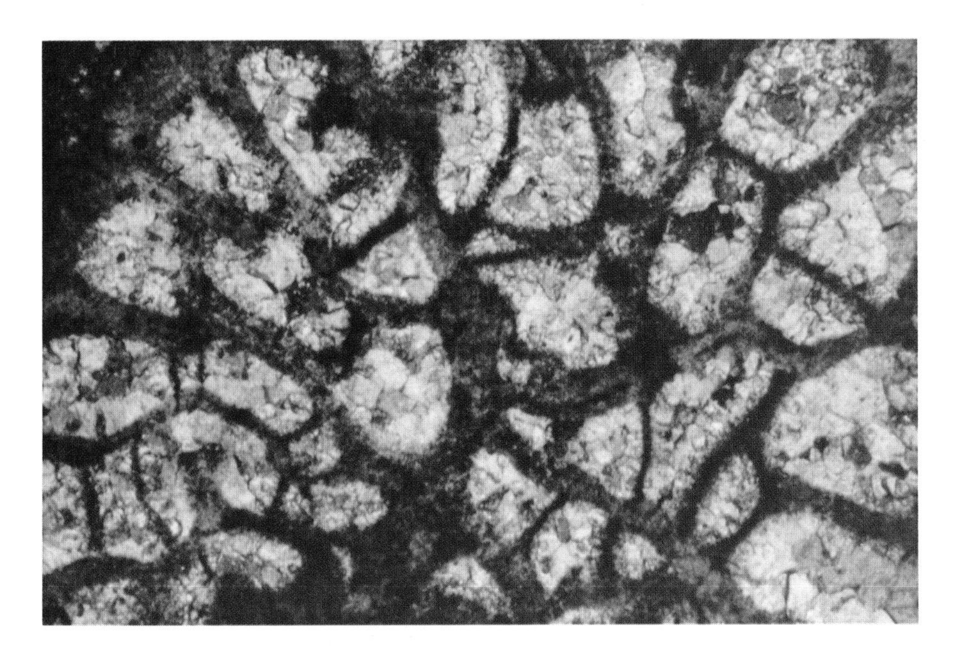

Lo. Carboniferous Coral Ls.
England

Closeup of large solitary coral showing irregular chamber shapes and dissepiments of varying thickness. Sparry calcite fills chambers.

X.N. 0.38 mm

Bryozoans

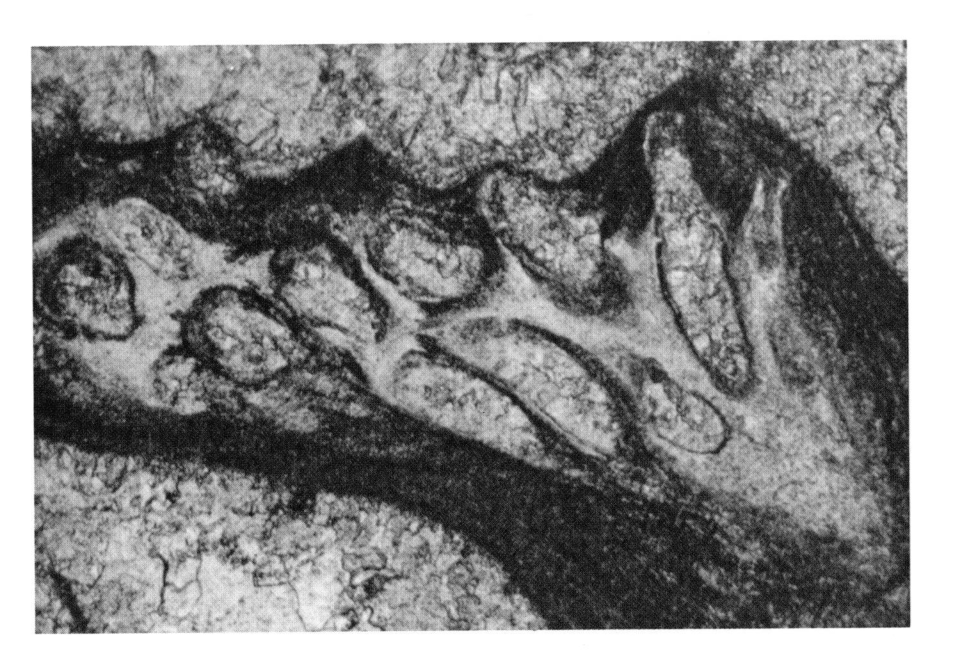

Up. Permian Capitan Ls.
New Mexico

Large bryozoan frond with regular circular to elongate holes (zooecia) and finely fibrous wall structure.

0.38 mm

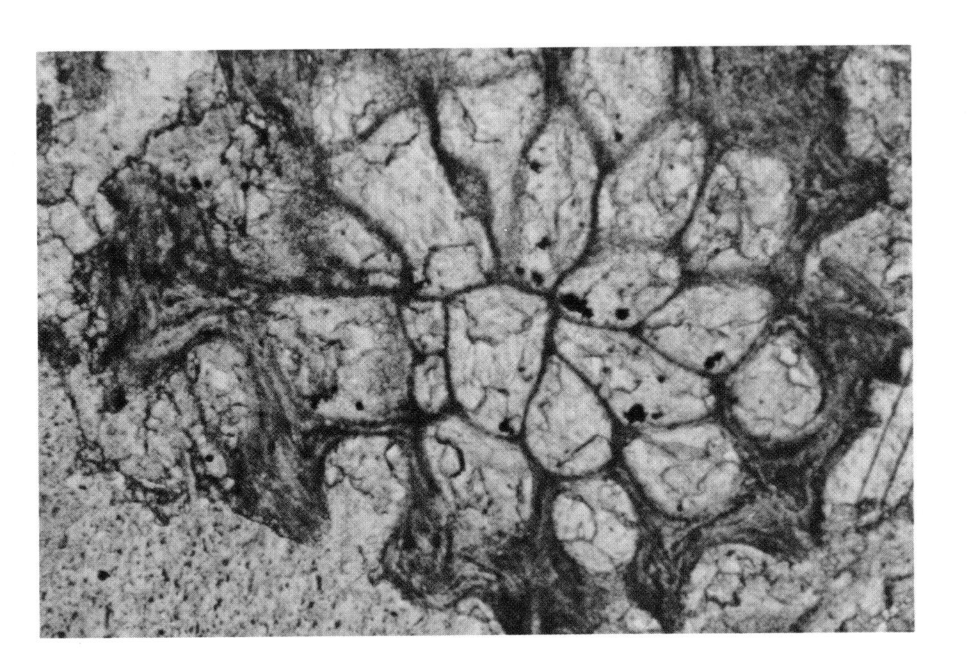

Up. Silurian part of Tonoloway-Keyser Ls.
Pennsylvania

Cross section of a bryozoan stem showing fairly regular rounded chambers and thick, fibrous wall structure.

0.10 mm

Up. Silurian part of Tonoloway-Keyser Ls.
Pennsylvania

Section of long axis of a stick-shaped bryozoan with characteristic inclined cellular structure and regular, fibrous walls.

X.N. 0.38 mm

Up. Pennsylvanian Winchell Ls. of Canyon Gp.
Texas

Cross section of a fenestrate bryozoan with typical skeletal structure (note especially the thick fibrous walls).

P.X.N. 0.30 mm

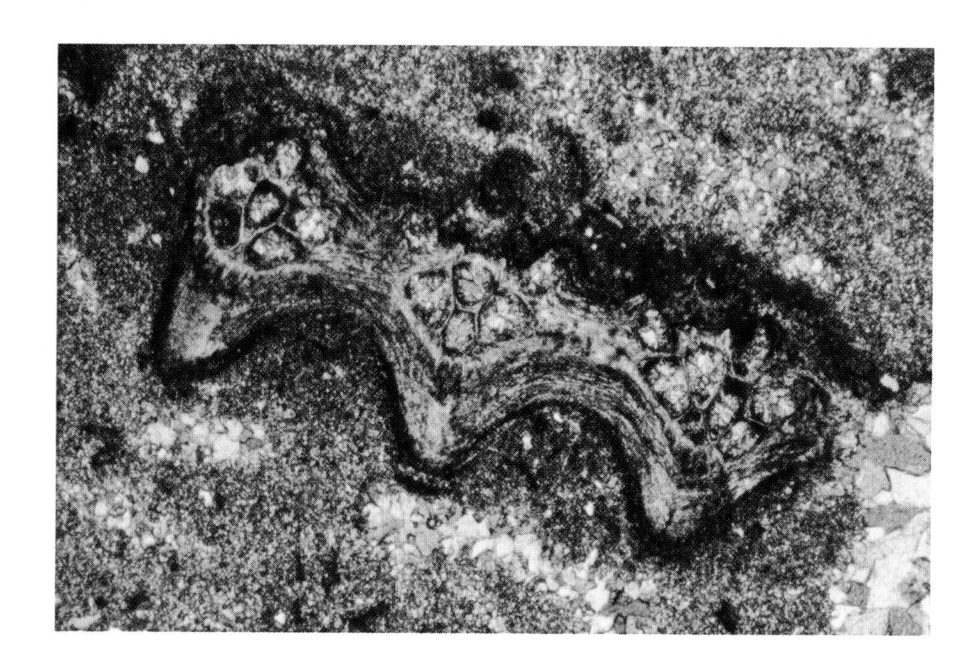

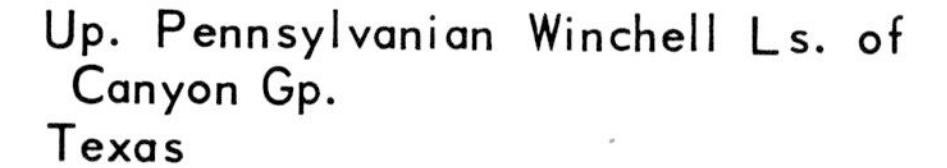

Up. Pennsylvanian Winchell Ls. of Canyon Gp.
Texas

Fenestrate bryozoan cut in a section showing separation of individual segments.

0.38 mm

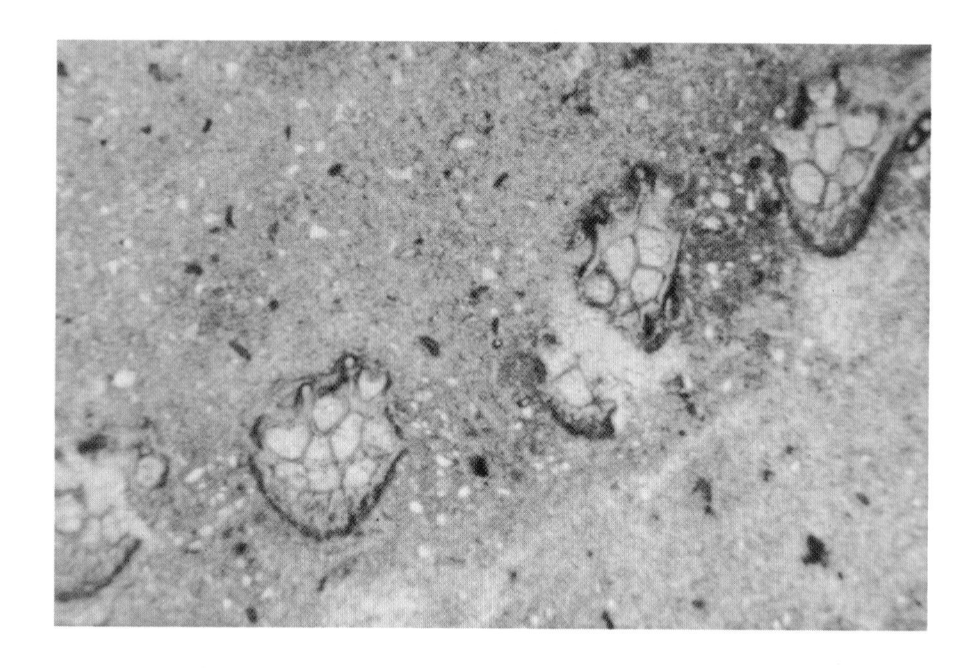

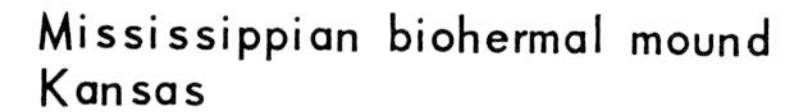

Mississippian biohermal mound
Kansas

Fenestrate bryozoan in a spar matrix. Shows little structure except for the characteristic even spacing of isolated segments. This is a common texture in Mississippian reefs.

0.38 mm

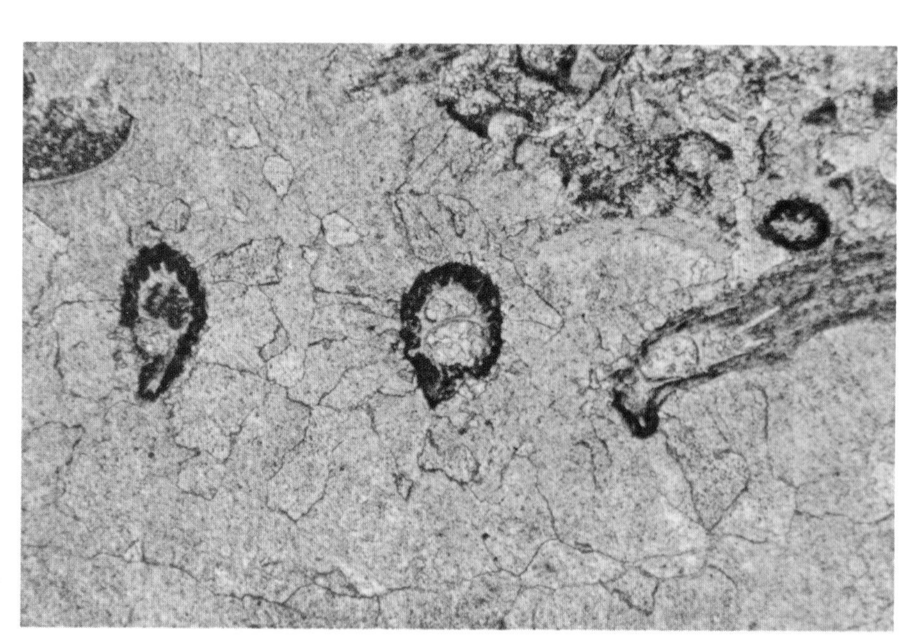

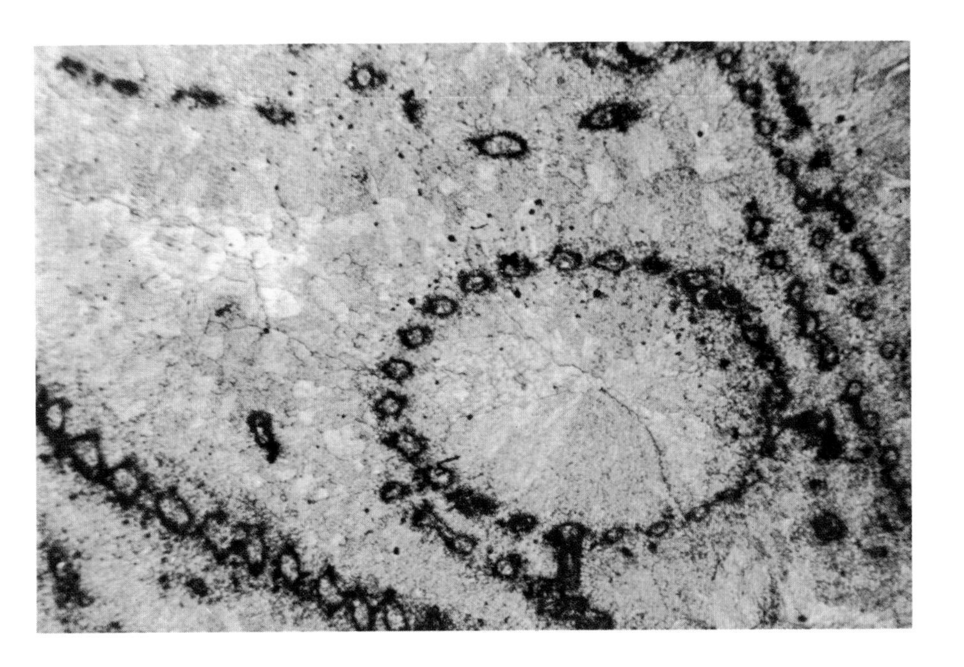

Mississippian Waulsortian (reef-flank) Ls.
Ireland

Abundant circular and elongate fenestrate bryozoans compose entire rock fabric. Shows the large amount of pore space (now filled with sparry calcite cement) which can be present in loosely packed bryozoan sediments.

1.24 mm

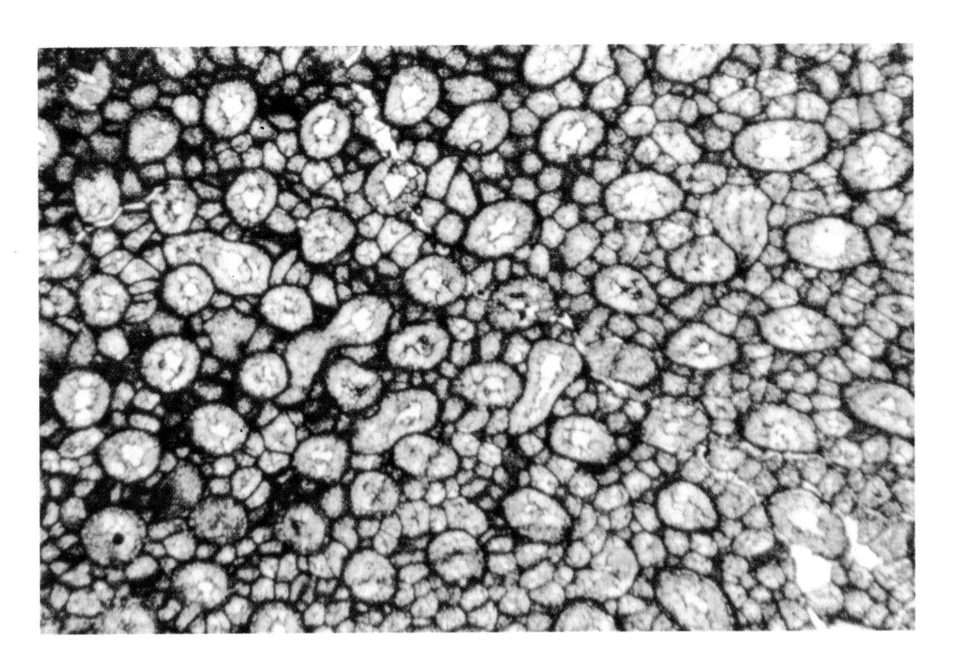

Up. Permian Bell Canyon Fm.
Texas

Large bryozoan fragment in reef talus. Shows unusually good preservation of zooecia because of early, finely crystalline dolomitization. Zooecia tend to be smaller and more uniform than corallites.

0.64 mm

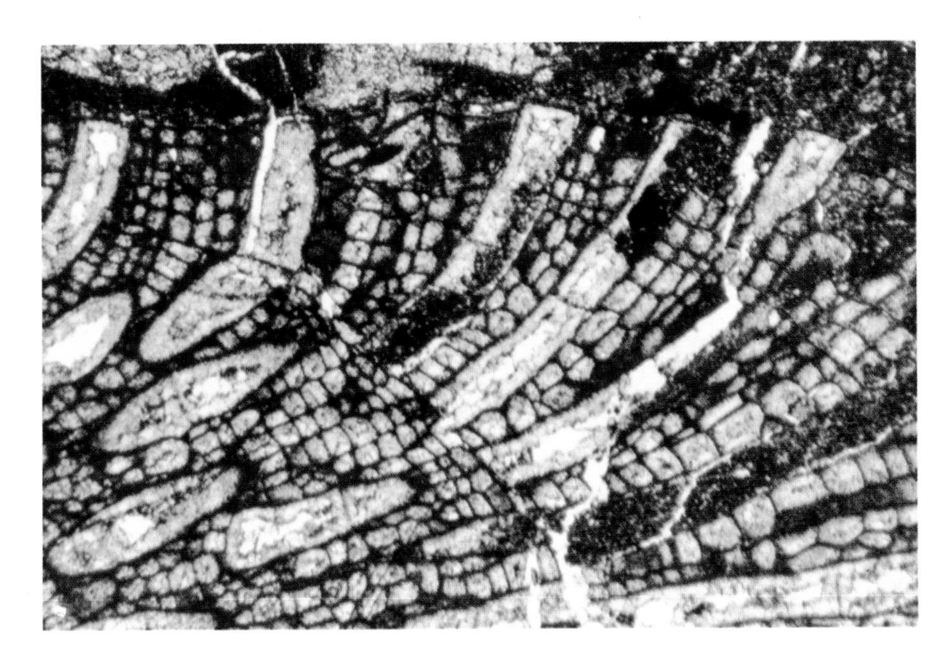

Up. Permian Bell Canyon Fm.
Texas

A different cut through the same specimen as above. Note the characteristic shape of the zooecia and their curvature toward the margins of the colony.

0.64 mm

Up. Pennsylvanian Graham Fm.
Texas

A series of three sections through a cryptostome bryozoan, *Megacanthopora* sp. This transverse cross section through the stick-shaped colony shows the radial distribution of zooecia and the thickened wall structure at the exterior margins.

0.64 mm

Up. Pennsylvanian Graham Fm.
Texas

A longitudinal section through the same specimen as above. Shows curvature of zooecia toward outer margins and again illustrates the thickening of wall structure toward the exterior.

0.64 mm

Up. Pennsylvanian Graham Fm.
Texas

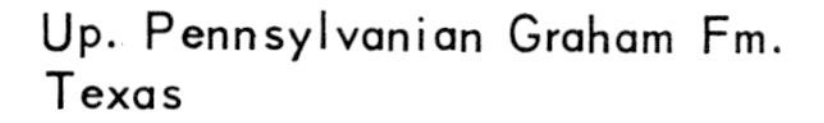

An oblique section through the same specimen as above. Shows uniform to slightly elliptical zooecia and fibrous calcitic wall structure. These 3 photos illustrate the range of different structures which are possible in a single specimen given various angles of sectioning.

0.64 mm

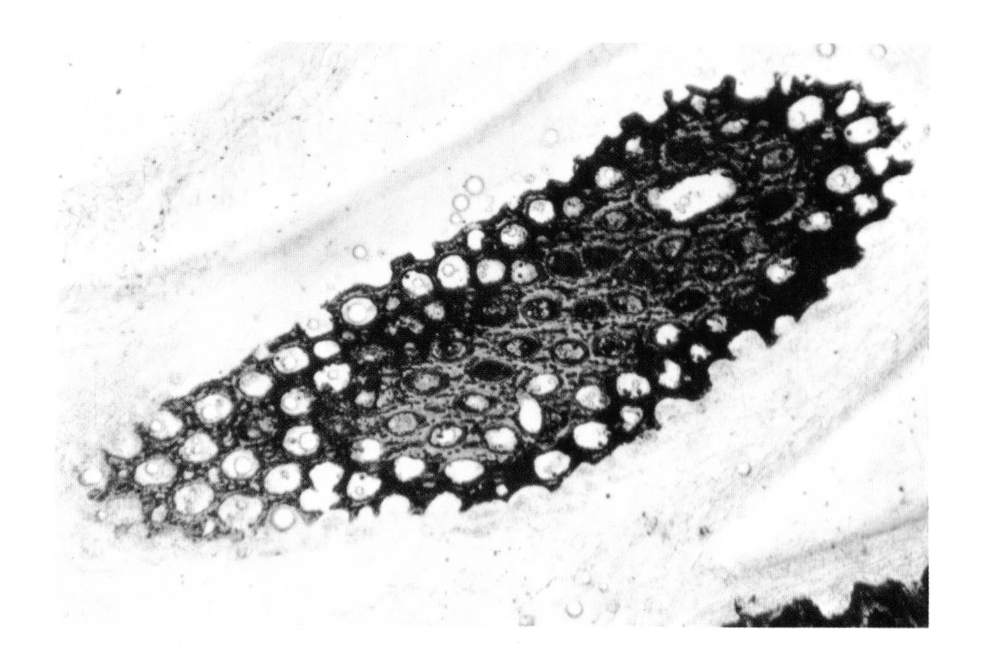

Up. Silurian part of Tonoloway-Keyser Ls.
Pennsylvania

Large bryozoan frond showing typical shape and structure.

0.30 mm

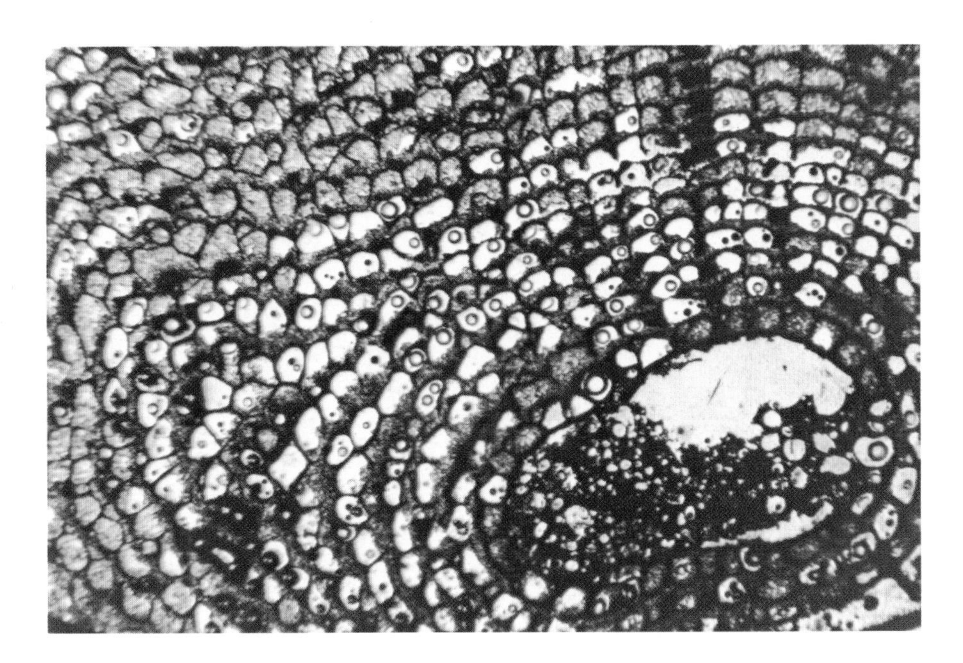

Pleistocene Key Largo Ls.
Florida

Section of a single colony of the modern encrusting bryozoan *Schizoporella floridana.* The central area, now partly filled with infiltrated sediment, once held the encrusted organism (probably some Thalassia grass). These encrusters are most abundant today in the somewhat restricted bay and lagoonal environments.

1.24 mm

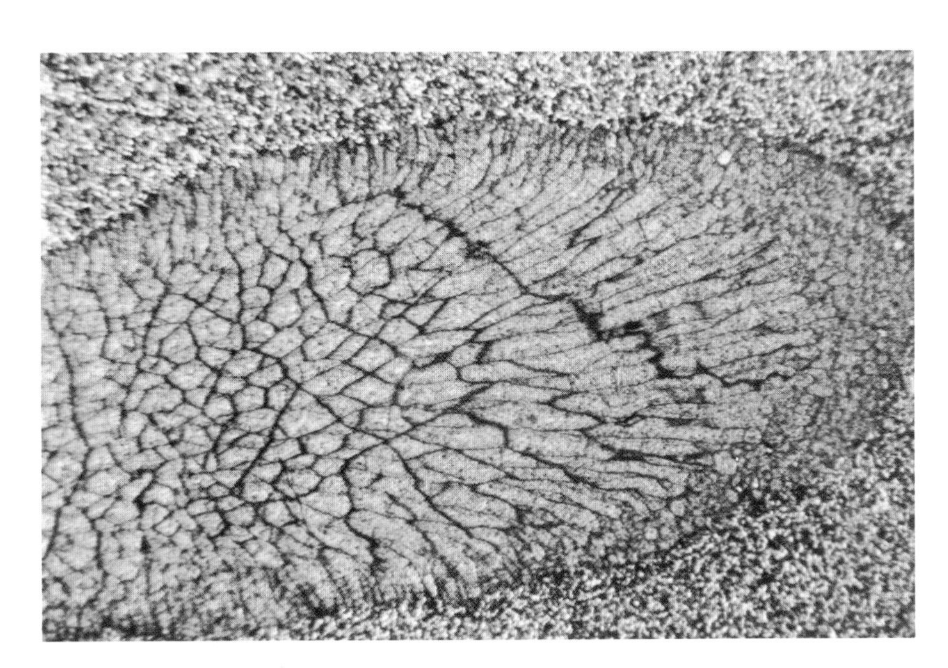

Mid. and Up. Ordovician Martinsburg Fm.
Virginia

A large bryozoan colony with irregular zooecial shapes but characteristic outward rotation of zooecia at margins.

0.10 mm

Brachiopods

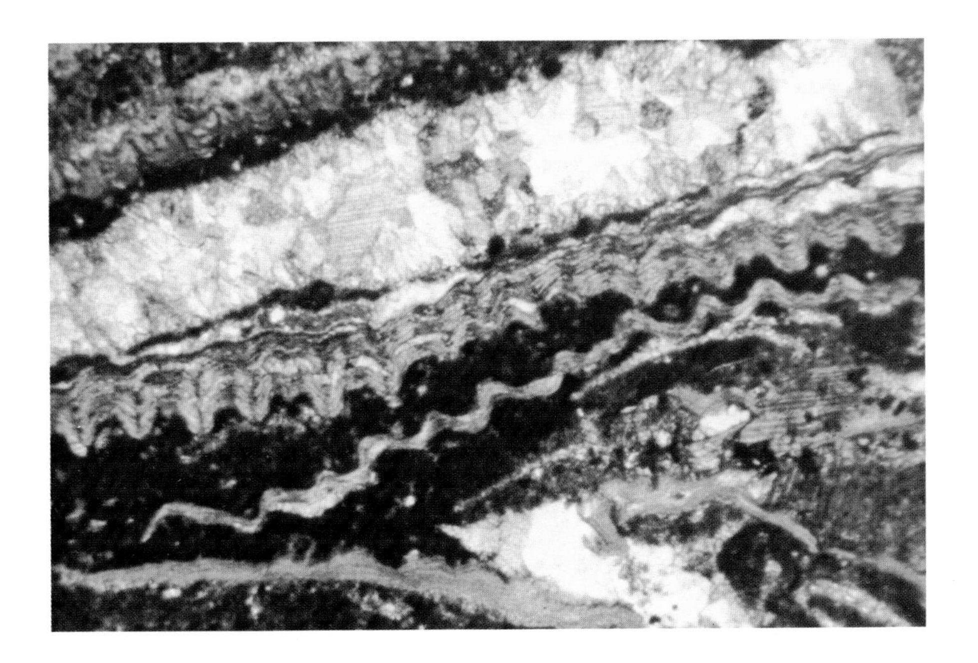

Mid. Ordovician Black River Ls.
Pennsylvania

Several elongate brachiopod shells (plus other grains) showing crenulate shape of many brachiopod shells.

X.N. 0.38 mm

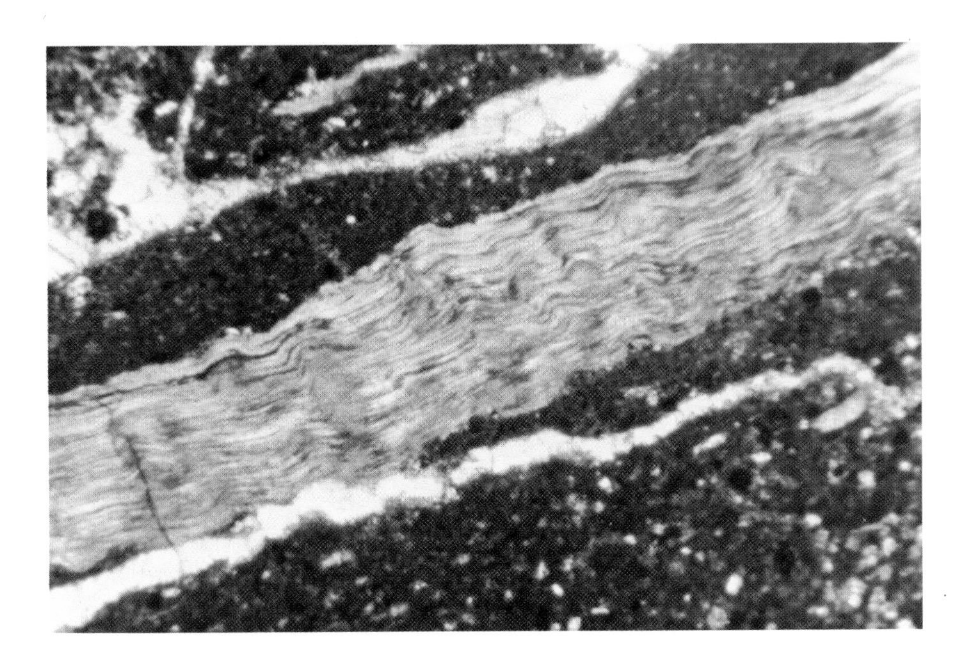

Mid. Ordovician Black River Ls.
Pennsylvania

Large brachiopod shell with wavy, parallel laminated shell structure sub-parallel to shell edge.

0.38 mm

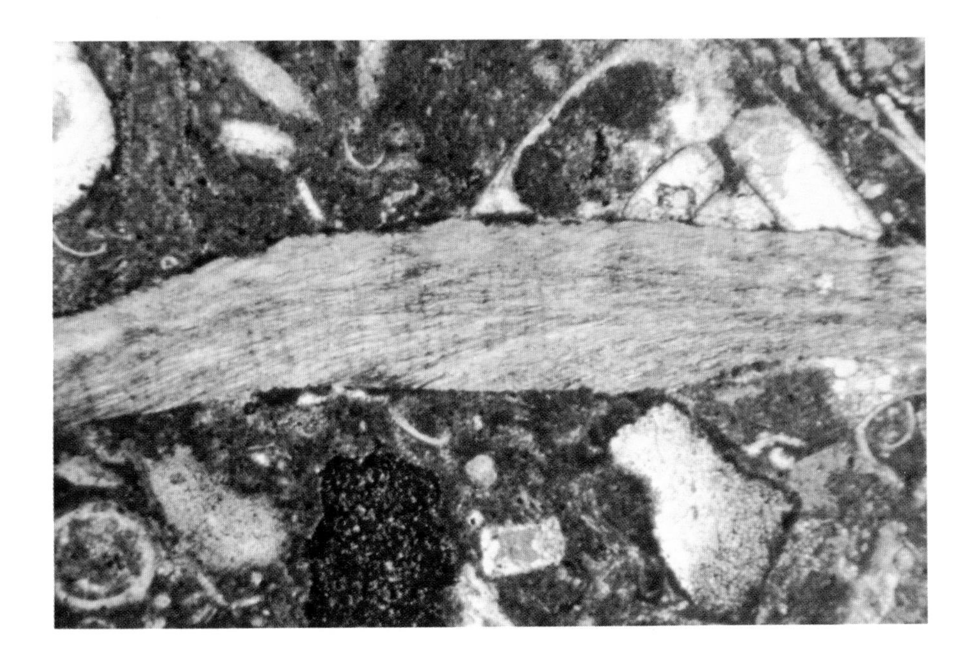

Up. Silurian part of Tonoloway-Keyser Ls.
Pennsylvania

Brachiopod shell with parallel fibrous structure running sub-parallel (inclined about 15 degrees) to the outer margin of shell. Shell is a simple one or two layered structure, unlike most mollusk shells.

X.N. 0.38 mm

Up. Mississippian Hindsville Ls.
Oklahoma

Shows a number of coated brachiopod grains (as well as crinoid plates). All the brachiopods have low-angle fibrous wall structure. One clearly shows the additional outer wall (with perpendicular arrangements of fibers). Another shell shows micrite-filled punctate structure, and a third shows impunctate structure with wavy shell contortion.

0.25 mm

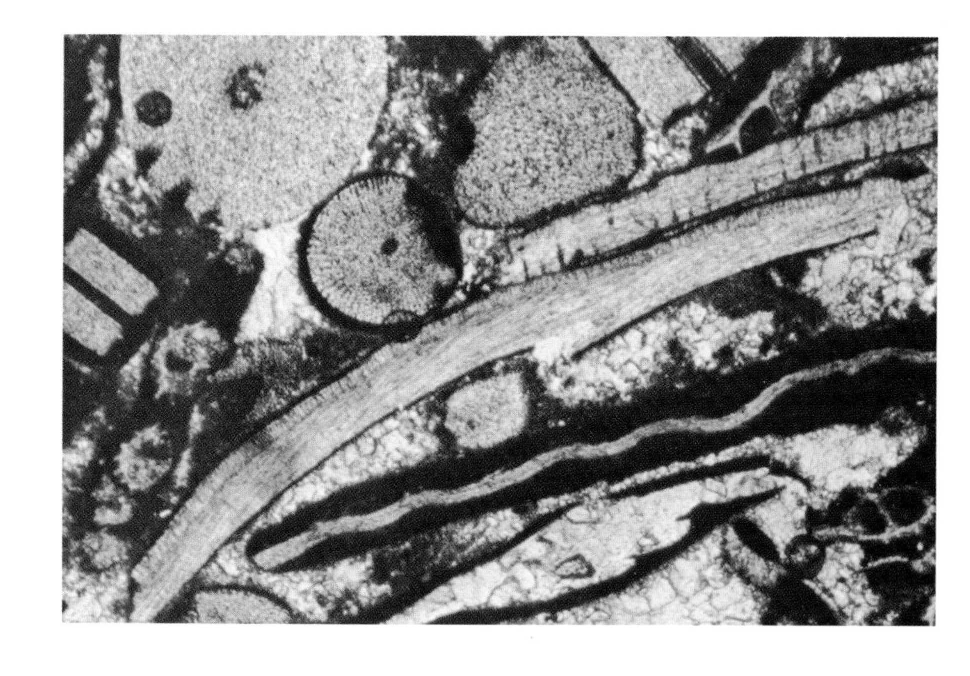

Up. Mississippian Hindsville Ls.
Oklahoma

A punctate brachiopod. Has clearly defined pores or punctae which completely penetrate the shell and here are filled with micritic sediment. Sediment also contains micrite coated echinoderm and bryozoan fragments.

0.25 mm

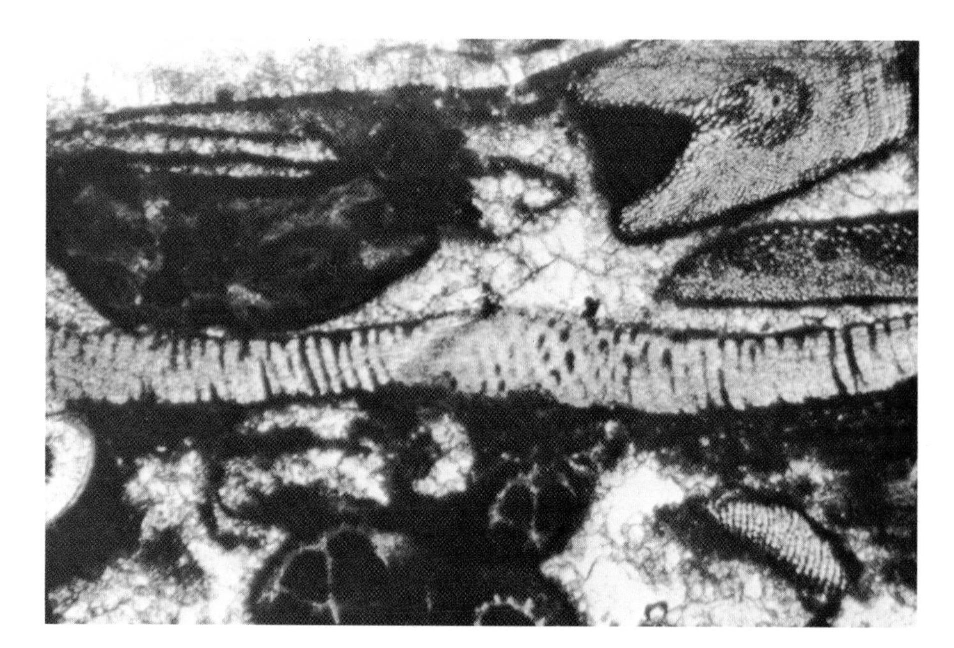

Up. Ordovician Reedsville Shale
Pennsylvania

Brachiopod shell with parallel fibrous structure. Also shows some oblique pseudopunctae.

X.N. 0.30 mm

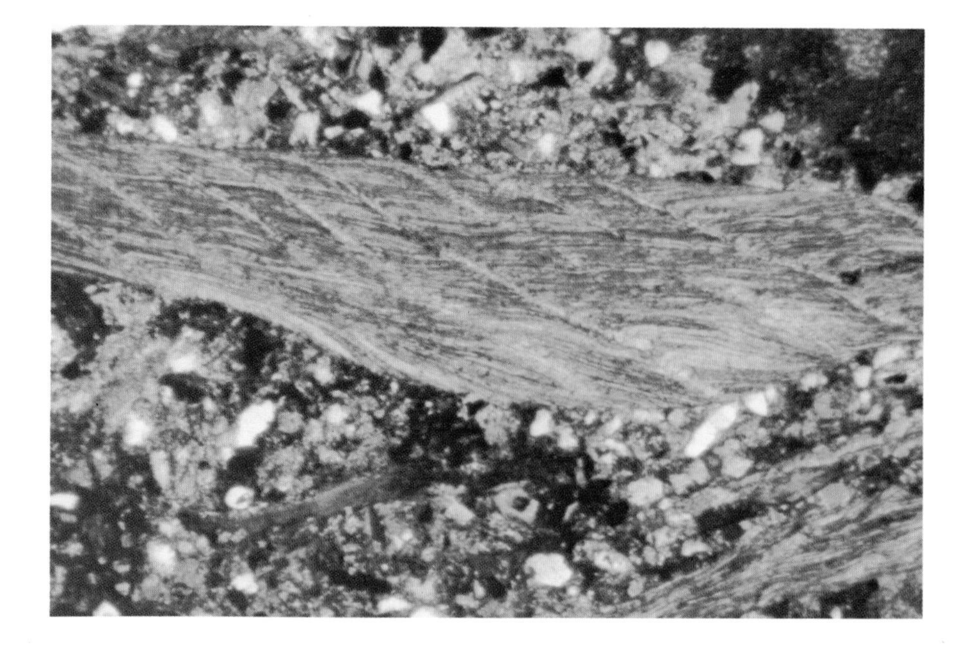

Lo. Devonian Becraft Ls.
New York

Section through a pseudopunctate brachiopod shell. Note the parallel fibrous wall structure with small plications which run vertically through the shell. These are the pseudopunctae; they do not represent actual openings or pores in the shell.

0.11 mm

Up. Permian Capitan Ls. (reef facies)
New Mexico

Section of a portion of the wall of *Composita* sp., a brachiopod with an atypical prismatic wall structure. This particular group of brachiopods is restricted to Carboniferous and Permian sediments. These shells can be differentiated from pelecypods on the basis of the presence of internal calcareous spires in the *Composita* (not always preserved).

X.N. 0.51 mm

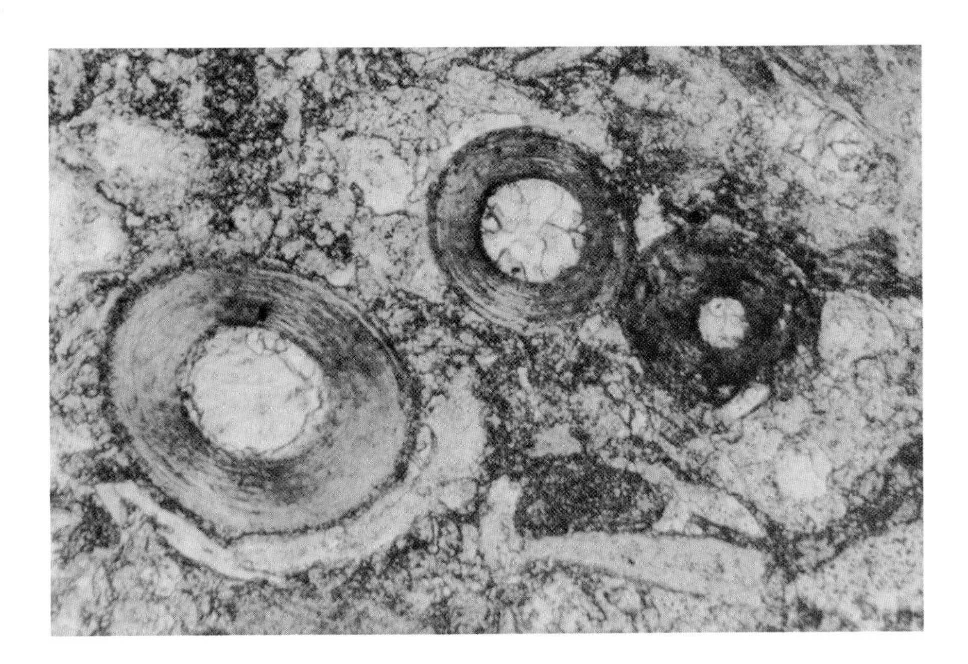

Mississippian Glencar Ls.
Ireland

Three brachiopod spines with concentric parallel fibrous inner, and radial-fibrous outer wall layers, as well as hollow central canal.

0.09 mm

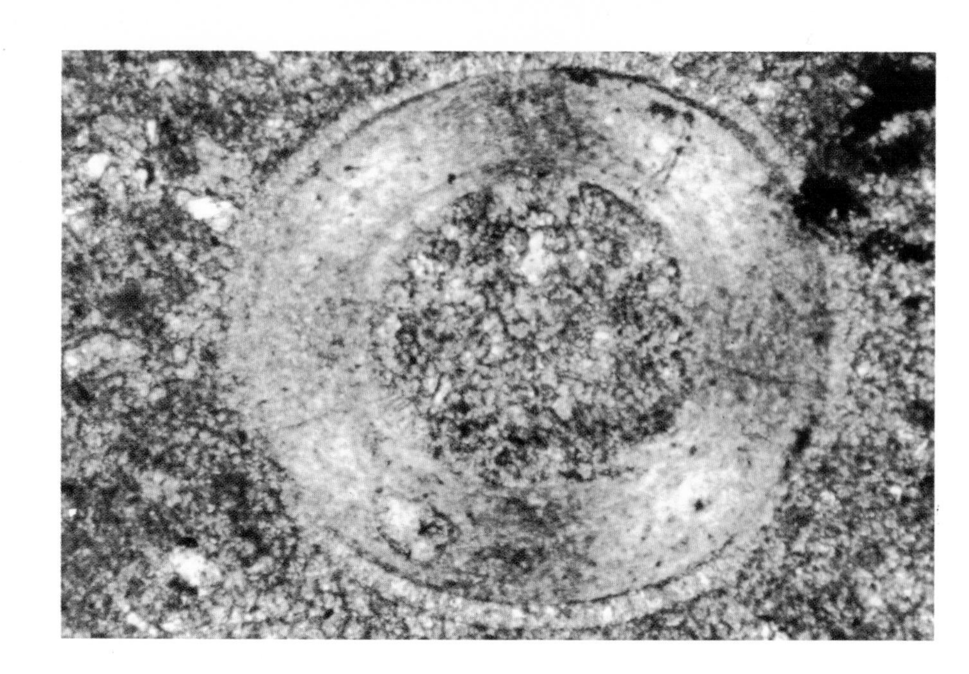

Lo. Permian Bone Spring Ls.
Texas

Transverse cross-section of a brachiopod spine. Note fibrous structure yielding pseudo-uniaxial cross, central canal, and outer wall.

X.N. 0.08 mm

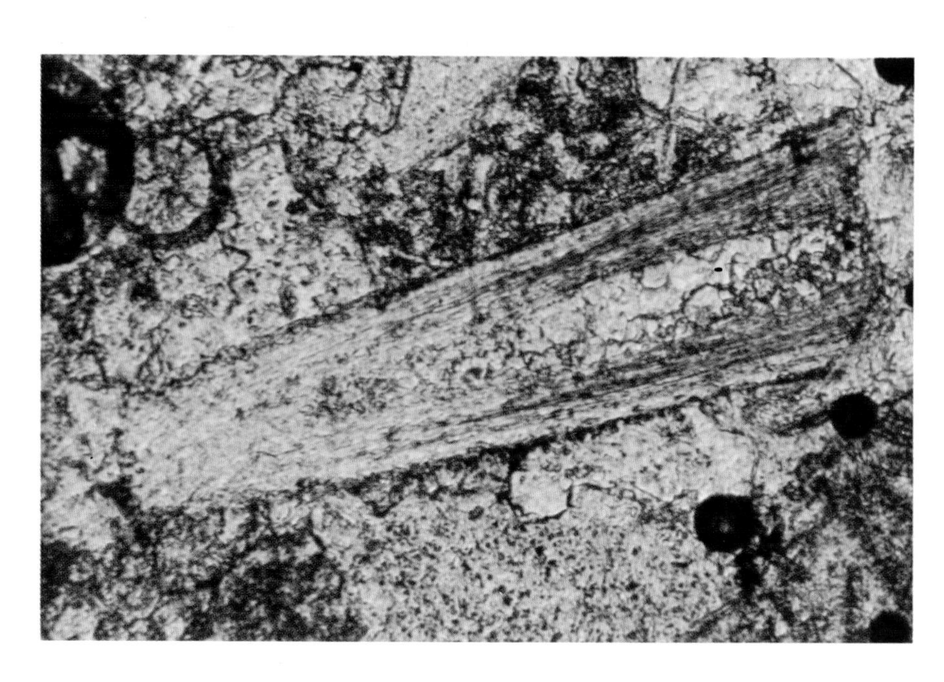

Mississippian Glencar Ls.

Ireland

Oblique longitudinal cross section of a brachiopod spine. Layered, fibrous wall structure is characteristic of such grains.

0.09 mm

Mollusks

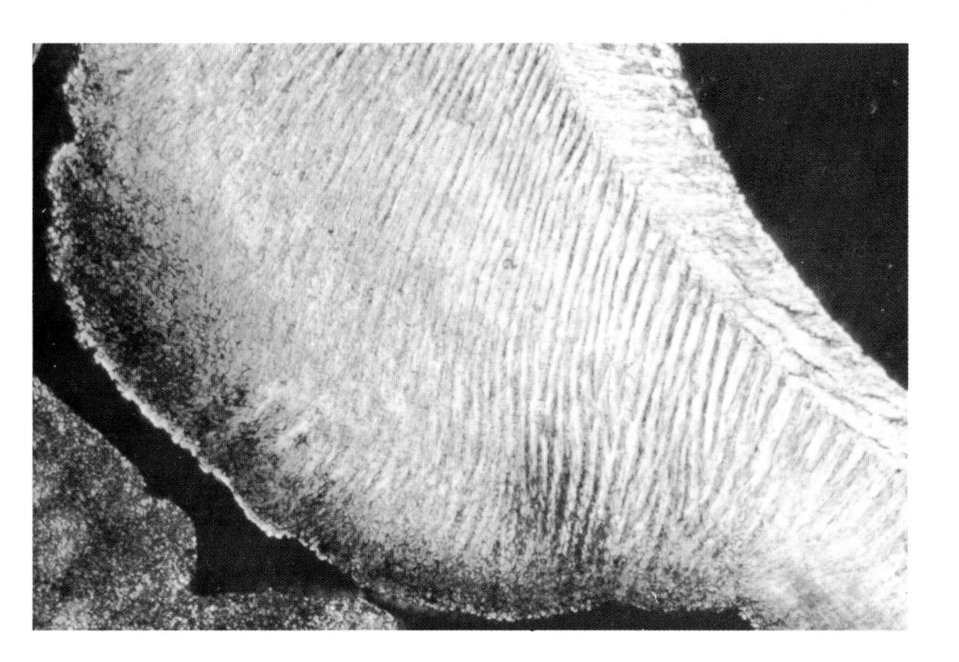

Holocene sediment
Bimini, Bahamas

An example of crossed-lamellar wall structure–one of the major types of aragonitic textures in mollusks. This example is from a gastropod. Note the narrow bands of alternating light and dark extinction which wedge out along axis.

X.N. 0.09 mm

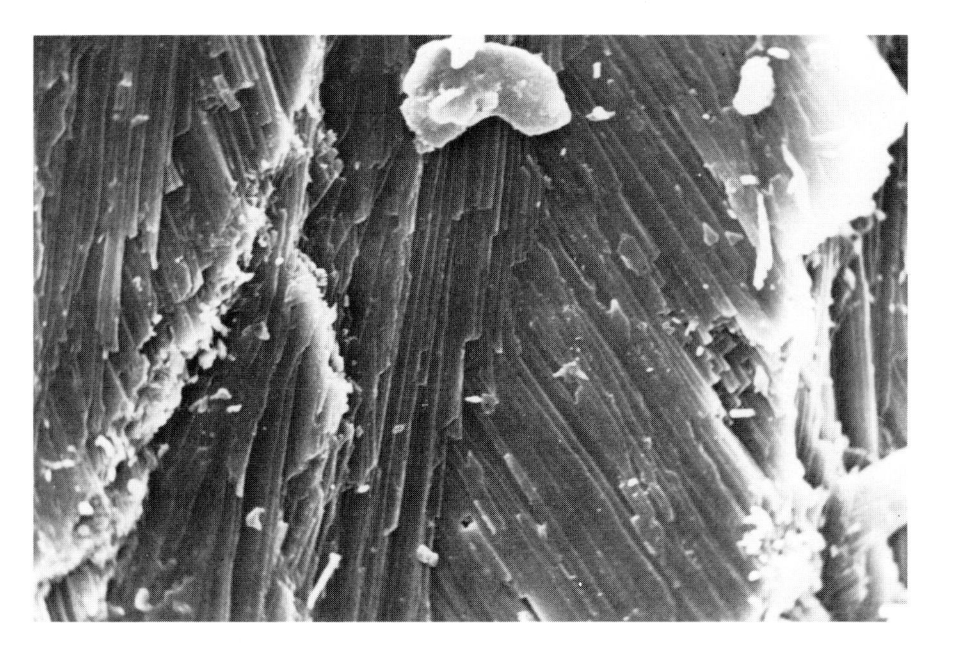

Holocene sediment
Belize (British Honduras)

SEM view of a broken shell of *Pecten (Lyropecten)* sp. Shell wall shows intersecting bundles of aragonite crystals characteristic of crossed-lamellar structure.

SEM Mag. 1000X 12.5 μm

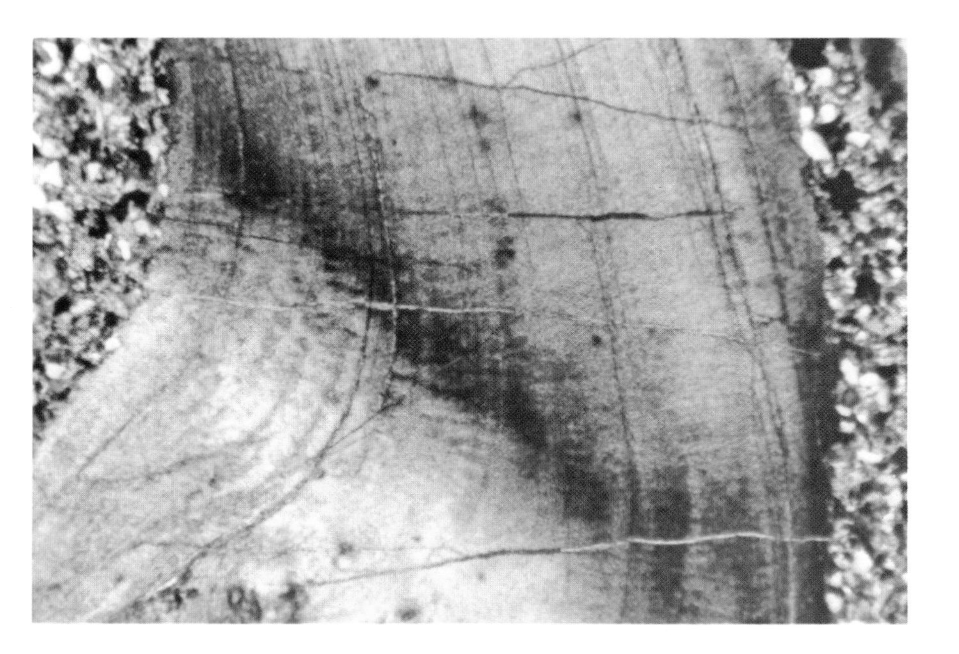

Up. Oligocene Molasse
Germany

Part of a large *Cyrenia* shell easily recognizable as a mollusk by its shell shape. Has homogeneous structure with extinction bands perpendicular to outer shell wall (and sweeping the length of the shell as stage is rotated).

X.N. 0.38 mm

Up. Cretaceous San Carlos Fm.
Texas

An *Inoceramus* shell with very distinctive prismatic structure. Often these shells break up into individual prisms which may constitute an important fraction of some sediments. Note boring in shell wall.

X.N. 0.38 mm

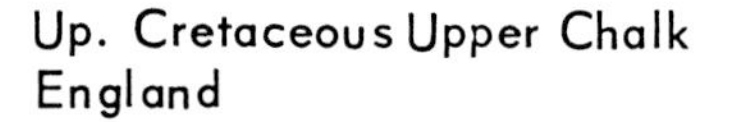

Up. Cretaceous Upper Chalk
England

Micritic sediment with numerous, scattered, individual *Inoceramus* prisms. The large variation in shape is due to different angles of cut through prisms, but all are recognizable by their size, shape, and characteristic extinction behavior.

X.N. 0.30 mm

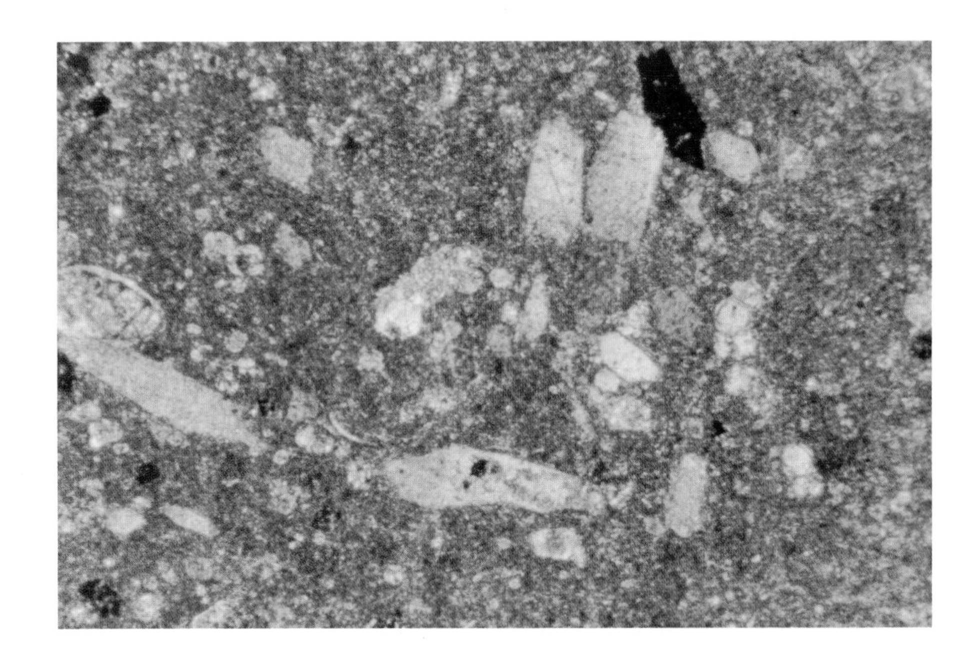

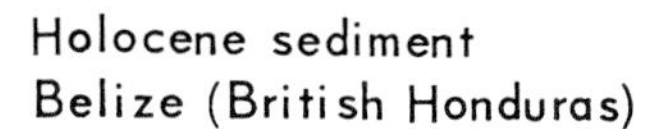

Holocene sediment
Belize (British Honduras)

SEM view of a broken shell of *Pinna* sp. Note the large prisms which are arranged perpendicular to the margin of the shell and which constitute the entire shell wall. Most organisms with prismatic-structured shell walls secrete primarily calcite.

SEM Mag. 100X 125 μm

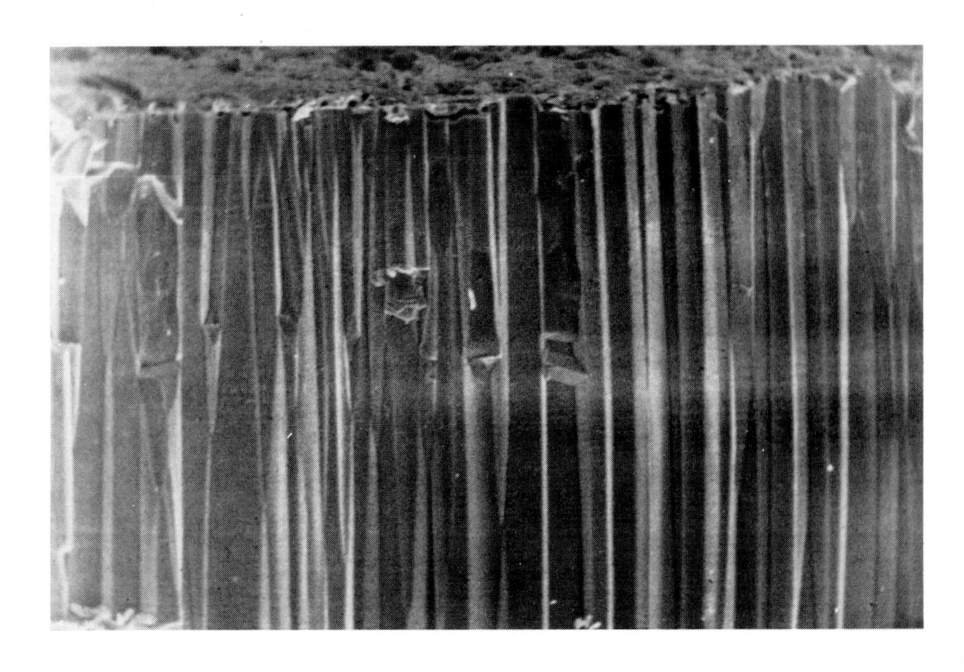

Pliocene and Pleistocene Caloosahatchee Marl
Florida

Cross section of an oyster shell showing alternating constructional layers of lamellar and vesicular calcite. Vesicular structure is common in ostreid mollusks.

X.N. 0.38 mm

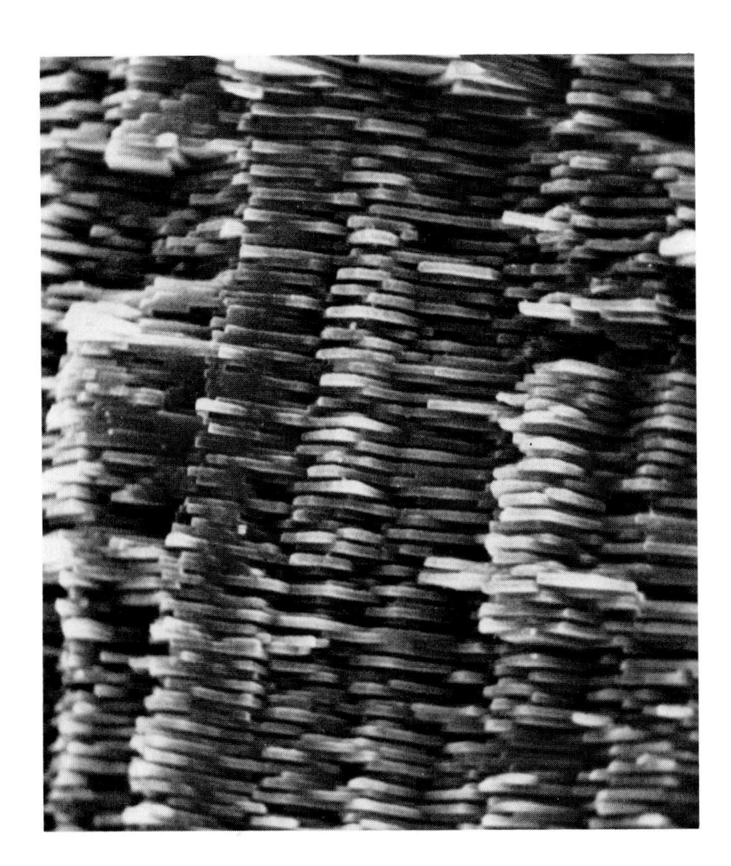

Holocene sediment
Belize (British Honduras)

SEM view of broken mollusk shell. The nacreous layer (shown here) consists of stacked, overlapping vertical columns of tabular aragonite crystals separated by thin sheaths of organic material.

SEM Mag. 2000X 6.2 μm

Holocene sediment
Belize (British Honduras)

SEM closeup of the nacreous layer of a mollusk shell showing detail of the interlocking aragonite tablets. Although this structure is lost during diagenesis, it is useful in identification of modern shell fragments.

SEM Mag. 10,000X 1.3 μm

Pliocene and Pleistocene Caloosahatchee Marl
Florida

Bored fragment of an oyster shell. Borings (dark patches) are filled with lime mud (micrite), and shell shows multilayered structure (alternating horizontal-foliated and vertical crossed-lamellar structure).

X.N. 0.38 mm

Up. Cretaceous Rio Yauco Fm.
Puerto Rico

Fragment of a rudistid mollusk. Rudistids have complicated and variable shapes but are often recognizable by the boxwork (vesicular) structure of their shells.

X.N. 0.38 mm

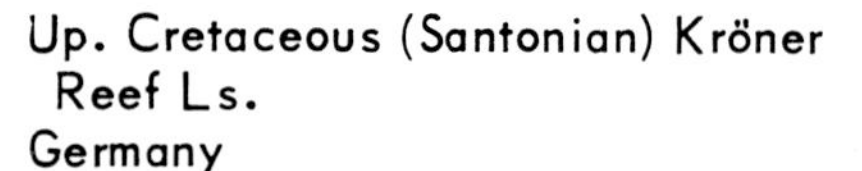

Up. Cretaceous (Santonian) Kröner Reef Ls.
Germany

A large fragment of the Chamacean pelecypod *Hippurites* sp. (a close relative of other rudistids). Again recognizable by the vesicular boxwork texture. Distinguished from corals, bryozoans, etc. by the size, shape, and details of vesicular structure.

0.64 mm

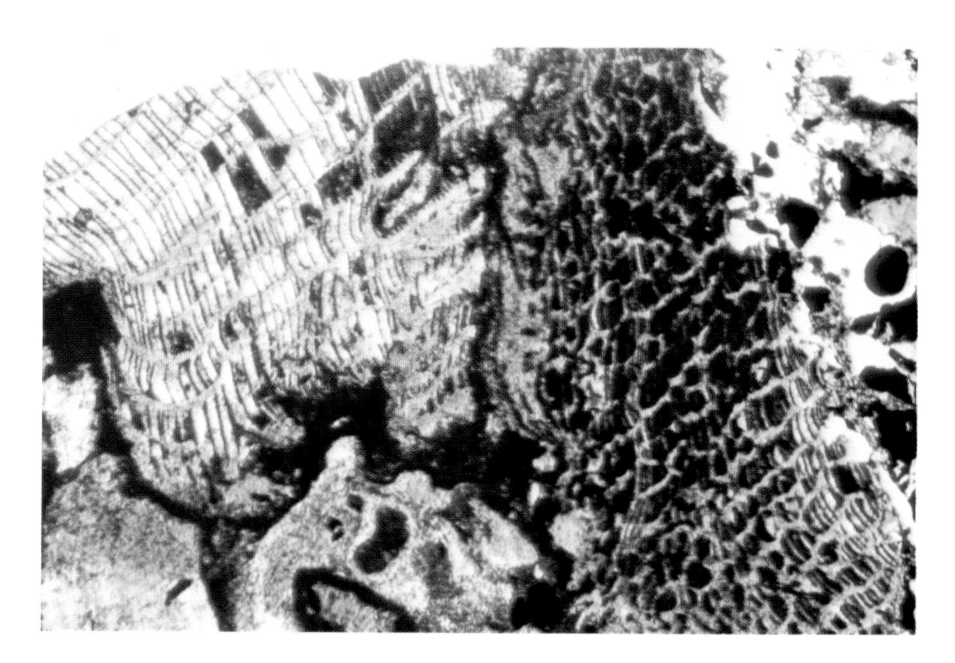

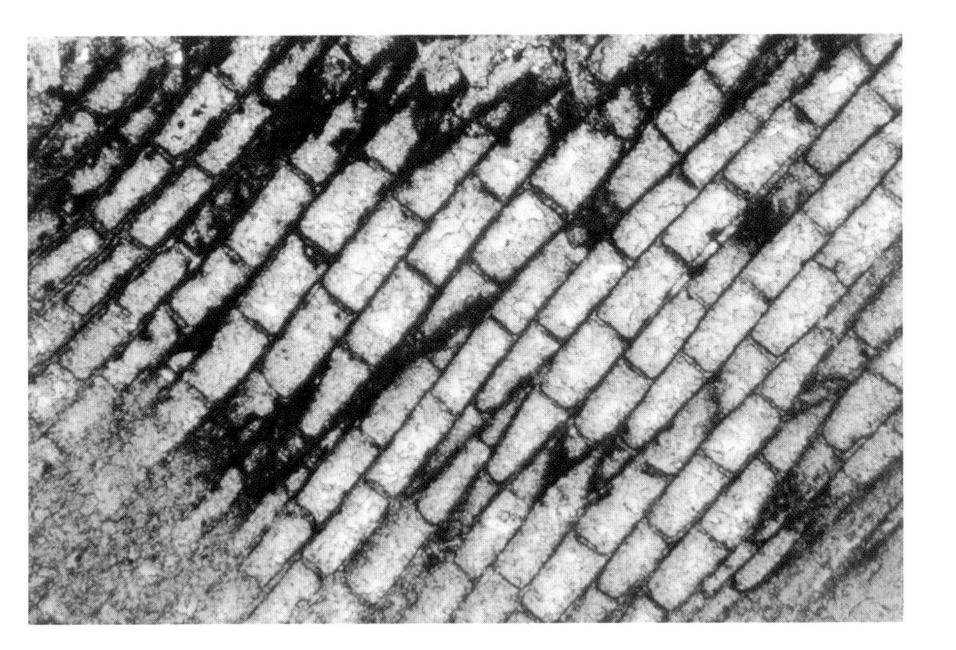

Cretaceous Tamabra Ls.
Mexico

Section of the wall of a probable radiolitid rudist illustrating the well preserved cellular prismatic outer layer of the shell. These rudists are major contributors to Cretaceous reefs and bioherms.

0.64 mm

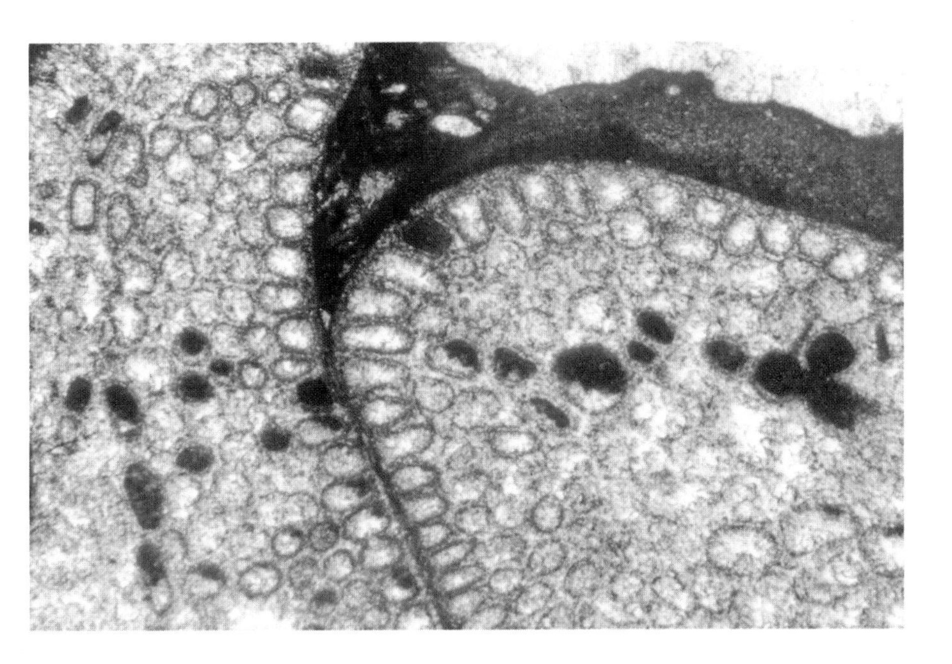

Cretaceous Tamabra Ls.
Mexico

Section of the wall of a probable caprinid rudist showing canals within the very poorly preserved (once, presumably, aragonitic) shell wall. These rudists were also important reef and bioherm builders in the Cretaceous. Micritic encrustation visible in upper part of photograph is probably algal.

0.64 mm

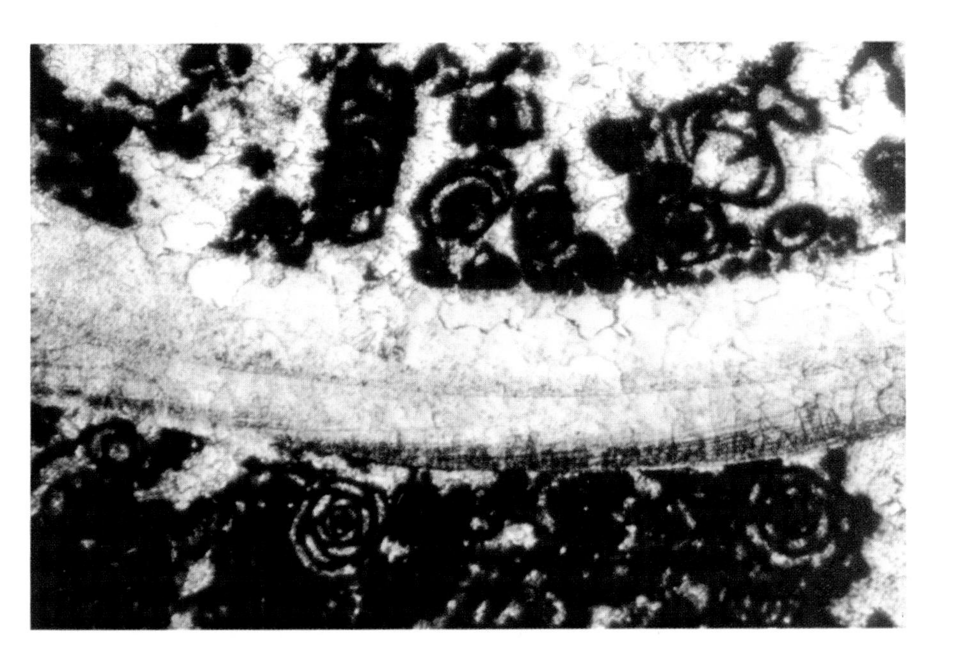

Cretaceous El Abra Ls.
Mexico

Section through a small portion of the wall of a toucasiaid rudist (Diceratidae) shows two distinct shell layers. One (exterior) retains original structure and pigmentation; the other is completely recrystallized to calcite and is marked by line of "perched" miliolid Foraminifera.

0.64 mm

Triassic (Karnian) Hallstatter Ls.
Germany

Numerous thin-walled shells of the Pectinoid pelecypod, *Halobia* sp. These very thin shells were adapted to a possibly motile existence on soft bottoms or may even have been pelagic, and are common in Triassic deeper-water sediments. Recognizable by shape.

0.64 mm

Lo. and Mid. Pennsylvanian Bloyd Fm.
Oklahoma
A foram-encrusted, partly recrystallized pelecypod with very large external shell plications. This grain is clearly recognizable as molluscan on the basis of shape alone.

X.N. 0.22 mm

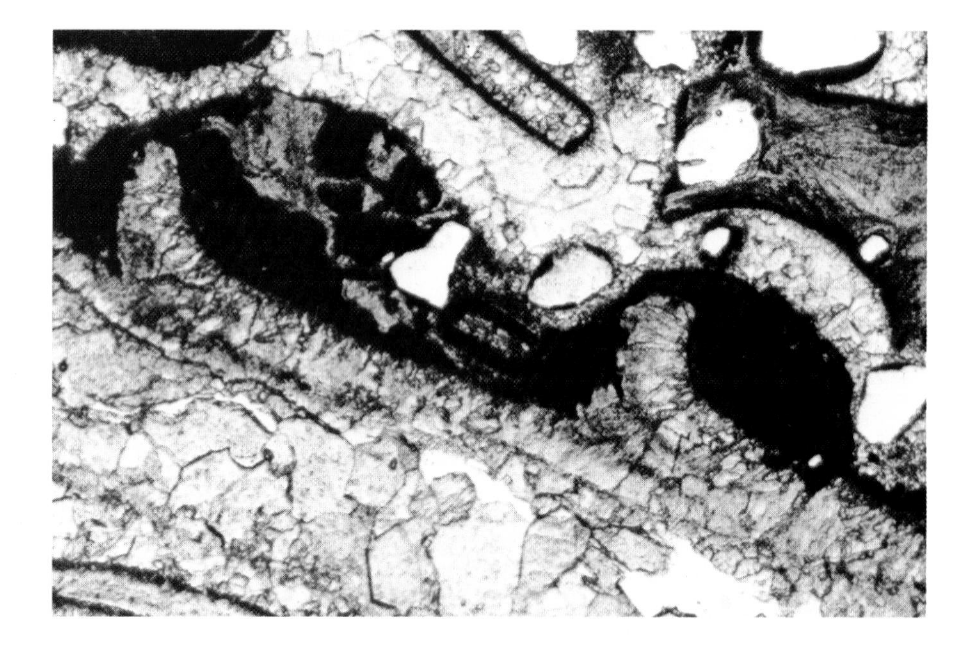

Lo. Cretaceous Cupido Fm.
Mexico

Two pelecypods showing common mode of alteration. These were originally aragonitic shells which inverted to calcite by a dissolution-precipitation mechanism. This obliterated all relict internal shell structure. Grains are identifiable only on the basis of characteristic shapes outlined by micrite envelopes.

0.22 mm

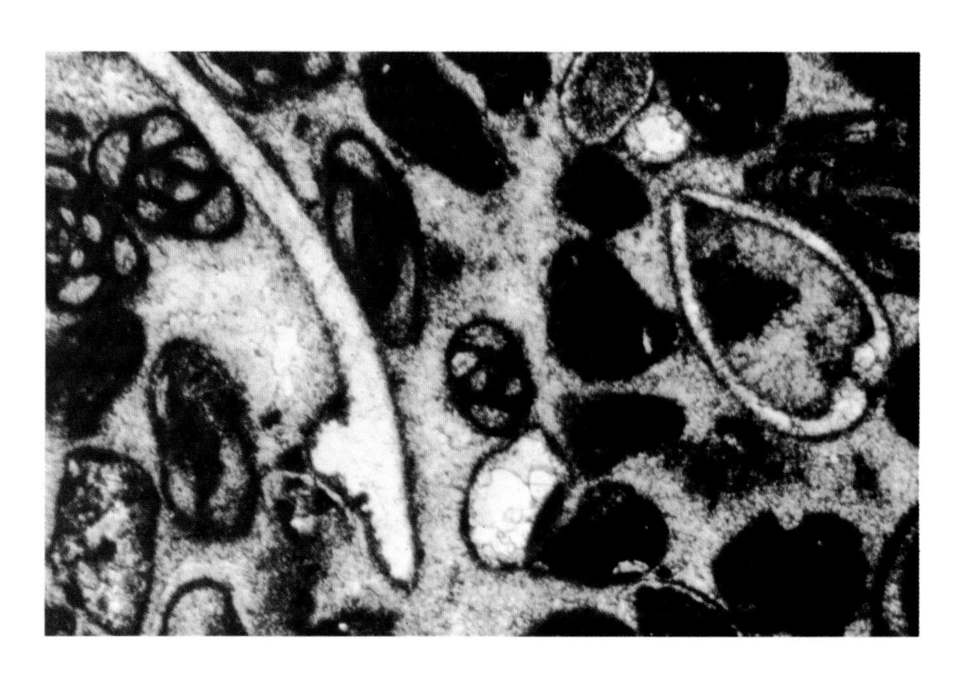

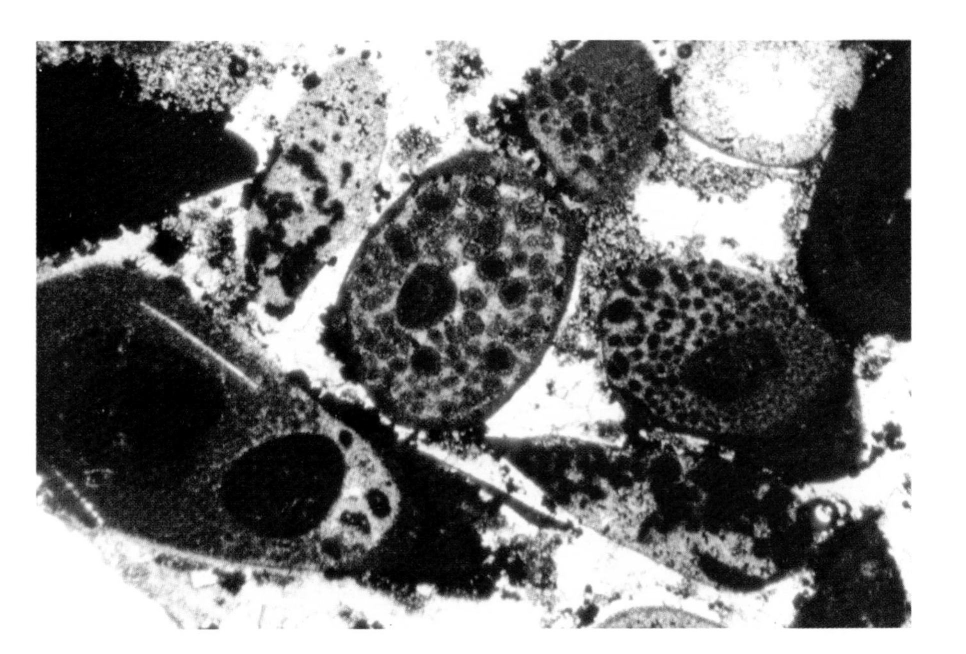

Lo. Cretaceous Glen Rose Ls.
Texas

These are the internal fillings of pelecypods--*Corbula* sp. All trace of the shell wall is gone and the only way to distinguish these complex pelletal grains from intraclasts is by their consistent "tear-drop" shape. The shells were rolled during the infiltration of sediment, yielding complex fillings.

0.64 mm

Up. Miocene Hydrobien Ls.
Germany

A well-sorted gastropod limestone composed entirely of the freshwater meso-gastropod prosobranchs, *Hydrobia* sp. Note the variety of shapes produced by different angles of section through these gastropods. Shells are aragonitic and have crossed-lamelar structure.

0.64 mm

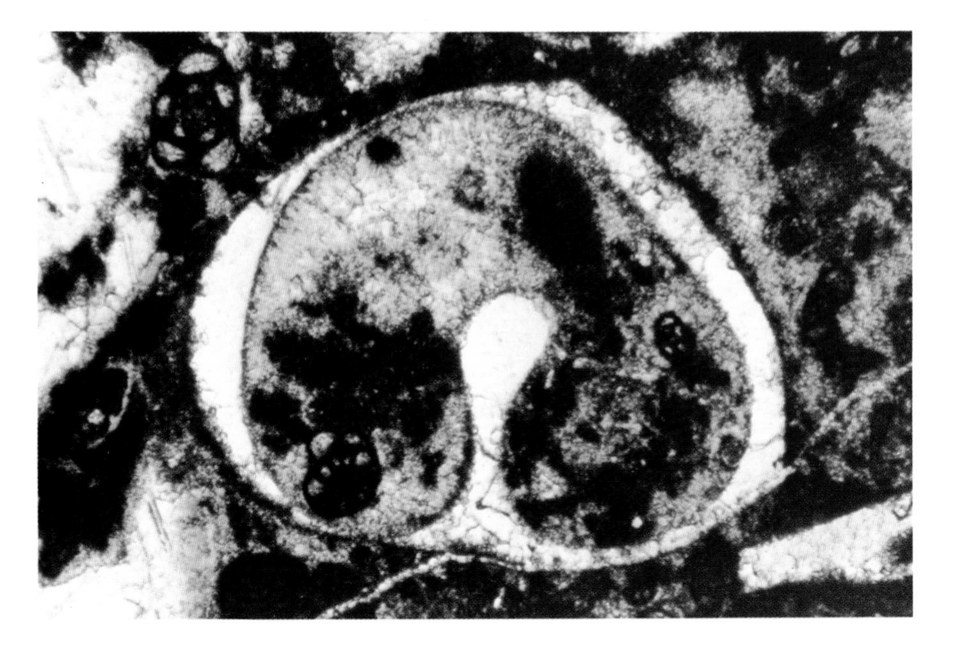

Lo. Cretaceous Cupido Fm.
Mexico

Transverse section through a single, originally aragonitic, gastropod. All trace of original wall structure has been obliterated during inversion to calcite, but recognizable outline is preserved by internal and external sediment plus cément.

0.22 mm

Lo. Cretaceous Cupido Fm.
Mexico

Longitudinal section of an originally aragonitic gastropod. All wall structure obliterated but recognizable outline is preserved, due largely to early diagenetic (probably synsedimentary) infilling of chambers with fibrous cement crusts. Shell also is outlined by micritization of outer margin.

0.25 mm

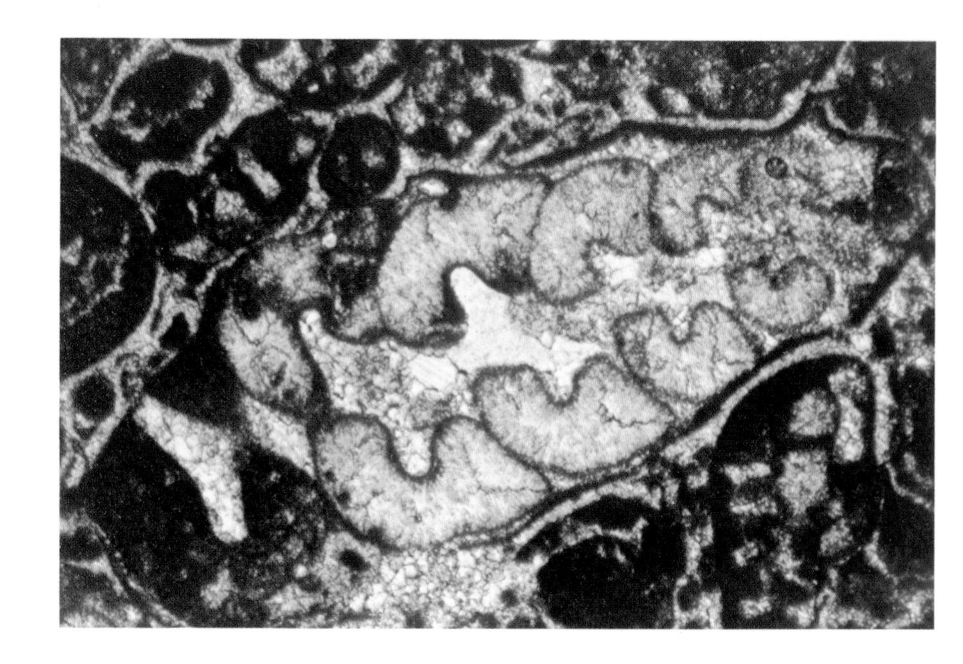

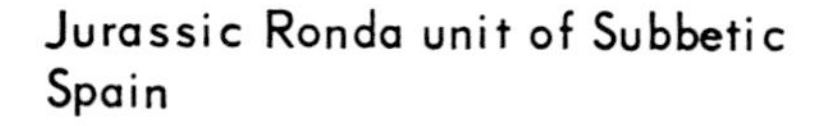

Jurassic Ronda unit of Subbetic
Spain

Longitudinal section of a high-spired gastropod. Original aragonite wall has inverted to calcite with loss of internal detail, yet original internal and external outlines are faithfully preserved by micritic sediment.

0.25 mm

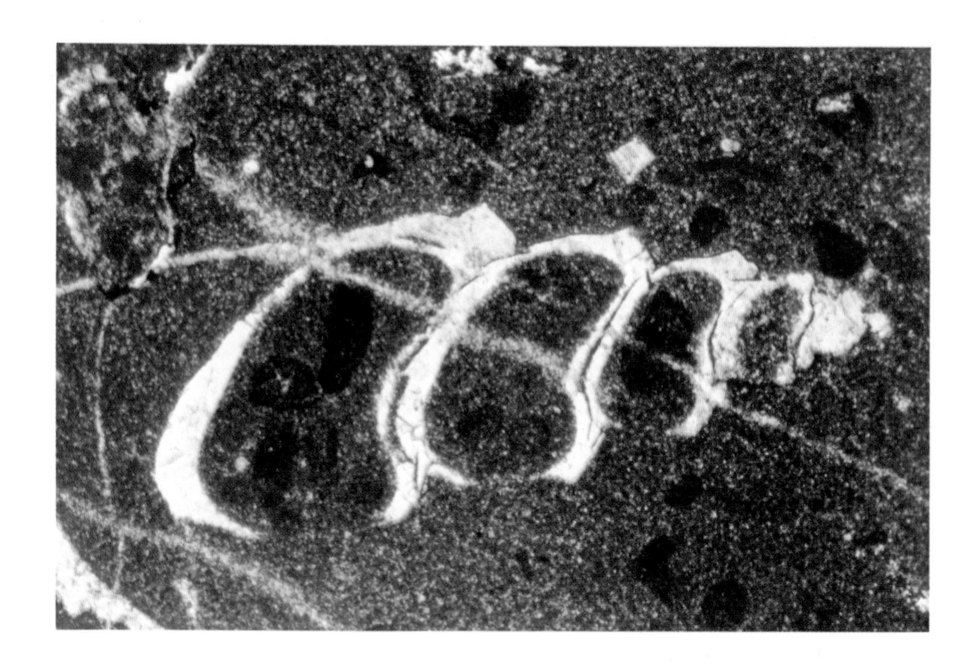

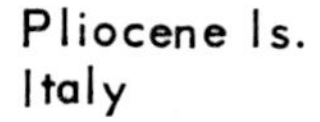

Pliocene ls.
Italy

Section through part of the wall of a scaphopod, *Dentalium sexangulare.* This group of mollusks is characterized by its conical shape, aragonitic shell, and concentrically laminated crossed-lamellar wall structure. One can also distinguish numerous dark growth bands.

0.64 mm

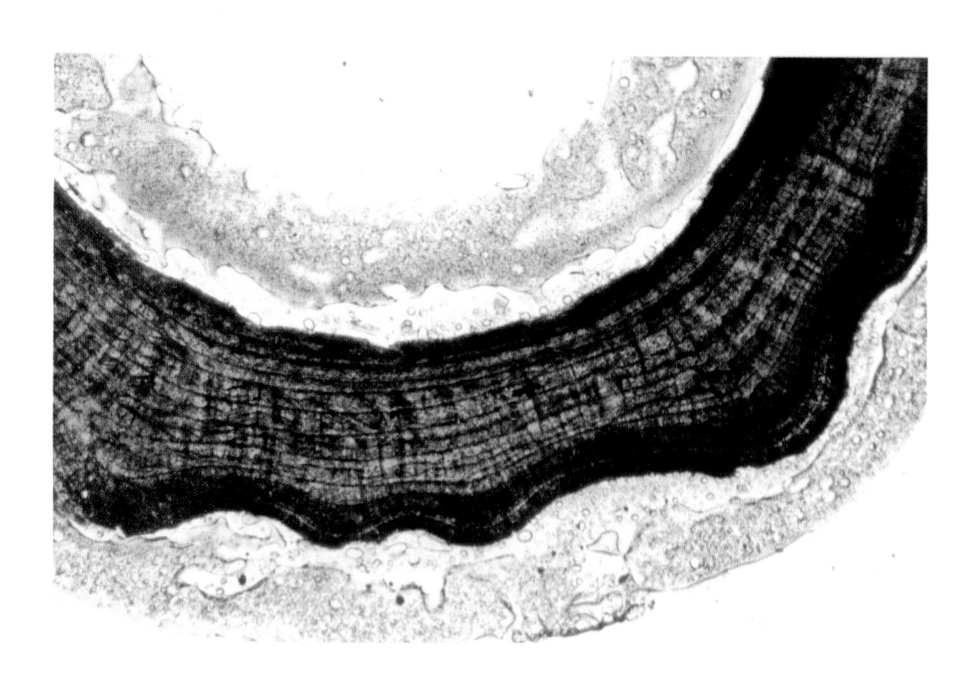

Pliocene ls.
Italy

Close-up of wall structure of the scaphopod *Dentalium sexangulare*. External plications are visible, and concentrically laminated crossed-lamellar structure is displayed along with growth bands.

X.N. 0.22 mm

Mississippian Dartry Ls.
Ireland

Cross section of a single cephalopod (nautiloid?). Size and shape are the primary distinguishing characteristics. In this case, all chambers are filled with sparry calcite cement.

1.24 mm

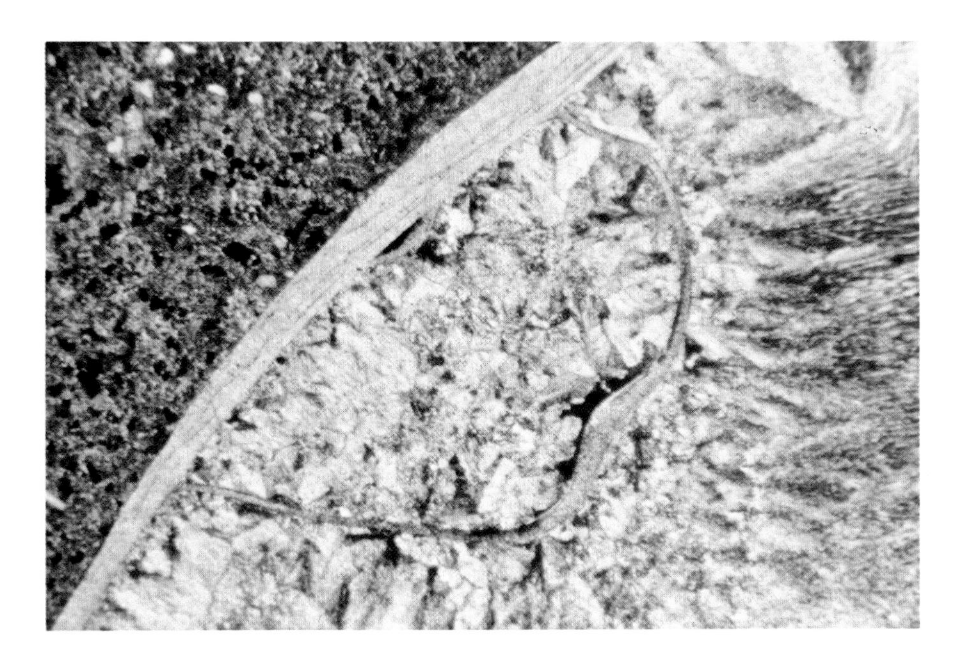

Up. Cretaceous Fox Hills Ss.
South Dakota

Thin-walled ammonite with complete fibrous spar infill of chambers. Ammonites are best recognized by their shapes--they are rarely a major rock-forming element. This shell has homogeneous structure (prismatic and nacreous walls are also known).

X.N. 0.38 mm

Pennsylvanian asphalt
Oklahoma

Transverse section of a nautiloid cephalopod. Note the extensive crushing of these uncemented and unaltered (still aragonitic) shells. Shows original homogeneous prismatic and nacreous structures.

0.38 mm

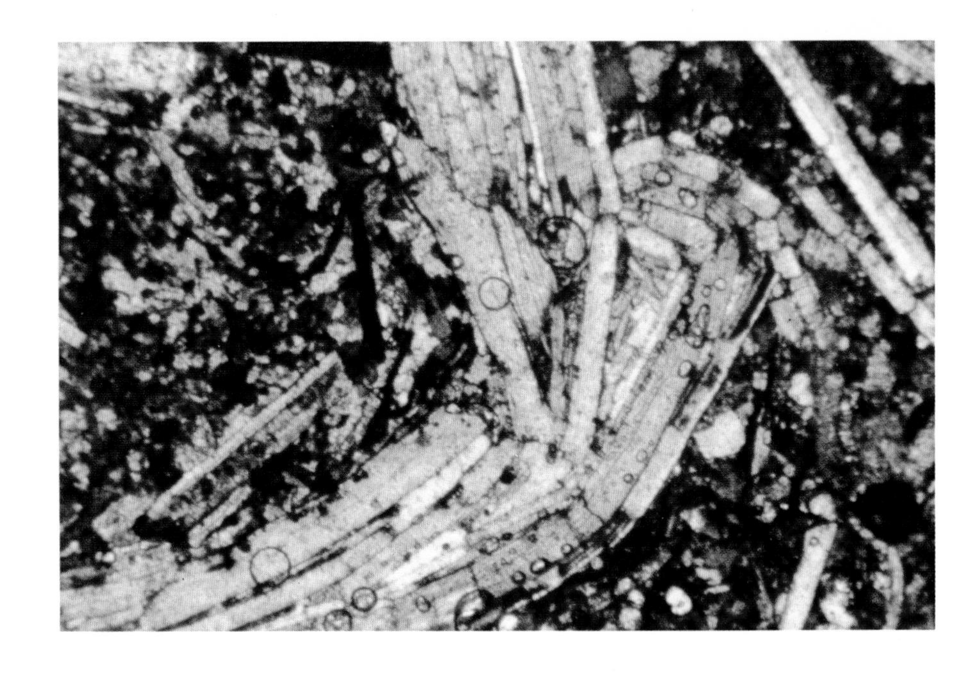

Up. Cretaceous sediment
New Jersey

Cross section of central part of a belemnite rostrum showing radially arranged prismatic structure and massive construction.

X.N. 0.38 mm

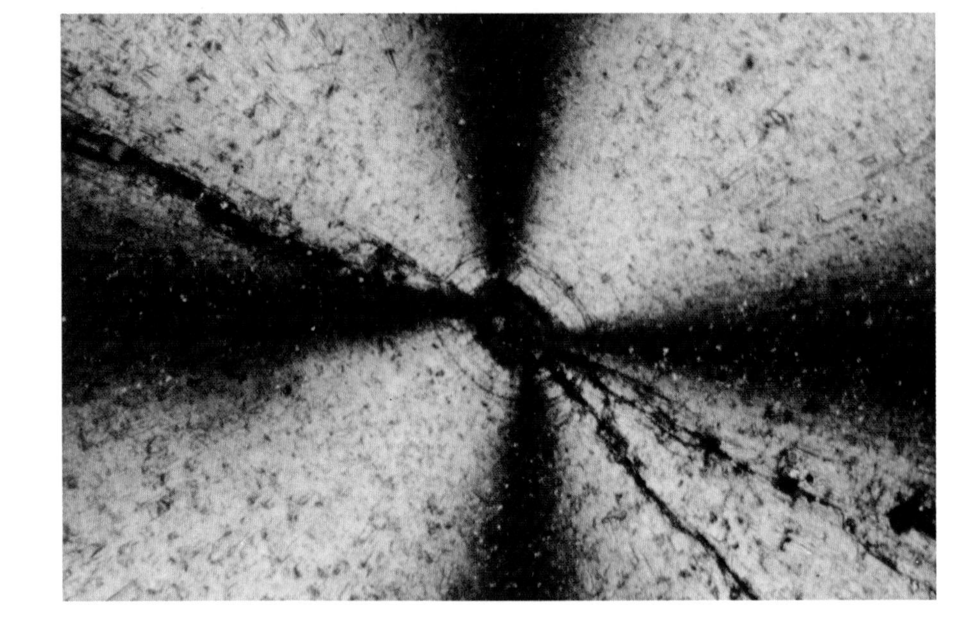

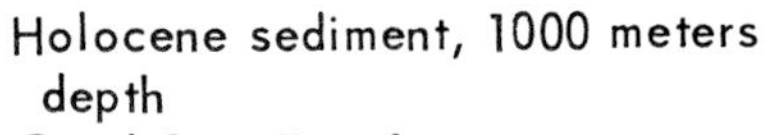

Holocene sediment, 1000 meters depth
Coral Sea, Pacific

Transverse and oblique sections of modern pteropods (generally classed with the gastropods). Pteropods have aragonitic shells with conical shape and homogeneous prismatic wall structure.

X.N. 0.17 mm

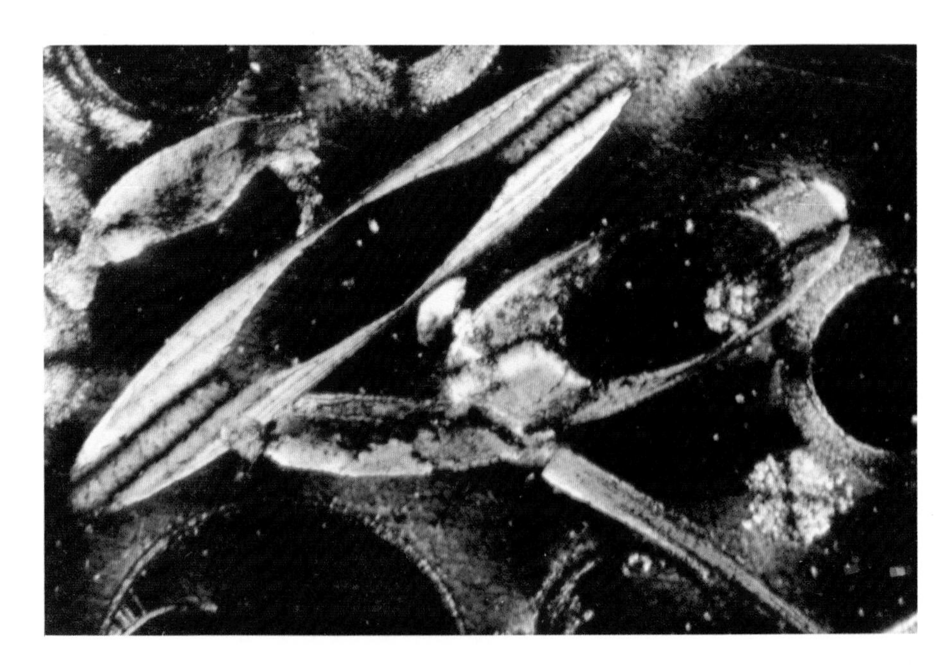

Echinoderms

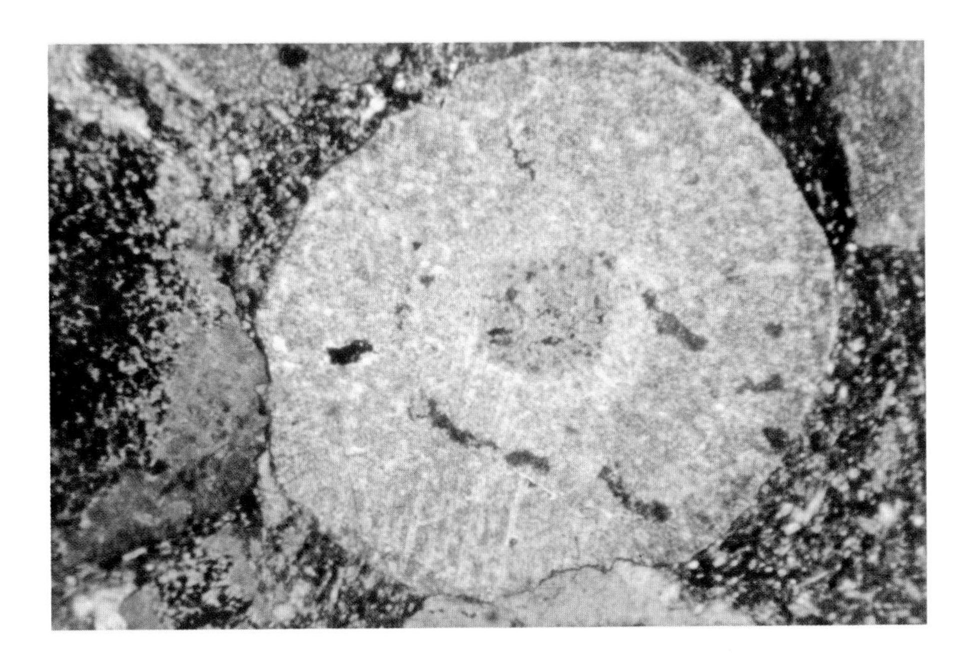

Lo. and Mid. Pennsylvanian Marble Falls Ls.
Texas

Large crinoid ossicle showing typical characteristic of echinoderm grains--single-crystal extinction. Circular shape and central canal are common in many crinoidal grains. Note embayment by adjacent grains.

X.N. 0.38 mm

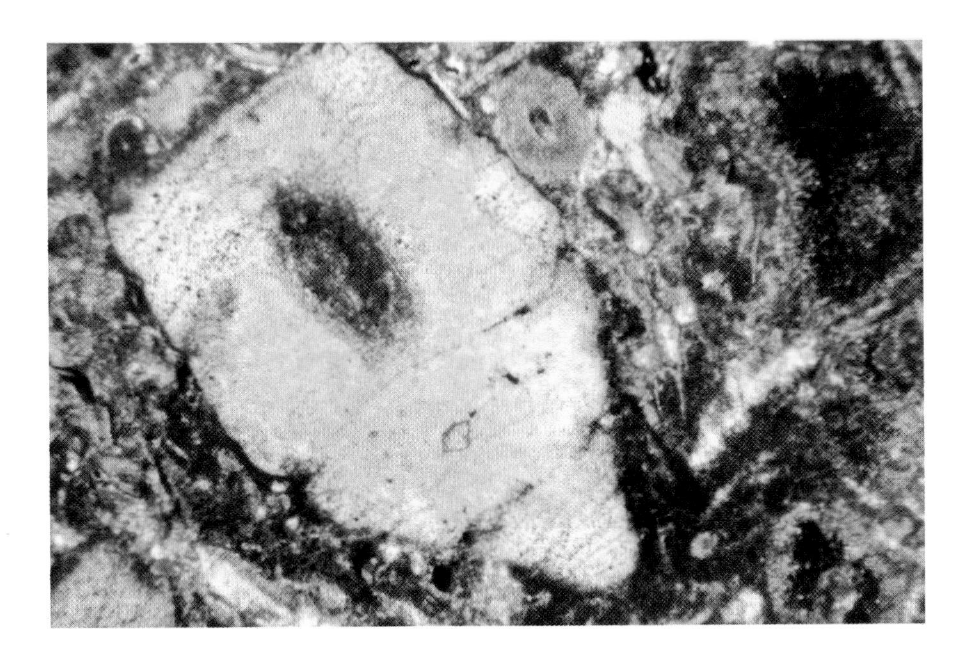

Up. Silurian part of Tonoloway-Keyser Ls.
Pennsylvania

Large and small crinoid fragments. Grains still retain single-crystal structure but microtexture is obliterated. Note central canal.

X.N. 0.38 mm

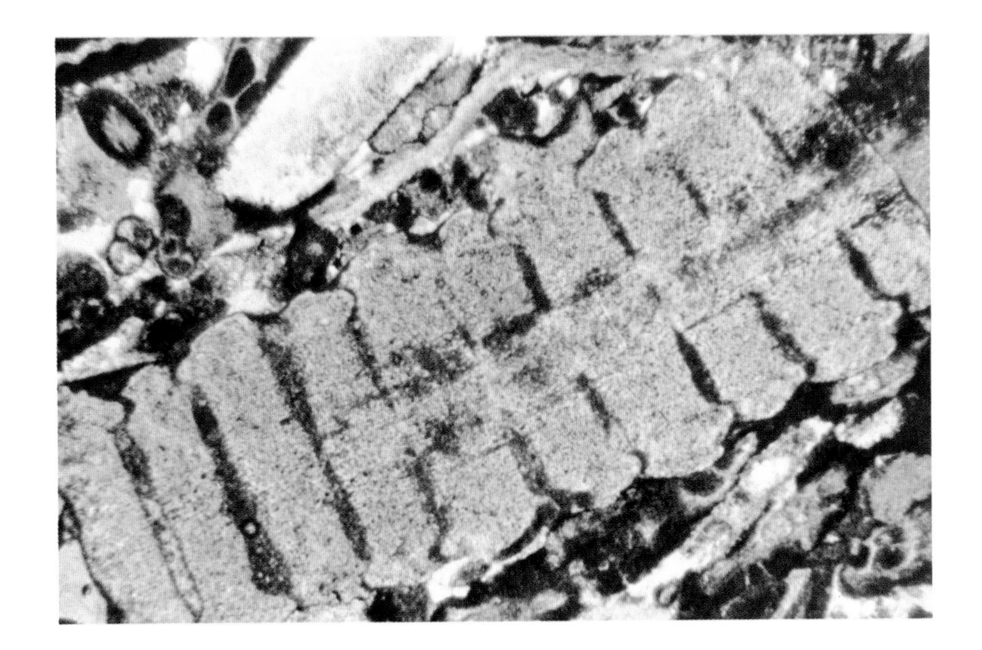

Up. Mississippian Salem Ls.
Indiana

Longitudinal section of a crinoid stem showing a series of connected stem segments (columnals). It is quite unusual to find crinoid columnals still articulated. Also note axial canal in center, and unit extinction.

X.N. 0.37 mm

Paleozoic crinoidal ls.
United States

Crinoid arm plate shown as the nucleus of an ooid. The U-shape is characteristic of arm plates, and the single-crystal extinction and coarse porous structure identify the grain as being echinodermal.

0.06 mm

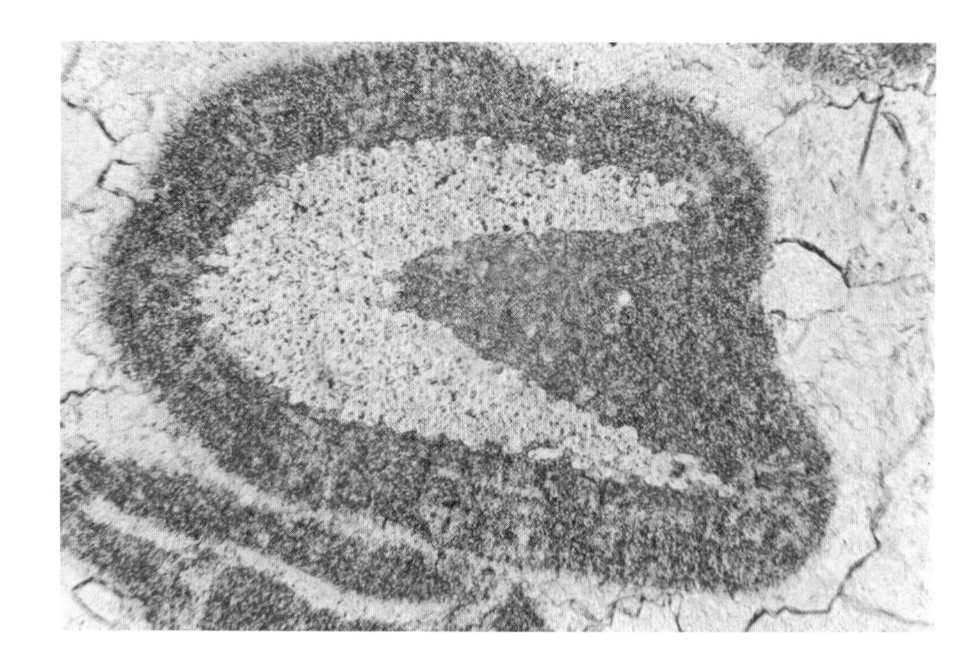

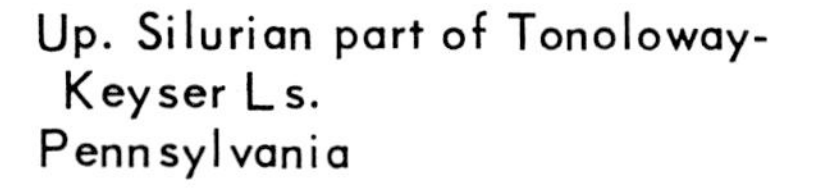

Up. Silurian part of Tonoloway-Keyser Ls.
Pennsylvania

Echinoderm fragment with single-crystal structure and large overgrowth in optical continuity (syntaxial rim cement). Twin lines of calcite crystal are continuous from grain to cement.

X.N. 0.24 mm

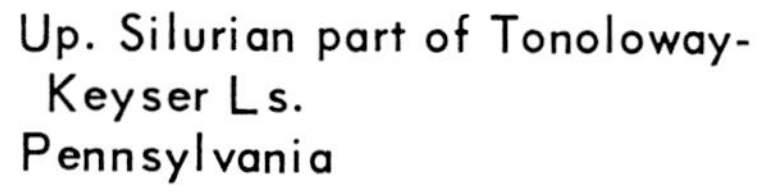

Up. Silurian part of Tonoloway-Keyser Ls.
Pennsylvania

Crinoid fragment with single-crystal structure, and much of the internal part replaced by chert. Crinoid fragments are very susceptible to this type of replacement.

X.N. 0.38 mm

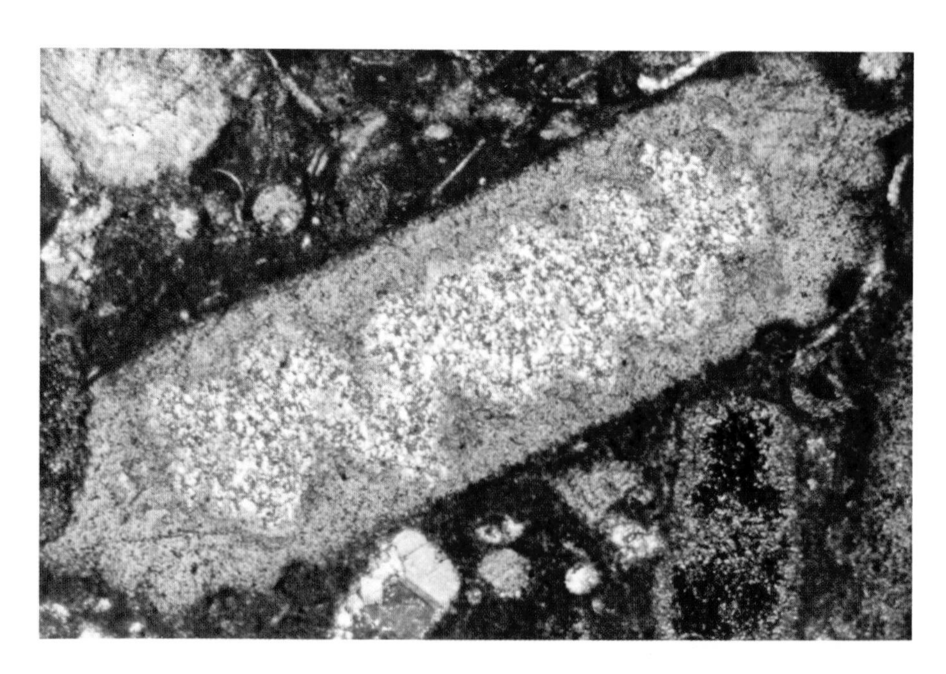

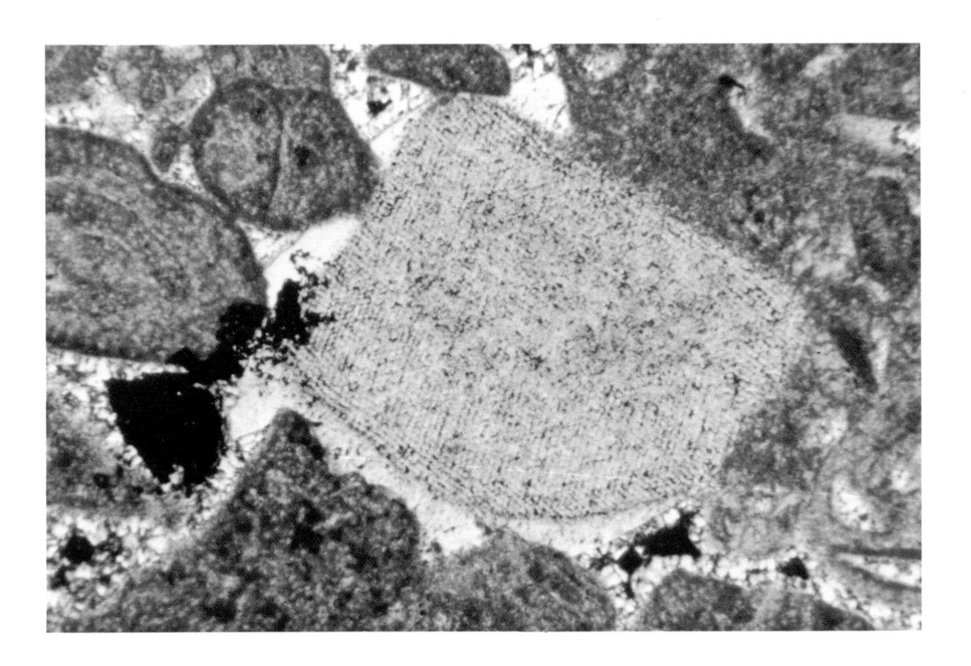

Up. Eocene Ocala Fm.
Florida

Large echinoderm fragment showing characteristic single-crystal extinction and uniform granular microtexture (small pores filled with "dirt"). Also note early calcite overgrowth in optical continuity with grain and later silica cement.

X.N. 0.30 mm

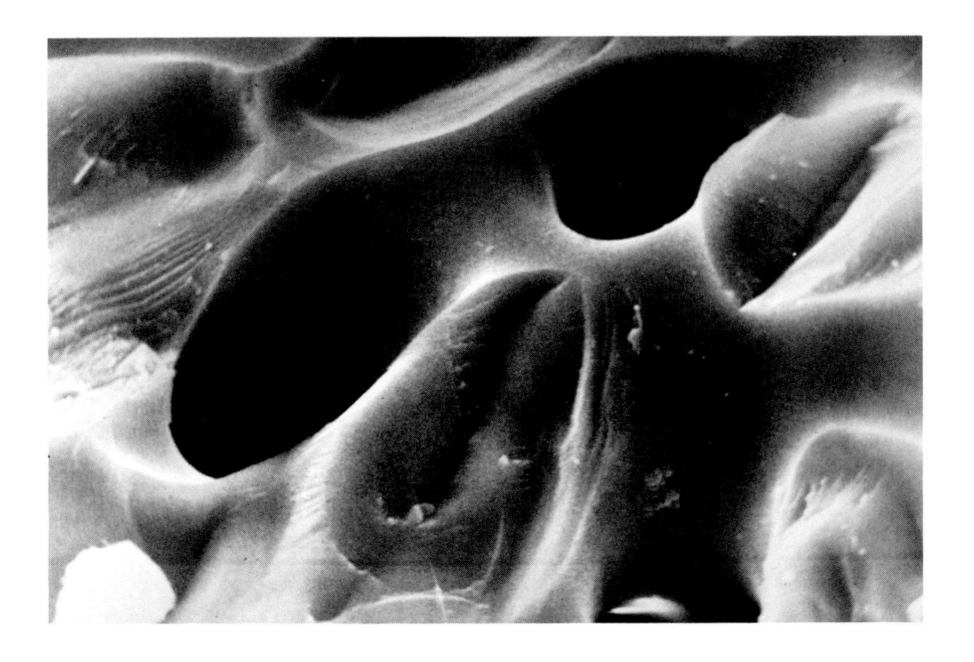

Holocene sediment
Belize (British Honduras)

SEM view of a fractured piece of echinoid wall. Illustrates the very solid, single-crystal construction and the network of large openings which transect the structure.

SEM Mag. 1600X 7.8 μm

Up. Cretaceous Upper Chalk
England

A large echinoid fragment in association with *Inoceramus* prisms and planktonic Foraminifera. The range of possible fragment shapes from the disintegration of an echinoid is enormous, but all are identifiable through their extinction behavior.

X.N. 0.30 mm

Lo. and Mid. Pennsylvanian Bloyd Fm.
Oklahoma

Longitudinal section of an echinoderm spine. Note the bulbous attachment area at one end and the elongate, ribbed, terminated spine itself. As with other echinoid grains, the entire spine is composed of a single calcite crystal with unit extinction.

X.N. 0.20 mm

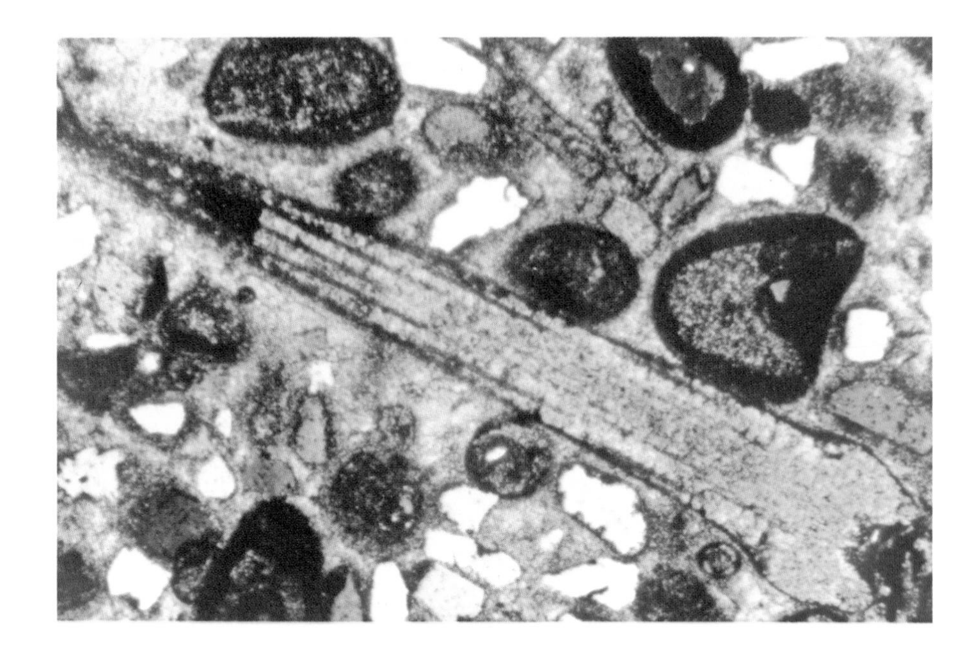

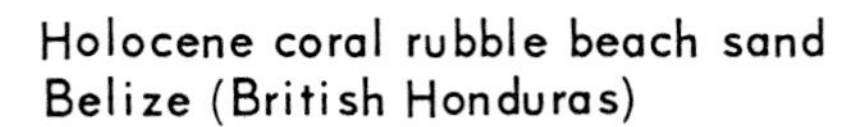

Holocene coral rubble beach sand
Belize (British Honduras)

Cross section of an echinoderm spine showing single-crystal structure and very characteristic lacy pattern.

X.N. 0.24 mm

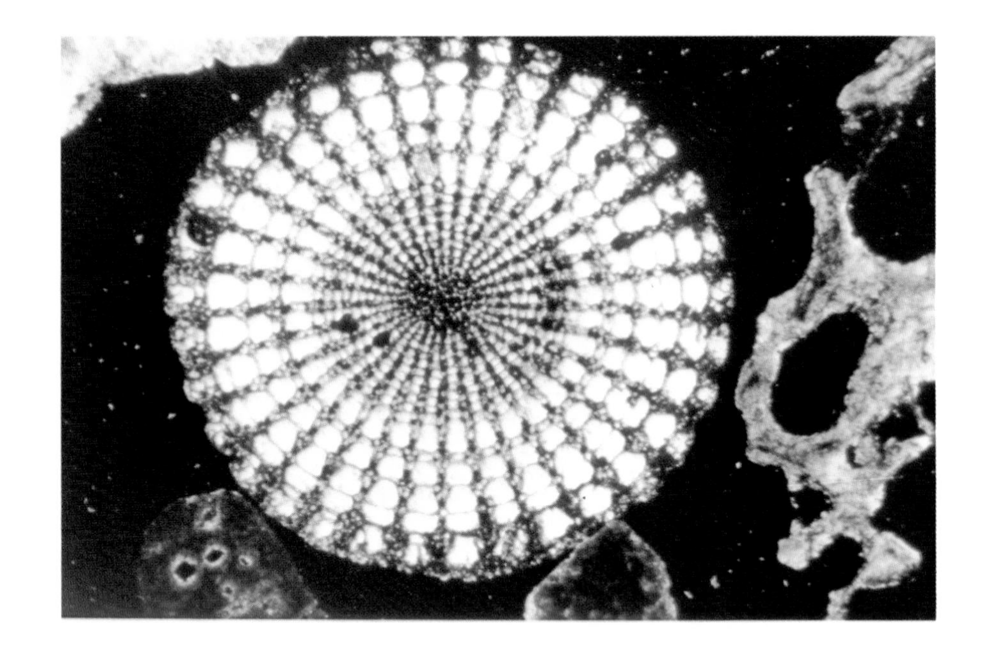

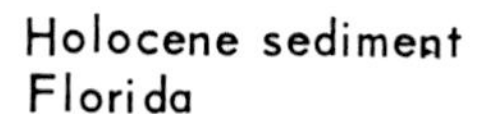

Holocene sediment
Florida

This is a clorox residue of a modern Holothurian (sea cucumber). Embedded within the fleshy tissue of these organisms are these small, irregular, perforated plates. Upon death and organic decay of the holothurian, these plates may be released to the sediment. Fossil examples are known, but are rare.

0.016 mm

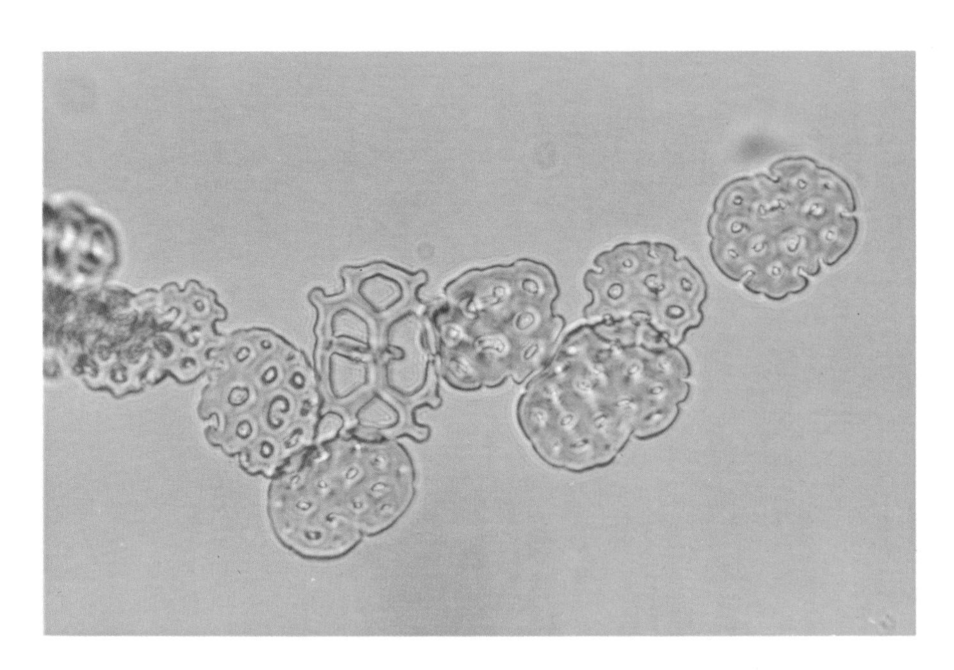

Trilobites and Ostracods

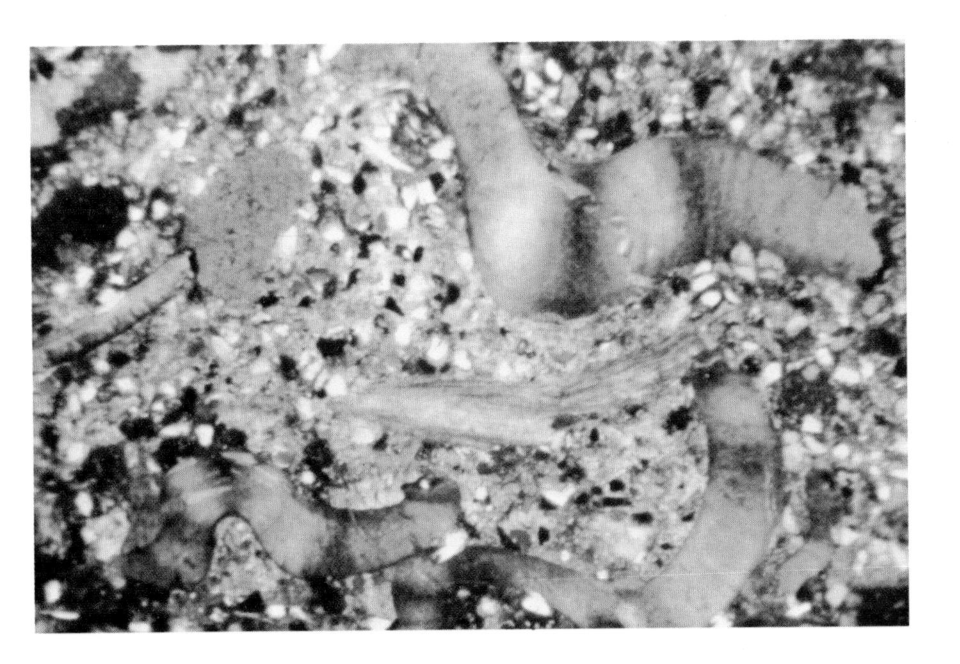

Up. Ordovician Reedsville Shale
Pennsylvania

Several trilobite fragments in a sandy matrix. Grains show fine, uniform structure with extinction bands (in polarized light) which sweep across the grain as the stage is rotated.

X.N. 0.38 mm

Mid. Silurian Rochester Shale
New York

Single trilobite fragment with typical convoluted shape (and turned-under end), uniform skeletal structure, and extinction bands. Some alteration apparent.

X.N. 0.30 mm

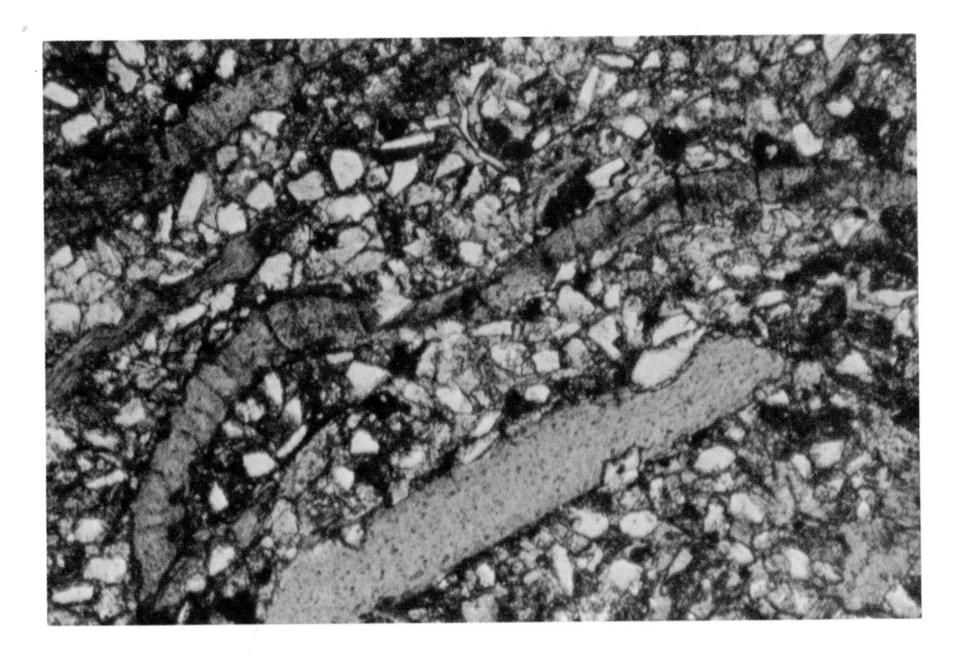

Up. Ordovician Reedsville Shale
Pennsylvania

Large trilobite fragment in center of photo shows fine perforations (canaliculi) filled with micritic sediment. Shell wall still shows uniform or homogeneous prismatic structure and sweeping extinction.

0.25 mm

Eocene Green River Fm.
Wyoming

Ostracodes. Note thin walls, small size, and shell shape. The thin-walled nature of the carapace is often a characteristic indicator of freshwater forms.

0.38 mm

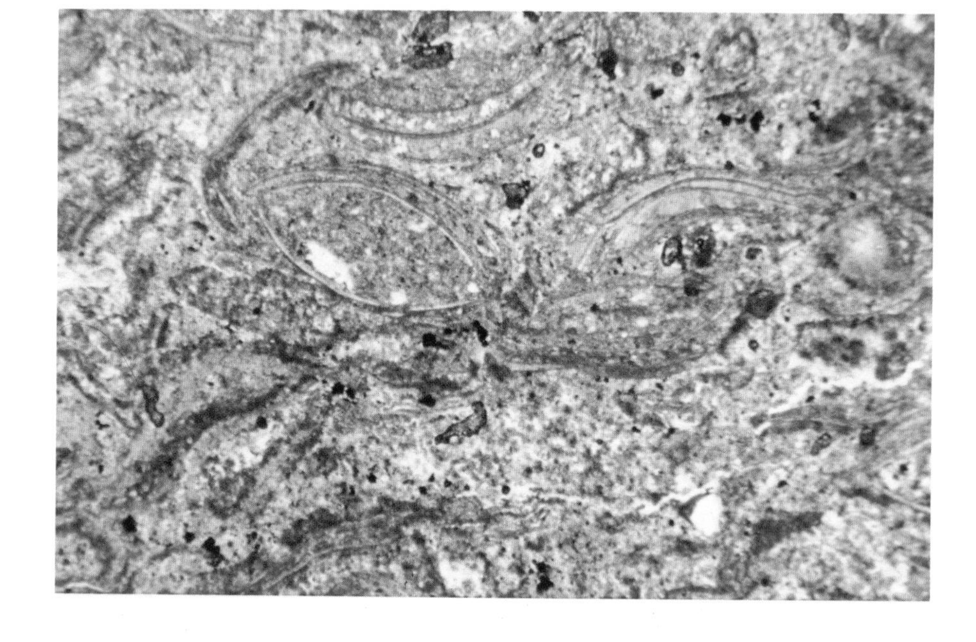

Up. Mississippian Hindsville Ls.
Oklahoma

Section of a single valve of an ostracode. This group is characterized by small size (average 0.5-0.75 mm in post-Paleozoic forms and 1-2 mm in Paleozoic ones), homogeneous prismatic wall structure with a calcite and chitin composition, and a shape like a pelecypod. Distinguished from pelecypods by smaller size and homogeneous wall structure.

0.07 mm

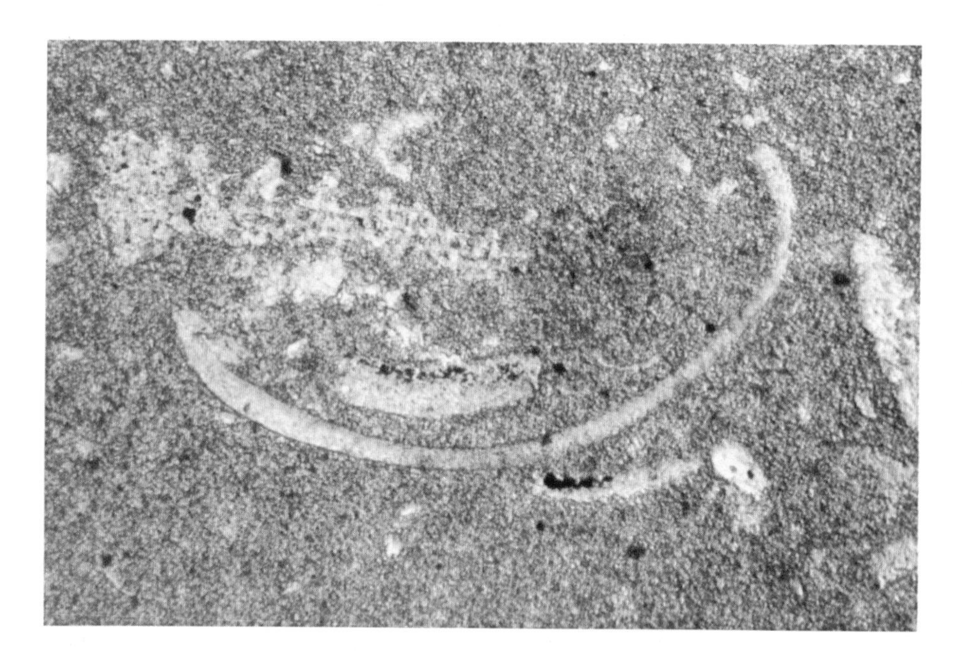

Up. Mississippian Hindsville Ls.
Oklahoma

Cross section of a single ostracode valve, illustrating the homogeneous prismatic wall structure with sweeping extinction, the small but thick shell, and the typical V-shaped carapace termination.

X.N. 0.05 mm

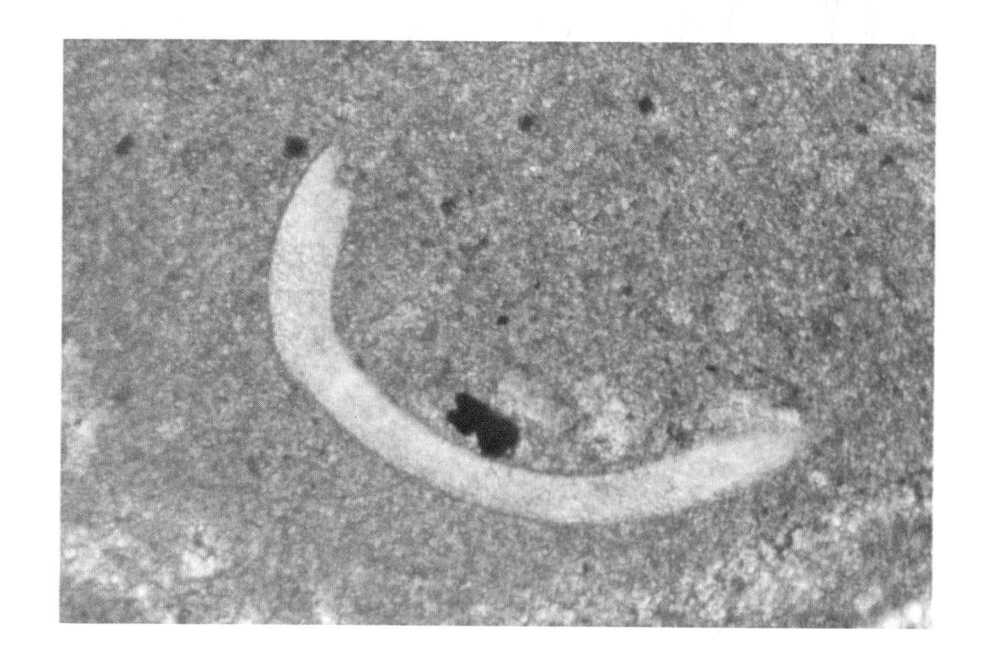

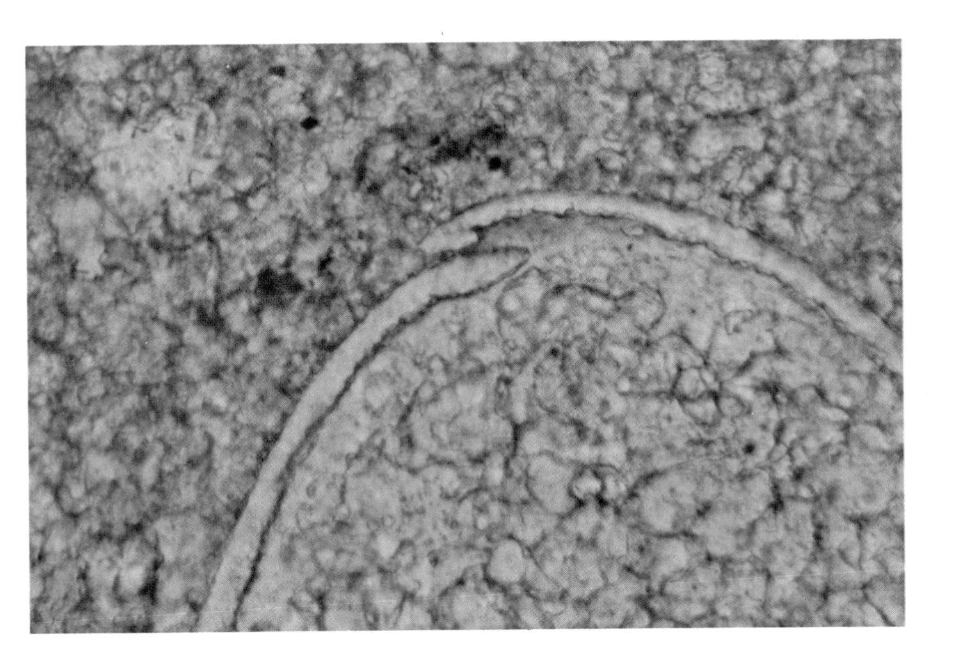

Mississippian ls.
Ireland

A pair of ostracode valves showing common overlap of one valve over the other, as well as the recurved termination of one valve.

0.027 mm

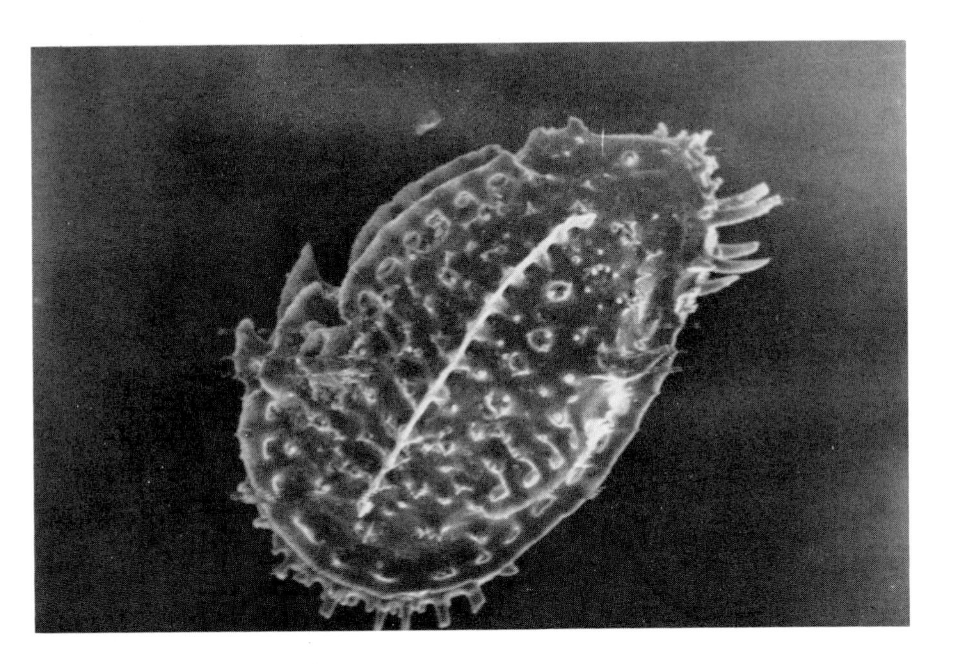

Holocene sediment
Belize (British Honduras)

SEM view of one surface of an ostracode, probably of the genus *Cativella*, showing irregular carapace outline with small perforations and spiny protrusions.

SEM Mag. 100X 125 μm

Plant Fragments

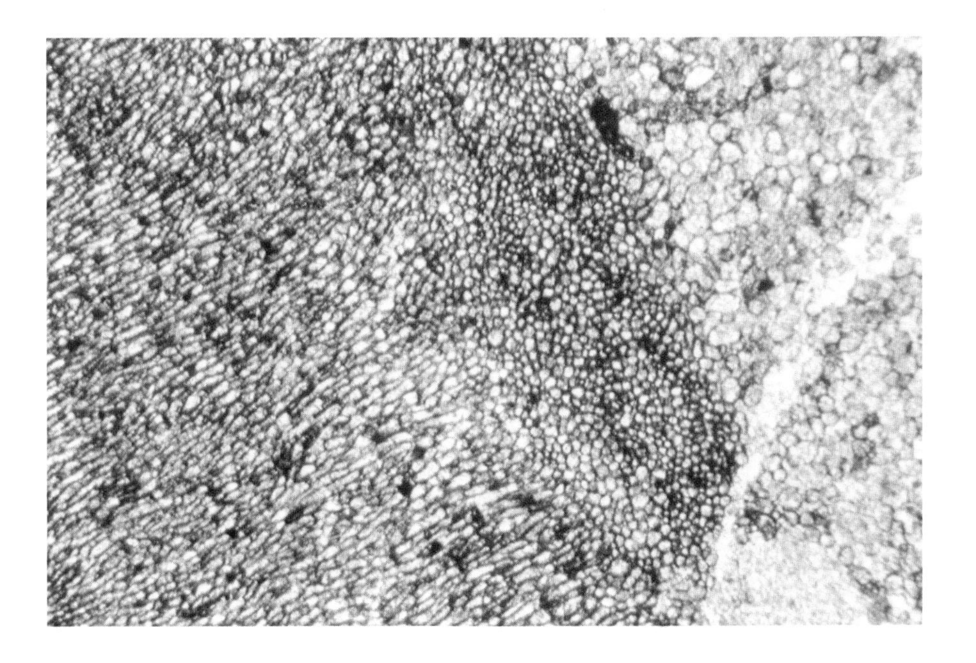

Pennsylvanian Mazon Creek coal balls
Illinois

Part of an HF acid acetate peel of a silicified bed of plant fossils. This is the seed (Pachytesta) of the Paleozoic pteridosperm Medullosa. Note the preservation of the cellular structure--distinguished from red algae by much coarser cell size.

0.25 mm

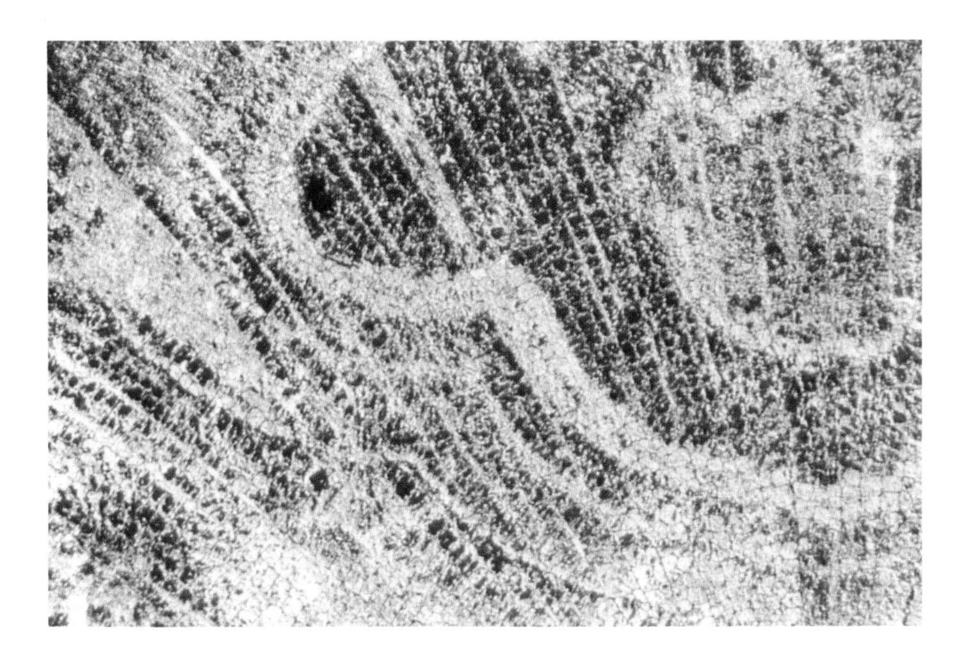

Pennsylvanian Mazon Creek coal balls
Illinois

Part of an HF acid acetate peel. Shows moderate preservation of cellular structure in Stigmaria, a root of Lepidodendrales. Note radiating rays of cellular tissue and curved, irregular lines which mark silica replacement fronts.

0.64 mm

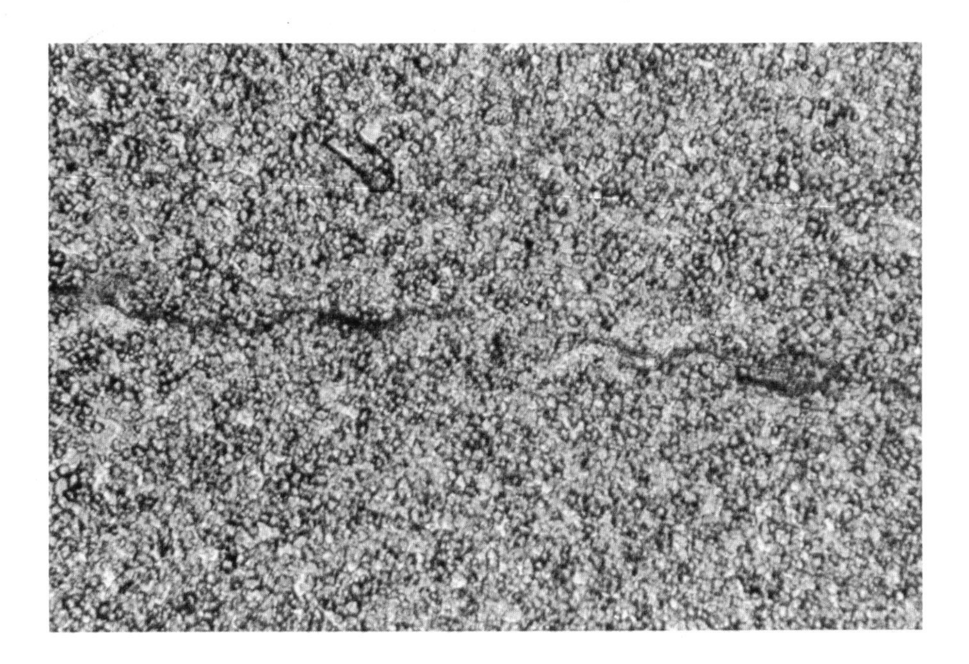

Oligocene Creede Fm.
Colorado

Brownish wisps in center of photo are small pieces of plant material lying parallel to bedding. The color of such organic matter depends upon the degree of thermal alteration the rock has seen (yellow at very low temperatures to black at very high temperatures).

0.04 mm

Up. Cretaceous Red Bank Sand
New Jersey

Stained palynological preparation showing a single trilete spore. Although these grains can be seen in thin sections of carbonate rocks, they are much more easily seen in stained insoluble residues. They can be used in stratigraphic as well as environmental studies.

0.018 mm

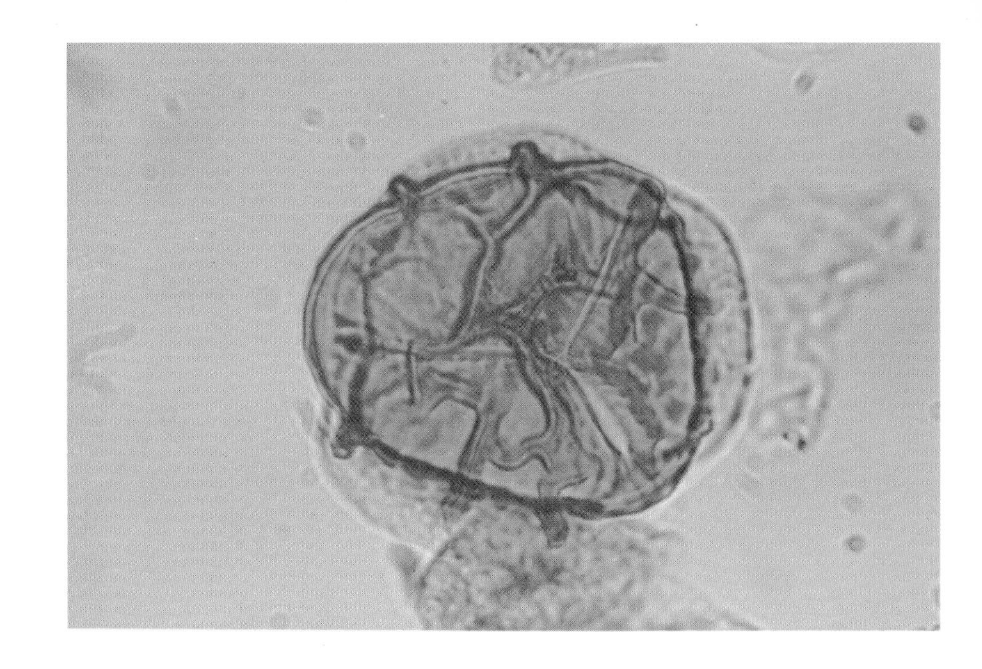

Up. Cretaceous Red Bank Sand
New Jersey

Stained palynological preparation showing a single spore of Sphagnum (a moss). These grains are noncalcified and consist entirely of organic matter.

0.018 mm

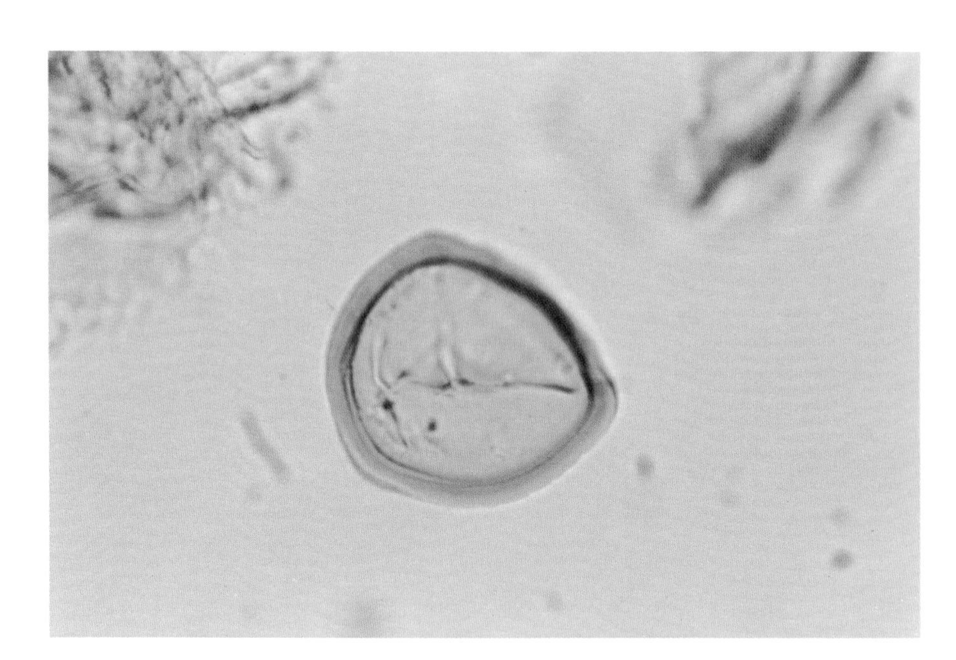

Up. Cretaceous Mount Laurel Sand
New Jersey

Stained palynological preparation showing a single pollen grain of Pinus. This is a typical bissacate form.

0.018 mm

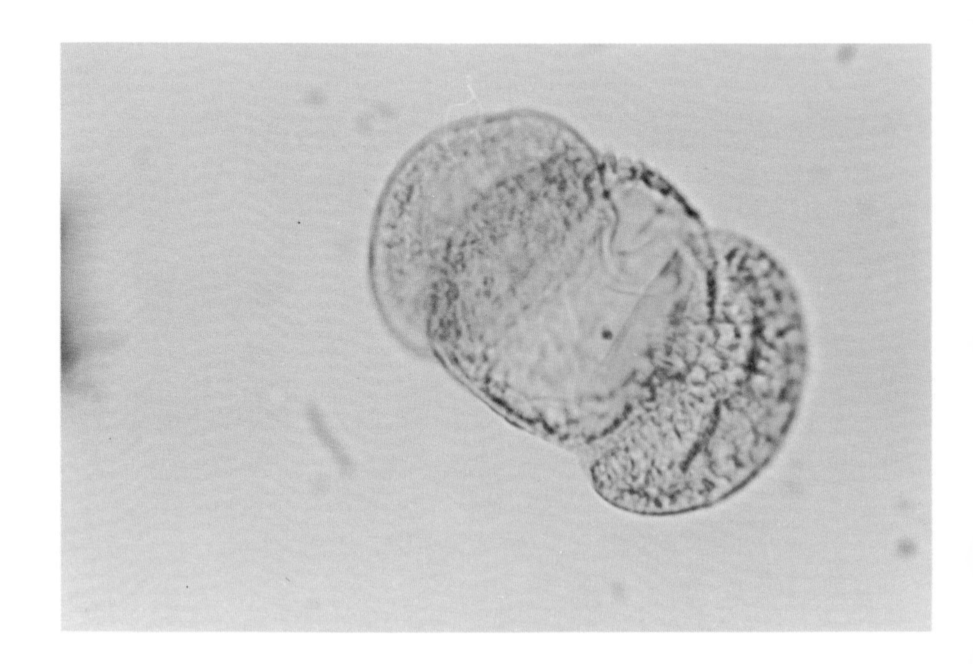

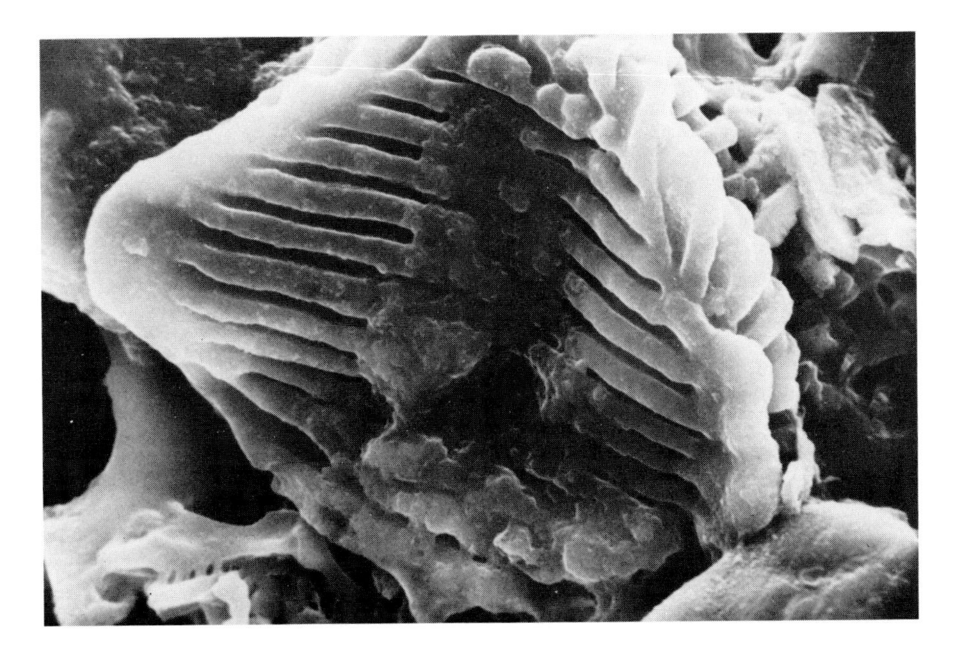

Lo. Cretaceous Dakota Group
Colorado

SEM view of a single trilete fern spore, *Appendicisporites* sp. Although these grains are hard to see in normal SEM photos of carbonates, they are frequently apparent in insoluble residues.

SEM Mag. ca. 2000X 6.2 μm

Lo. Cretaceous Dakota Group
Colorado

SEM view of a simple trilete spore, *Cyathidites* sp. which is partially collapsed. The 3 radiating tetrad scars are clearly visible.

SEM Mag. 2000X 6.3 μm

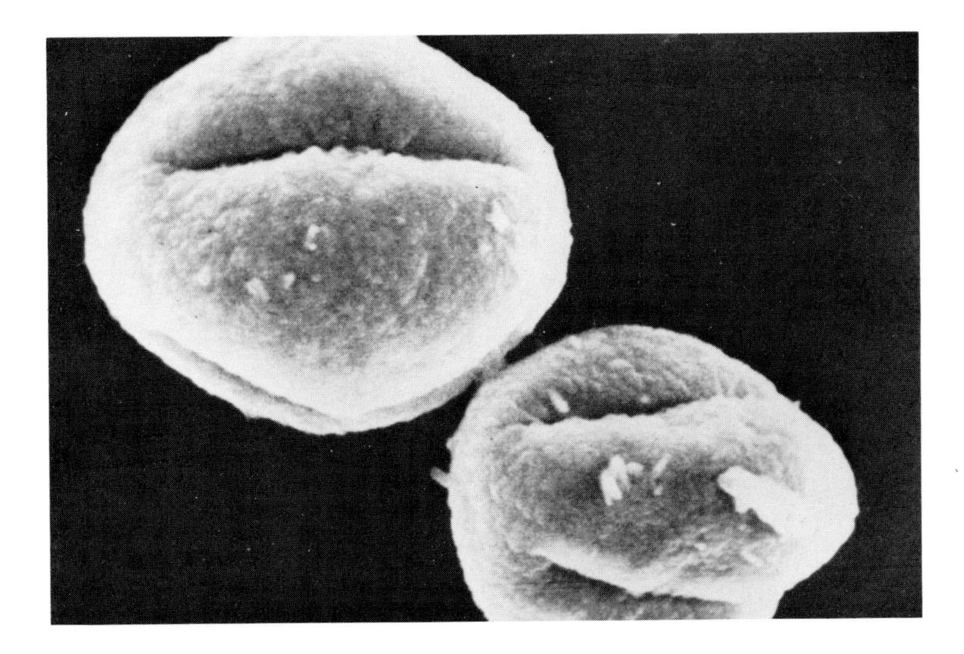

Lo. Cretaceous Dakota(?) Group
Colorado

SEM view of two tricolpate angiosperm pollen grains, *Tricolpites* sp. with the pollen apertures visible.

SEM Mag. ca. 1000X 12.5 μm

Lo. Cretaceous Dakota Ss.
Colorado

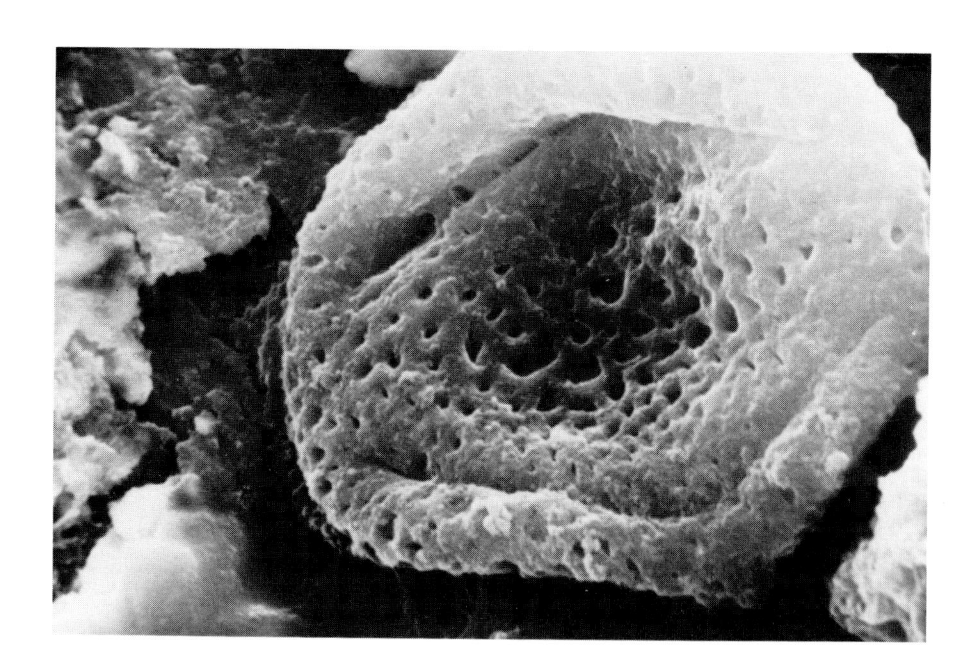

SEM view of a single tricolpate angiosperm pollen grain, *Retitricolpites* sp. It shows a pair of bowed pollen apertures as well as a reticulated surface of small pores.

SEM Mag. ca. 2000X 6.0 μm

Other Carbonate Grains

Pellets

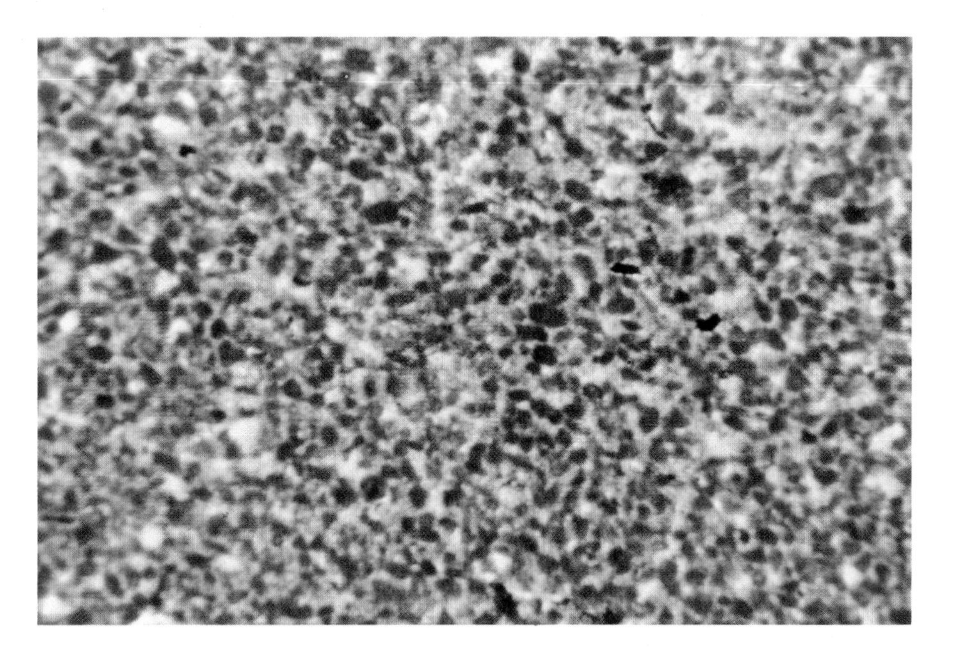

Lo. and Mid. Ordovician Deepkill Shale
New York

A pelsparite. Abundant pellets of unknown, but probably fecal, origin; cemented by sparry calcite. Note the uniformity of grain size and shape.

0.38 mm

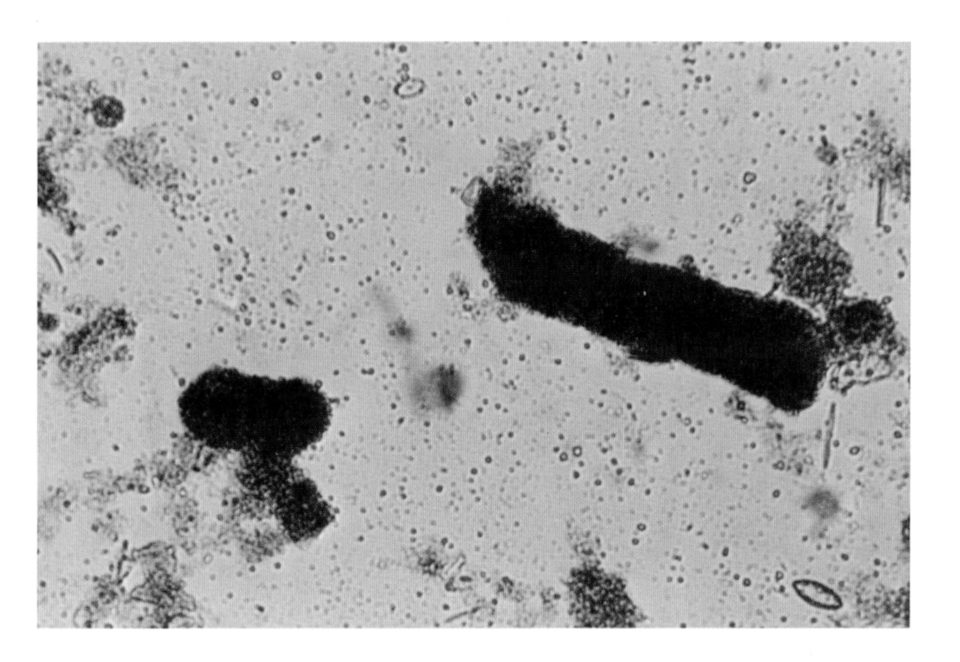

Holocene sediment
Mexico

Two pellets from a modern lagoonal setting. These are fecal pellets in all probability and show the common rounded, rod-shaped outline of fecal pellets. In this sediment, the uniformity of size and shape of pellets aids in their identification as being of fecal origin.

0.09 mm

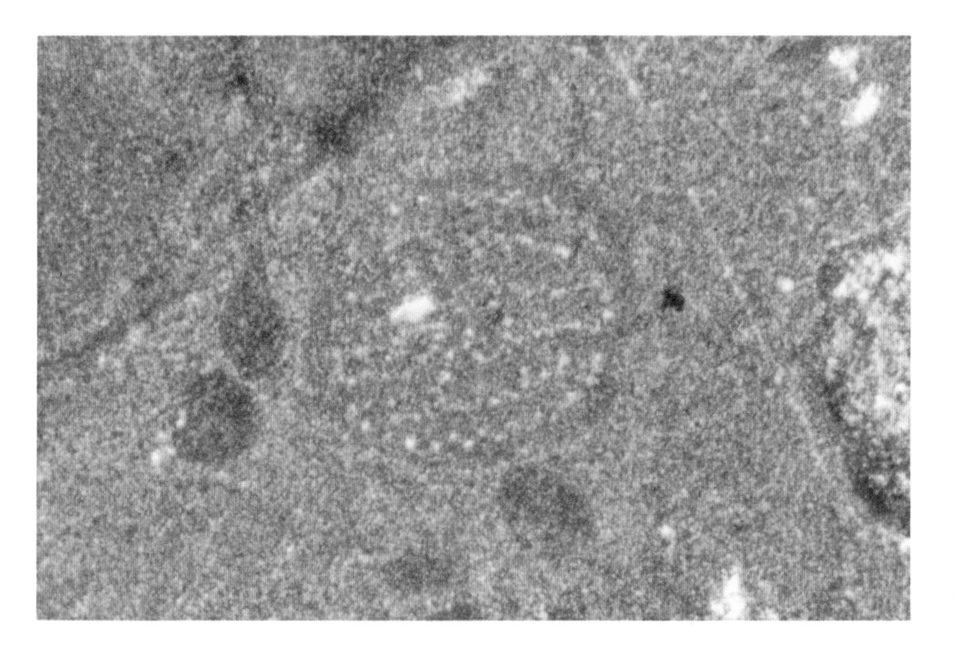

Jurassic Ronda unit of Subbetic
Spain

A *Favreina* sp. pellet. Note the regular arrangement of hollow tubules which are formed by elongate appendages from the intestinal walls. In the Holocene, such pellets are produced by anomuran crustaceans. These pellets can be found as major rock-forming elements in Jurassic and Cretaceous limestones.

0.17 mm

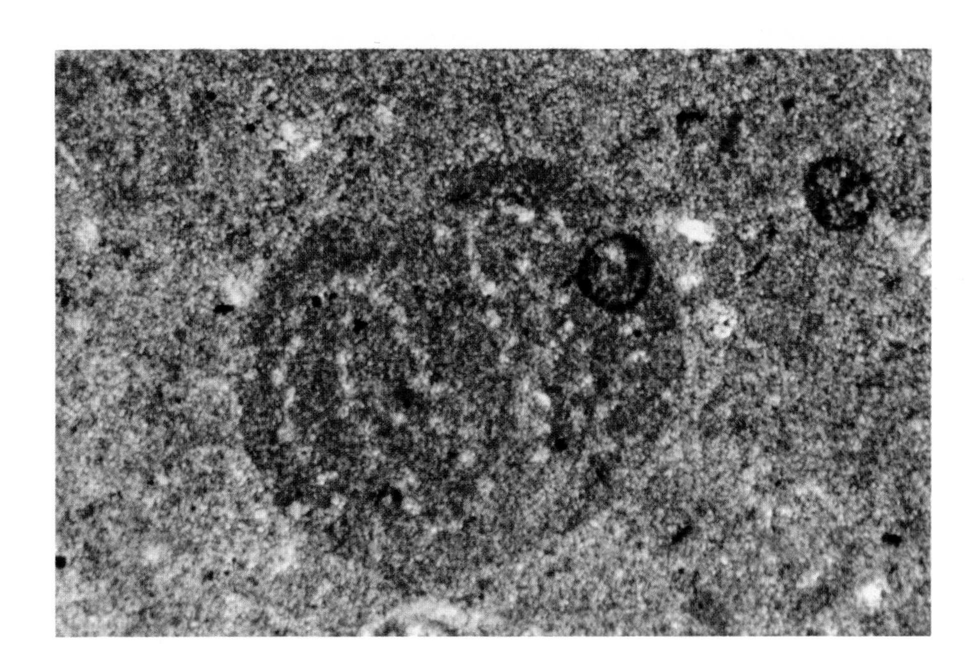

Up. Jurassic Smackover Fm.
Alabama

Another *Favreina* sp. pellet showing a different pattern of internal tubules These pellets have been subdivided into numerous species and have been used for stratigraphic zonation.

0.07 mm

Ooids and Pisolites

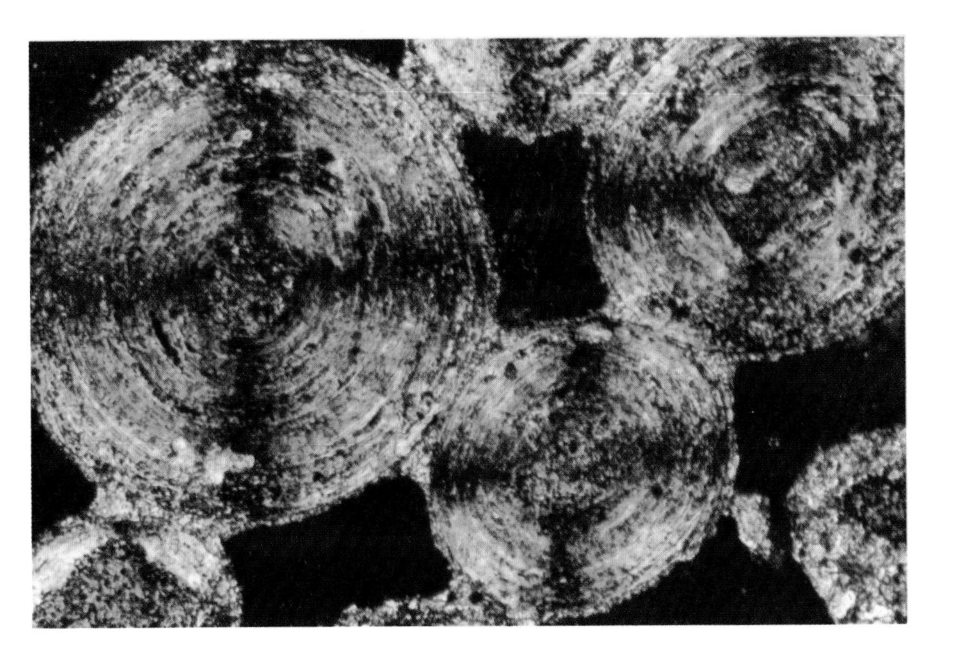

Pleistocene(?) dunes
Bahamas

Modern marine ooids are generally aragonitic, range in size from about 0.1 to 1 mm, and have a tangential arrangement of crystals which yield the type of pseudo-uniaxial cross seen here. Nuclei are varied and can include carbonate or noncarbonate grains. Deposits are often well sorted and form in highly agitated, bank margin environments in open marine, marginal marine, or lacustrine settings.

X.N. 0.11 mm

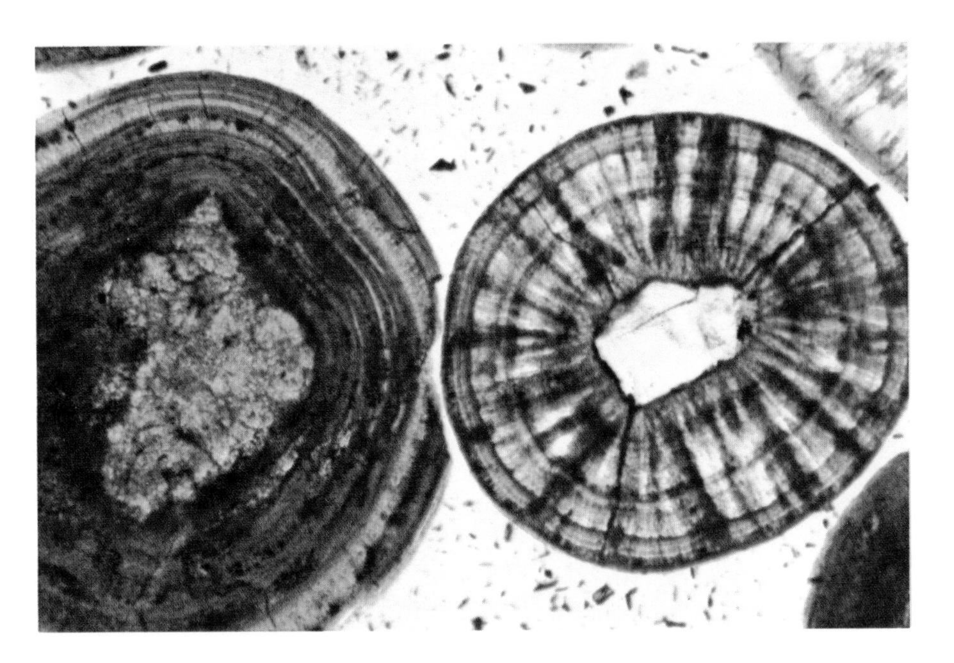

Holocene sediment
Great Salt Lake, Utah

These ooids show the textures commonly found in hypersaline lakes. Most grains have coarse, radiating crystals of bladed to fibrous aragonite interspersed with layers of tangential, very finely crystalline aragonite. The radial texture appears to be disruptive and therefore secondary. Such hypersaline lake deposits commonly have a very high percentage of broken grains.

0.10 mm

Holocene sediment
Florida

These ooids were formed synthetically in a water purification plant (thus without algal influence). Nuclei are quartz and glauconite while the oolitic layers are coarse, radially oriented blades of probable calcite. Note the similarity in texture to the Great Salt Lake ooids.

X.N. 0.25 mm

Lo. and Mid. Pennsylvanian Bloyd Fm.
Oklahoma

Superficial oolitic coatings on echinoderm fragments. These are thin (one to five layer) concentric coatings which must be carefully distinguished from micrite envelopes. They are important in echinodermal limestones as they isolate the grains from pores, preventing cementation by overgrowths. Common in moderately agitated waters. Note thickened coatings in embayed areas; this is typical of oolitic growth.

0.25 mm

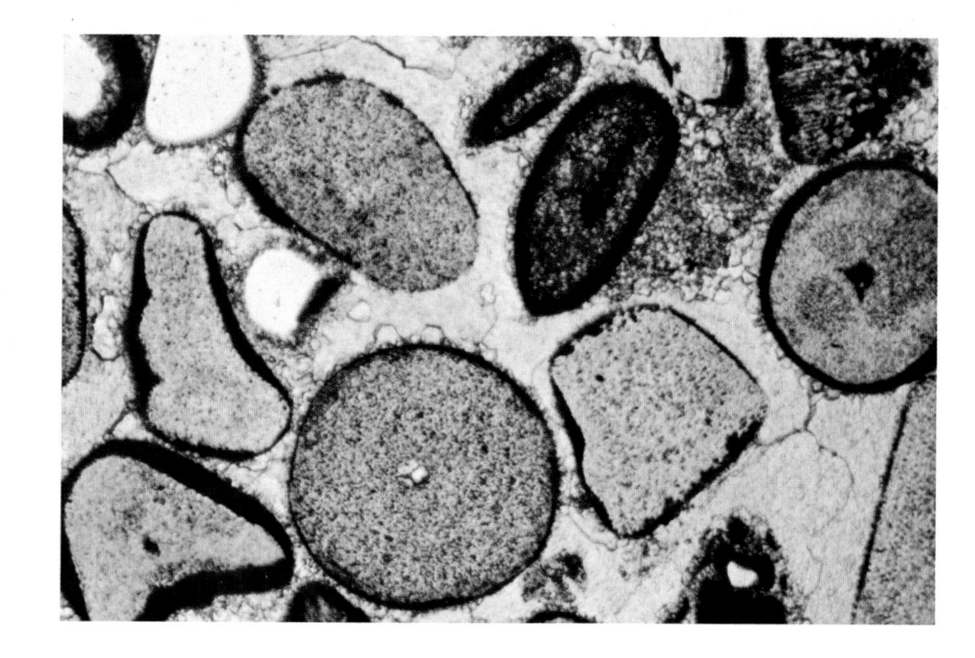

Holocene sediment
Laguna Madre, Texas

The relatively high salinity and moderately low energy environment of Laguna Madre and Baffin Bay, Texas produces superficial ooids with alternating layers of aphanocrystalline aragonite and lighter, coarser, radial high-Mg calcite (both seen here). The coatings are generally thin and often incomplete.

X.N. 0.07 mm

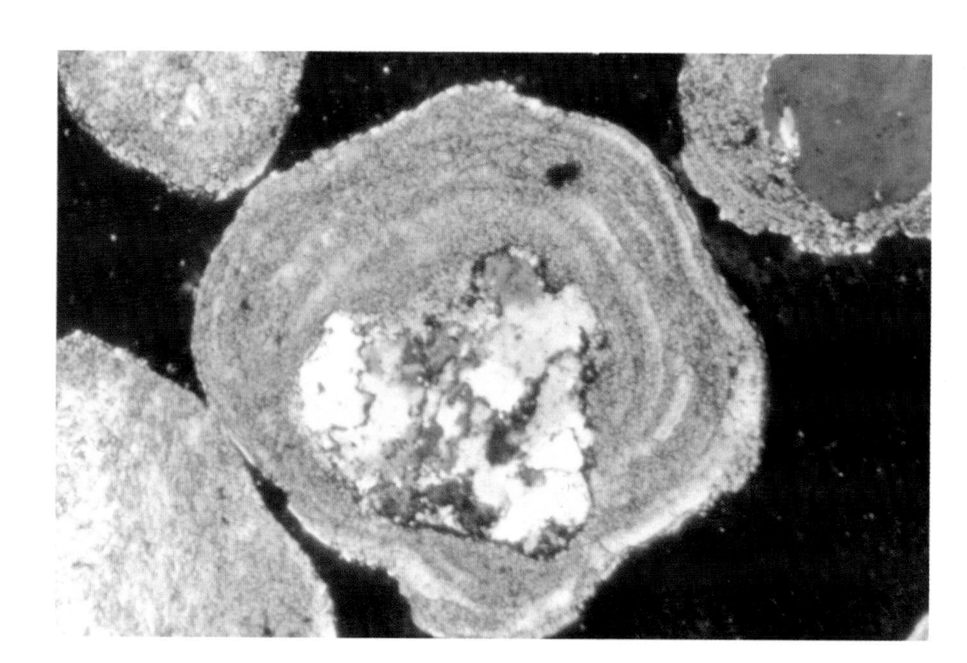

Holocene sediment
Laguna Madre, Texas

This illustrates an incomplete, eccentric oolitic coating which occurs commonly on grains from low energy oolite forming areas. The aragonitic coatings are quite thick, but only cover a portion of the grain. The rest of the grain was probably resting in the bottom sediment and agitation was insufficient to rotate the grain.

X.N. 0.09 mm

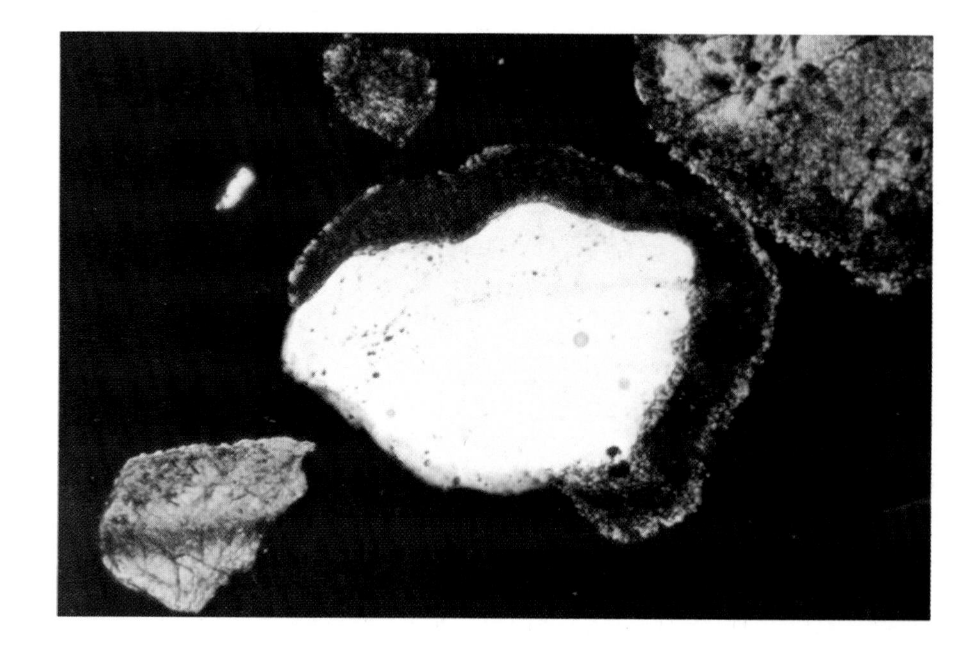

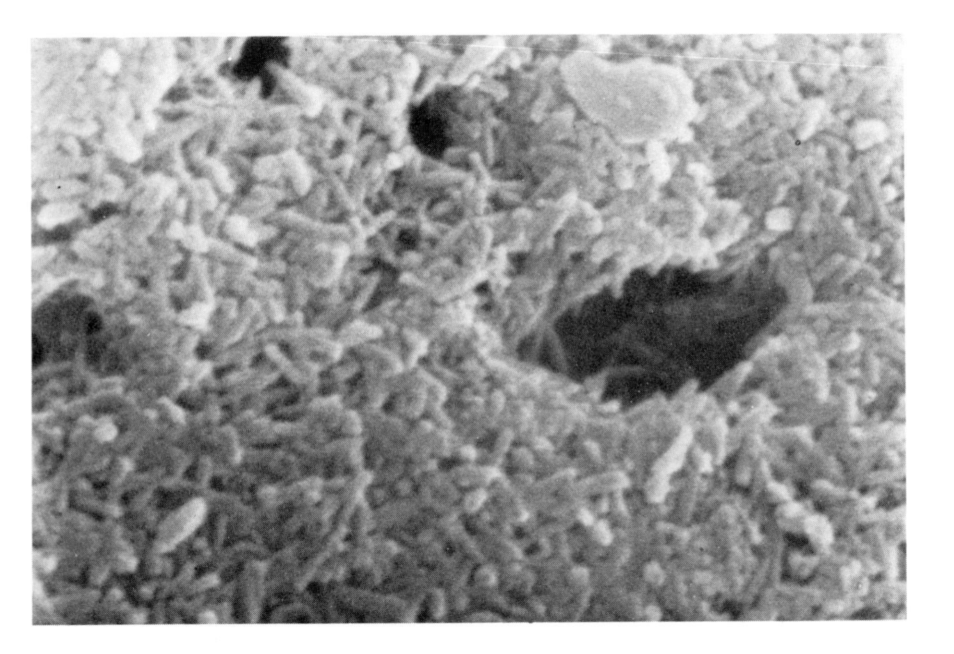

Holocene sediment
Bahamas

SEM view of the surface of a modern aragonitic ooid from an open-marine setting. The small aragonite needles which make up the ooid are seen to be oriented with their long axes tangential to the outer surface of the ooid, but randomly oriented within that curved plane. The large holes are probably algal borings.

SEM Mag. 10,000X 1.3 μm

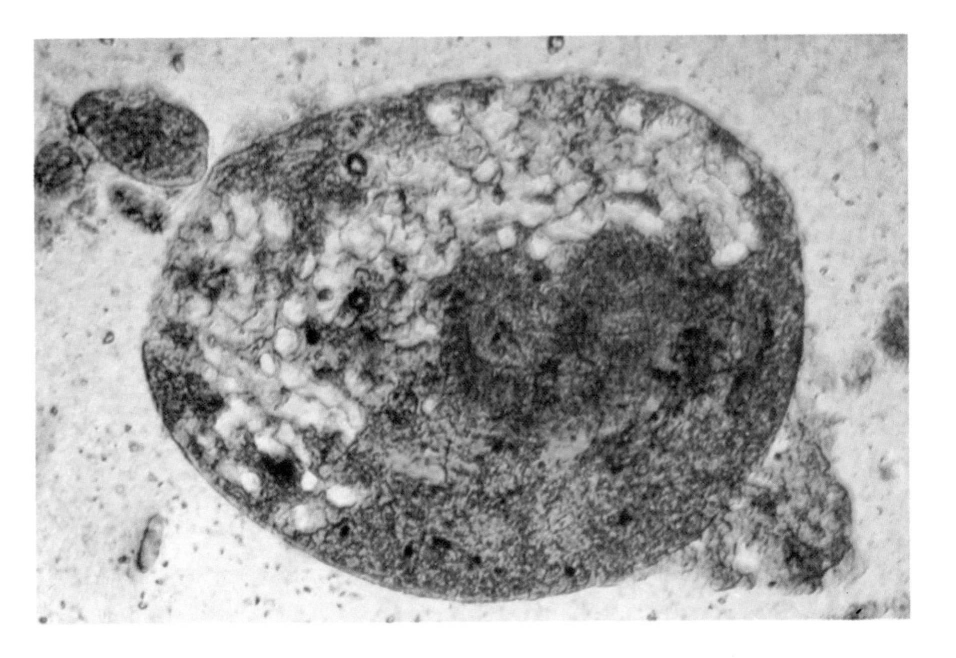

Holocene sediment
Joulters Cay, Bahamas

This is a common stage of alteration of modern ooids. Small portions of the original structure can be seen, but most of the grain has been obliterated by the formation of empty tubes by the boring of algae and other organisms. As these tubes are filled with fine-grained sediment or cement, the ooid takes on a micritized appearance (lower part).

0.05 mm

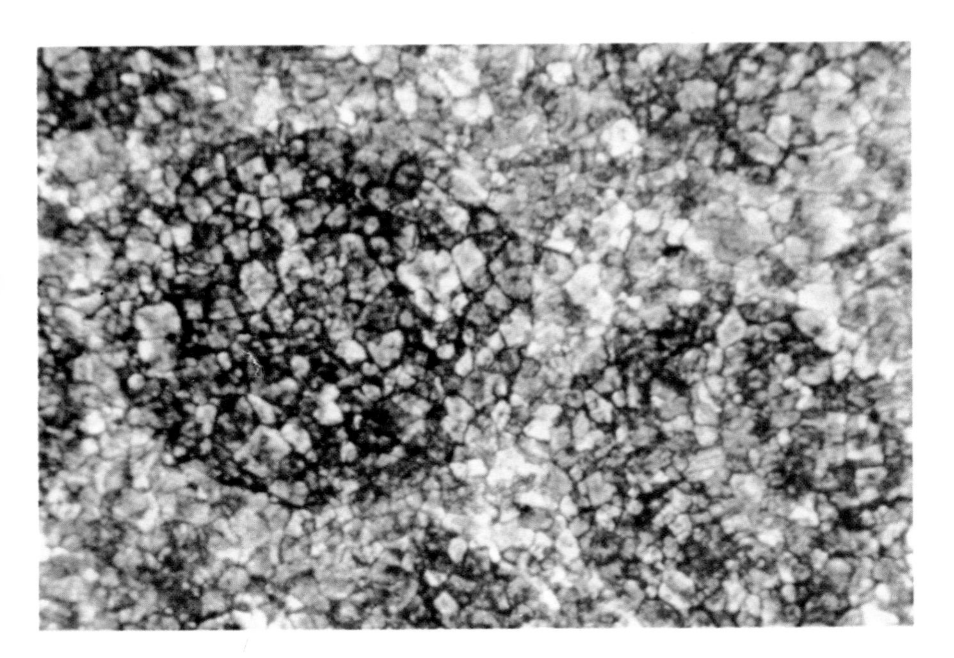

Up. Cambrian Kittatinny Ls.
New Jersey

The textural details of ooids can also be lost during later diagenesis, as in this case, where all internal structure has been obliterated by dolomitization. Grains are identified as ooids on the basis of size, shape, and sorting. Note lack of selectivity in replacement.

0.38 mm

Up. Mississippian Pitkin Ls.
Oklahoma

A very common texture in ancient ooids is the development of a radial fabric of the type seen here. Although the concentric laminations are still clearly visible, they are cut by occasional thin, disruptive radial crystals. This may be produced during alteration from original aragonite to later calcite. Also note spalling of outer cortex of one ooid.

0.09 mm

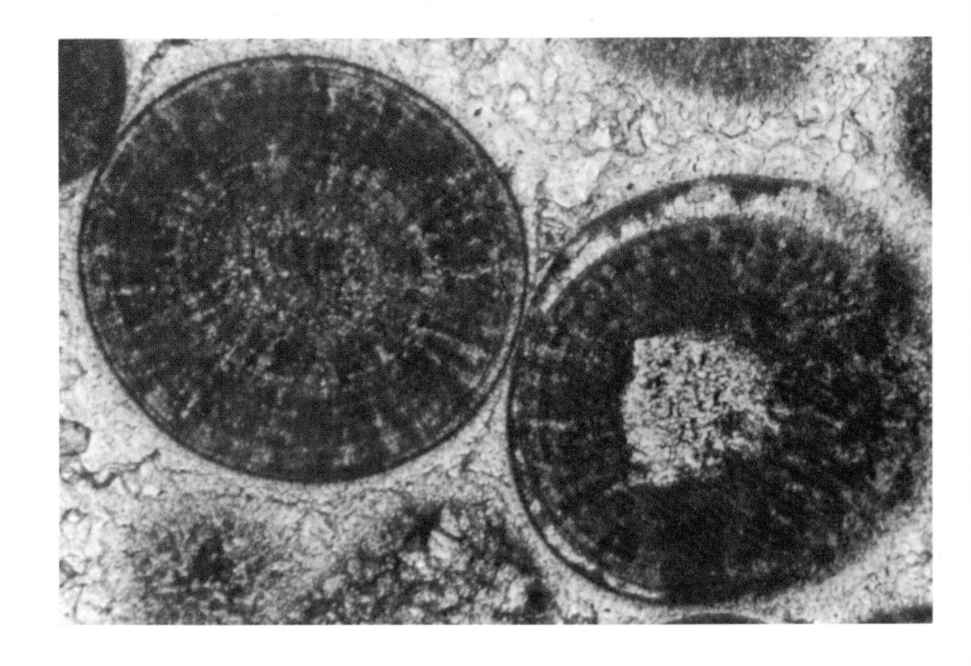

Up. Cambrian and Lo. Ordovician
Gallatin Fm.
Wyoming

These ooids show no internal structure and are classed as ooids only on the basis of shape, size, and sorting. They most probably underwent dissolution in which the ooids were removed (producing oomoldic porosity) and the resulting voids were later filled with coarse (often single-crystal) sparry calcite.

0.38 mm

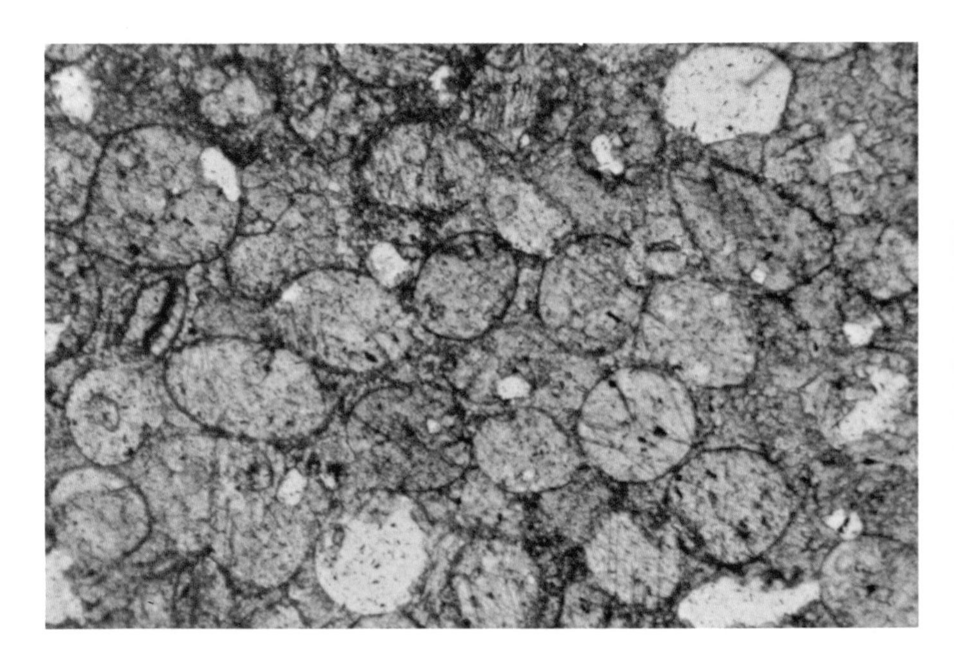

Lo. and Mid Pennsylvanian Bloyd
Fm.
Oklahoma

These ooids have suffered a complex sequence of alteration. They were completely dissolved during the alteration from aragonite to calcite and also have been strongly sheared, yielding a "spastolith" texture. The outer cortex layers have been sheared off to produce what looks like a series of linked grains.

0.04 mm

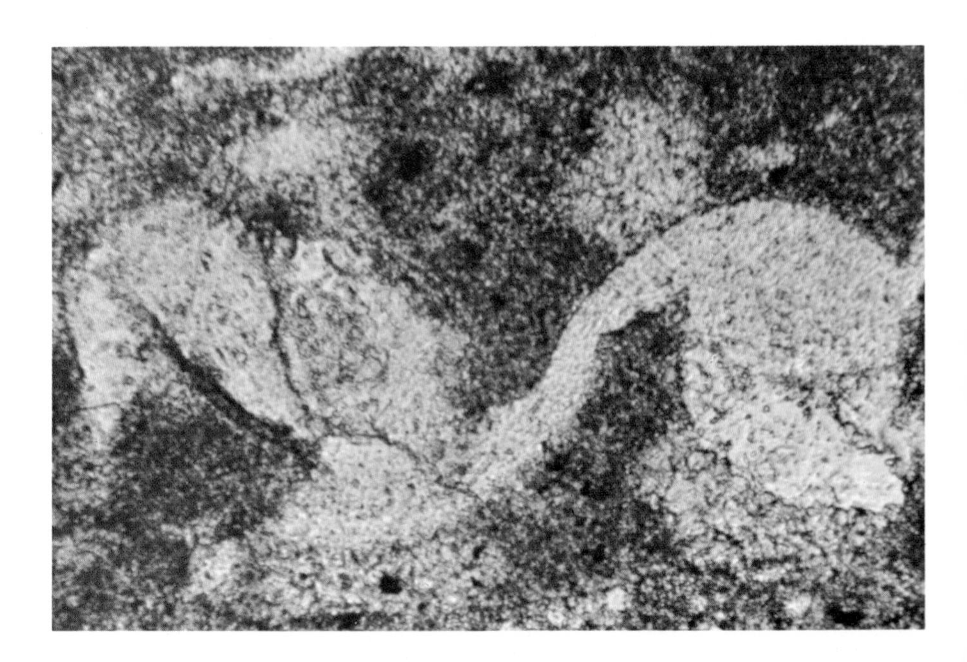

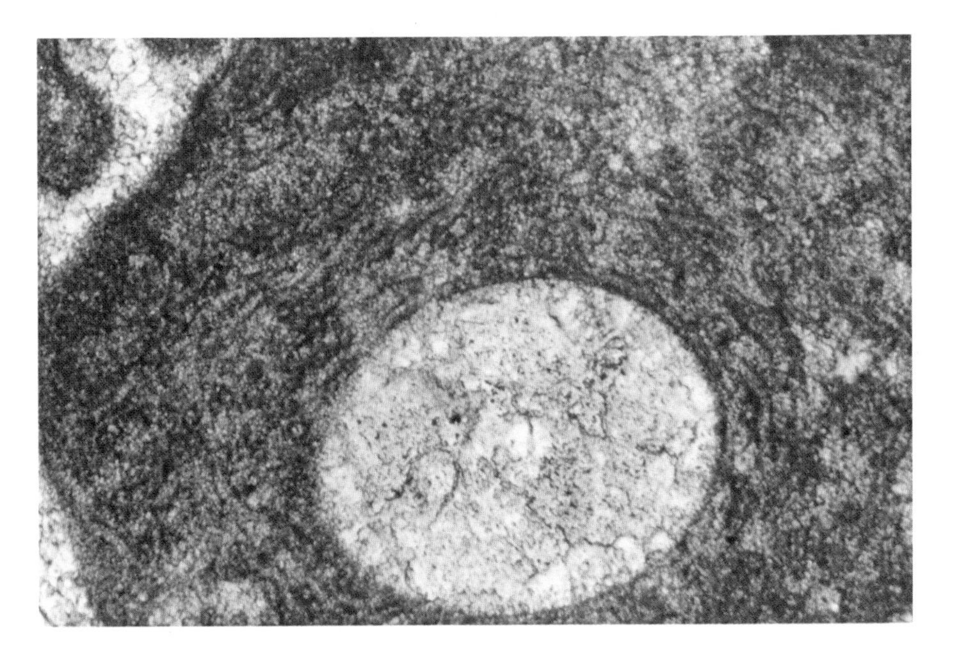

Lo. Ordovician West Spring Creek Fm.
Oklahoma

This pisolite (larger than 2 mm) has a nucleus and vague concentric laminations consisting of *Girvanella* algal tubules. The grain would thus be classed as an oncolite. Definite evidence of algal origin of pisolites is often considerably more difficult to see.

0.11 mm

Holocene sediment
Carlsbad Caverns, New Mexico

An example of a "cave pearl", an inorganically formed pisolite found on cave floors. It shows coarse, fibrous calcite crystals growing in radially oriented fan-shaped clusters interspersed with thin rinds of gypsum.

X.N. 0.25 mm

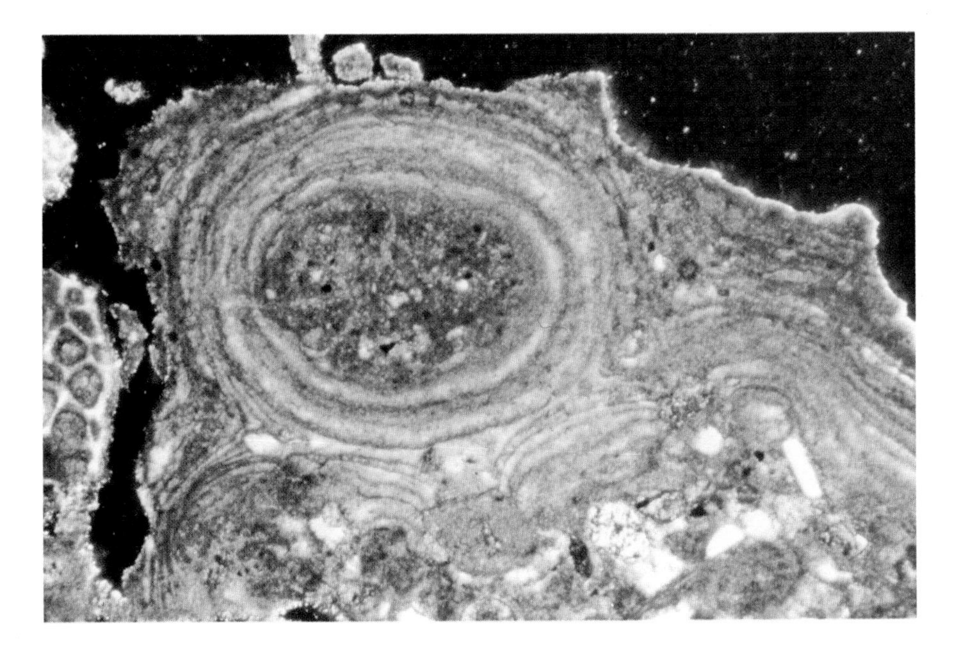

Quaternary sediment
Abu Dhabi

A portion of a coastal caliche deposit, this rock has large pisolites with well developed coatings set in a laminated crust. Yet, because of the marine and hypersaline-evaporitic pore waters in the Persian Gulf area, these vadose deposits are composed of aragonite and high-Mg calcite. Unlike other caliche, therefore, they are subject to extensive later alteration.

X.N. 0.25 mm

Quaternary caliche
Texas

A calcitic, irregularly laminated caliche pisolite with abundant quartz inclusions. Caliche pisolites are characterized by irregular lamination, inclusions, complex nuclei, autofracturing, microstalactitic textures, interlocking grains, and inverse grading of pisolite sizes within a single bed.

0.64 mm

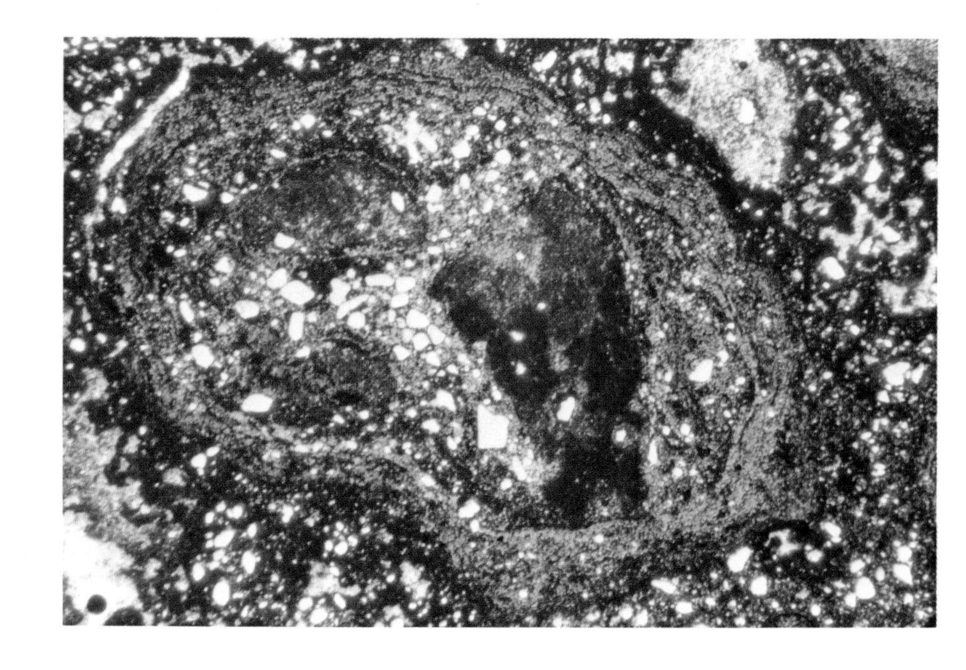

Lo. Pennsylvanian Hale Fm.
Oklahoma

This photo is included to show that some cementation and recrystallization features can produce textures at least superficially similar to oolitic and pisolitic ones. Here, early cement rims surround grains and contrast with later ankeritic cement.

0.25 mm

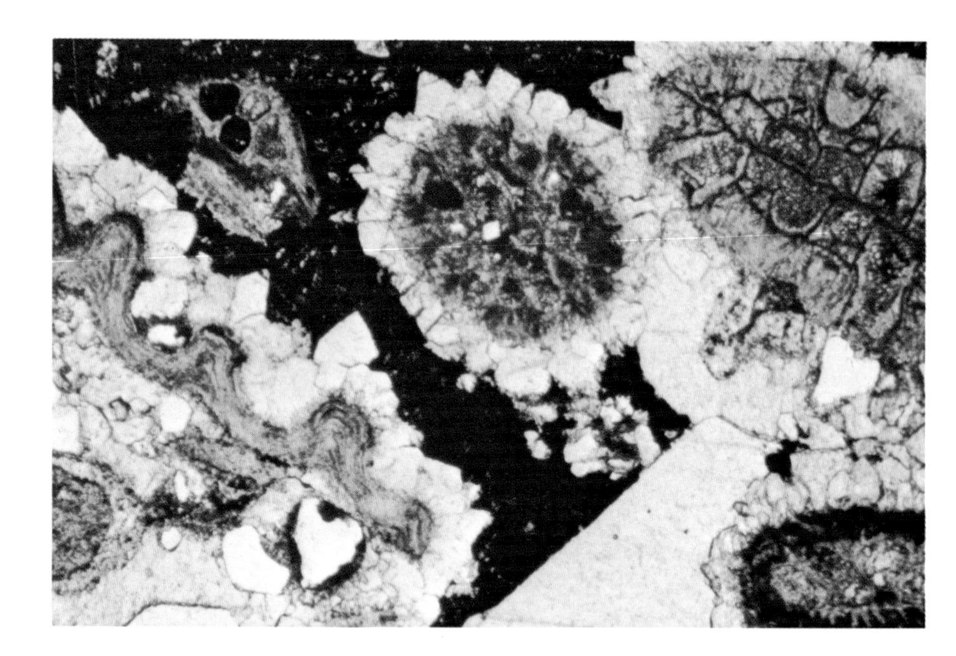

Permian Phosphoria Fm.
Idaho

All ooids and pisolites are not composed of carbonate. Phosphate, hematite, goethite, chamosite, barite and other ooids are known. Those shown here are phosphatic. In many cases, however, phosphatic and iron pisolites or ooids form by replacement of earlier carbonate ones.

0.38 mm

Intraclasts and Extraclasts

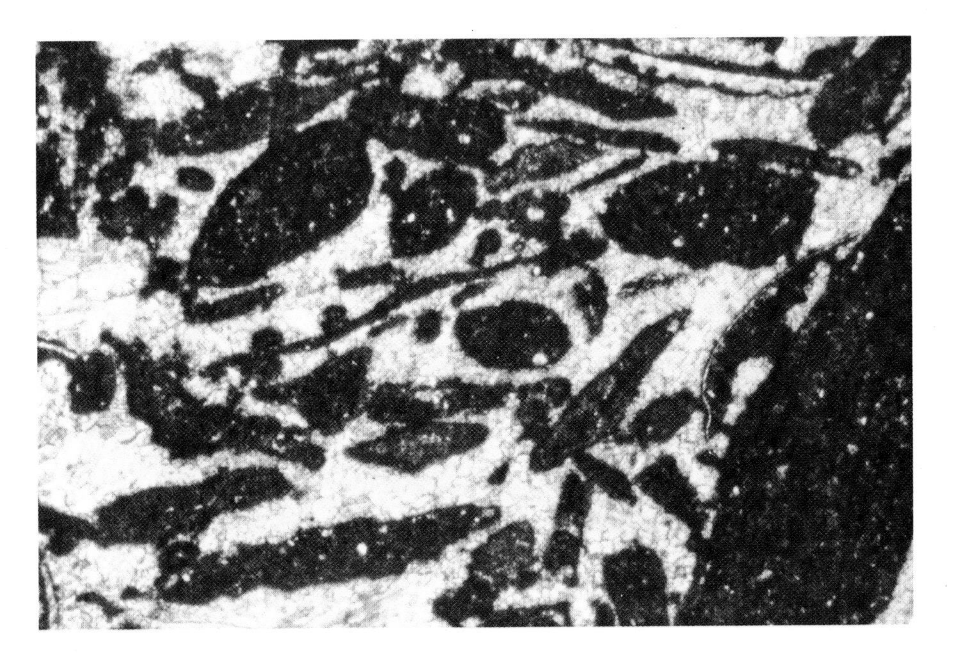

Mid. Ordovician (fm.?)
Virginia

Intraclasts. Large, rounded grains with internal structure (and encompassing fine grains) are clearly reworked carbonate sediments but probably were deposited in the same depositional cycle as the final sediment--thus they are intraclasts.

0.24 mm

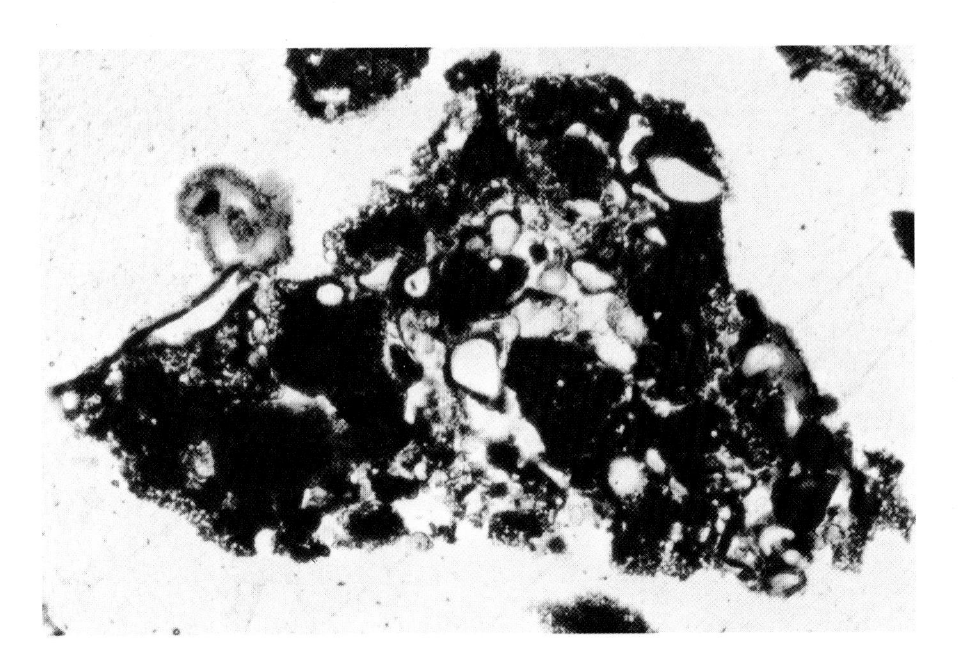

Holocene sediment
Berry Islands, Bahamas

A modern intraclast of the "grapestone" type. These are clusters or aggregates of other grains (skeletal fragments, ooids, pellets, etc.) held together by cement and algal, foraminiferal, and other encrustation. Grapestone forms in areas of moderate or intermittent agitation where cementation can take place but where reworking prevents formation of complete crusts.

0.22 mm

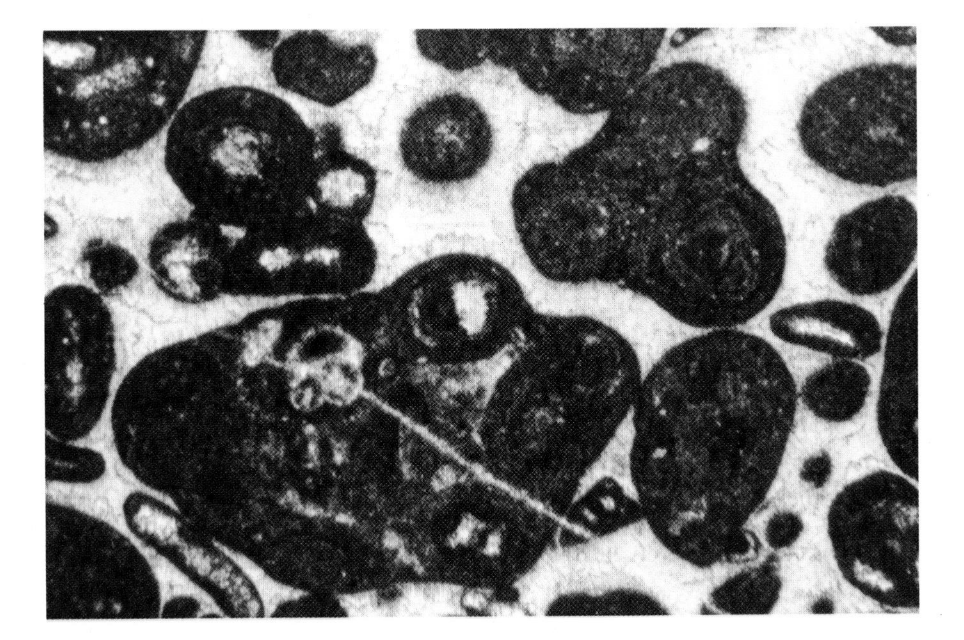

Lo. Cretaceous Cupido Fm.
Mexico

An ancient example of a grapestone deposit. Note the clusters of cemented ooids with a botryoidal external shape as well as the isopachous crust of submarine cement which surrounds all grains.

0.25 mm

Lo. Cretaceous Glen Rose Ls.
Texas

Although these grains look very much like intraclasts, they are actually the fillings (steinkerns) of *Corbula* sp., a pelecypod. Note the consistent teardrop shape. The complexity of internal filling textures is probably the result of rolling of the grains during infilling.

0.64 mm

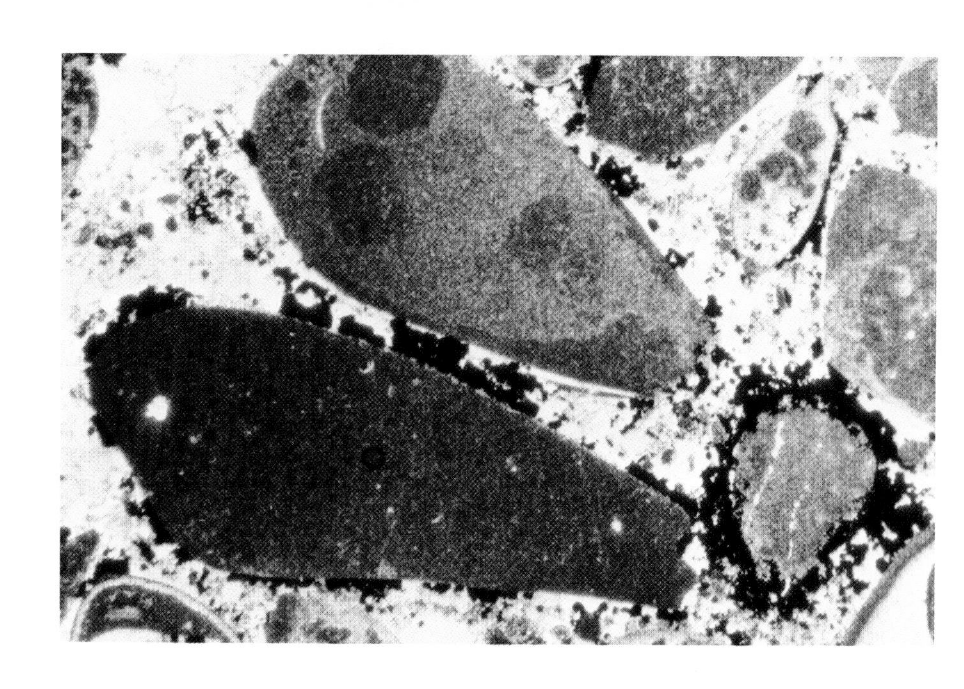

Eocene Oberaudorf Schichten
Austria

An example of a calclithite, a rock composed of extraclasts. These grains are detrital from older limestones within the rising Alpine region. Note the polymict character of the grains--they include spicular cherts, dolomites, and several types of limestone. Because of the solubility and low abrasion resistence of carbonate grains, such deposits are generally associated with major fault scarps or very arid climates.

X.N. 0.64 mm

Peloids

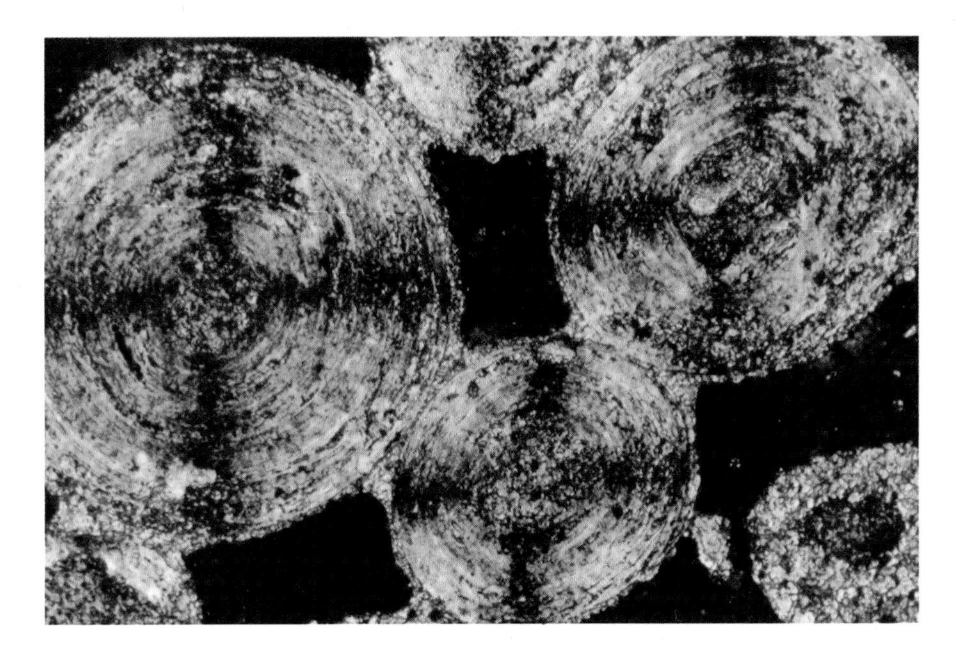

Pleistocene(?) dunes
Bahamas

This photograph and the following three show the progressive alteration of grains (with loss of characteristic textures) within the depositional sedimentary environment. These are relatively fresh and unaltered ooids illustrating strong birefringence and a pseudo-uniaxial cross.

X.N. 0.11 mm

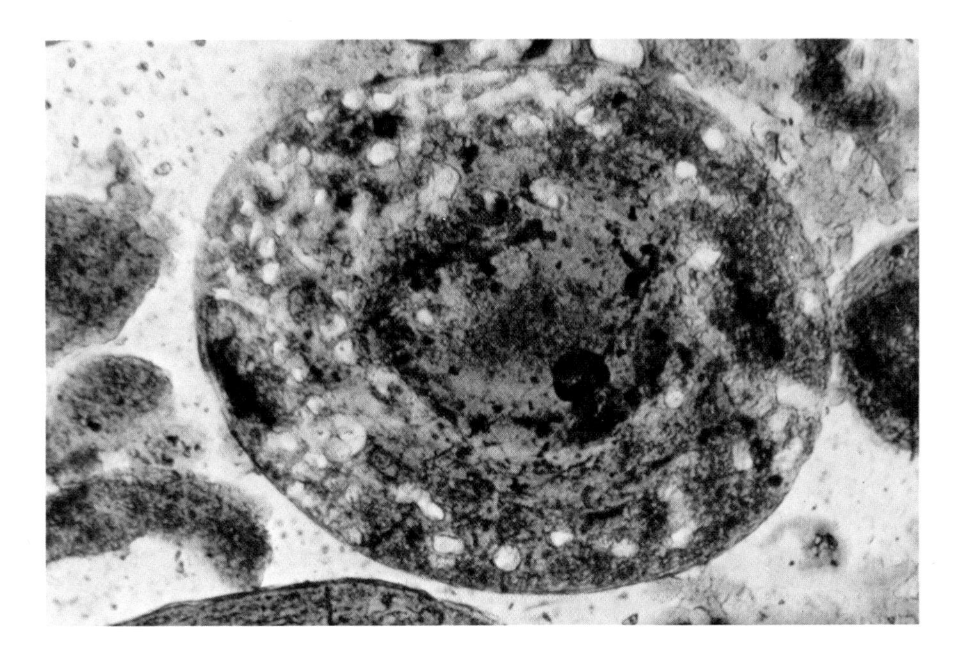

Holocene sediment
Joulters Cay, Bahamas

This is the same type of ooid as in the previous photo, but it has undergone considerable algal and fungal boring. Most of the borings are still unfilled, showing their tubular shape, but some have been filled and thus appear as diffuse micritic patches.

0.05 mm

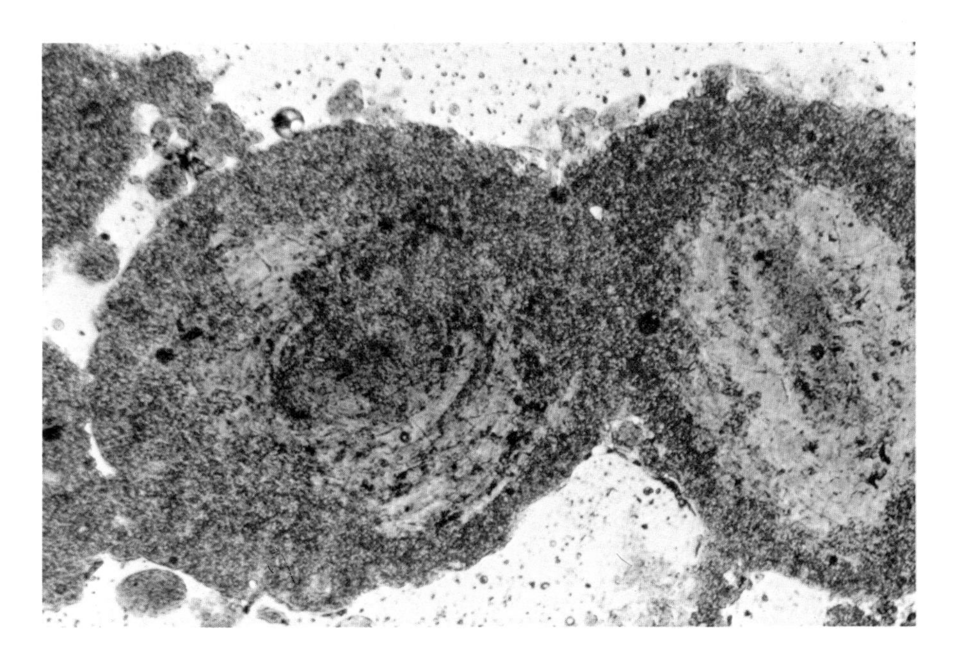

Holocene sediment
Joulters Cay, Bahamas

These ooids have undergone even more boring. Although a few tubules are still visible, most of the grain now has a uniform micritic texture with some areas of largely unaltered structure at the centers.

0.11 mm

Holocene sediment
Joulters Cay, Bahamas

An example of a probable ooid which has been virtually completely micritized. Although the section may be somewhat tangential to the ooid, no remnant structure is visible in this cut and the grain would have to be classed as a peloid. Other than size and shape, no textural characteristics would allow identification as an ooid.

0.09 mm

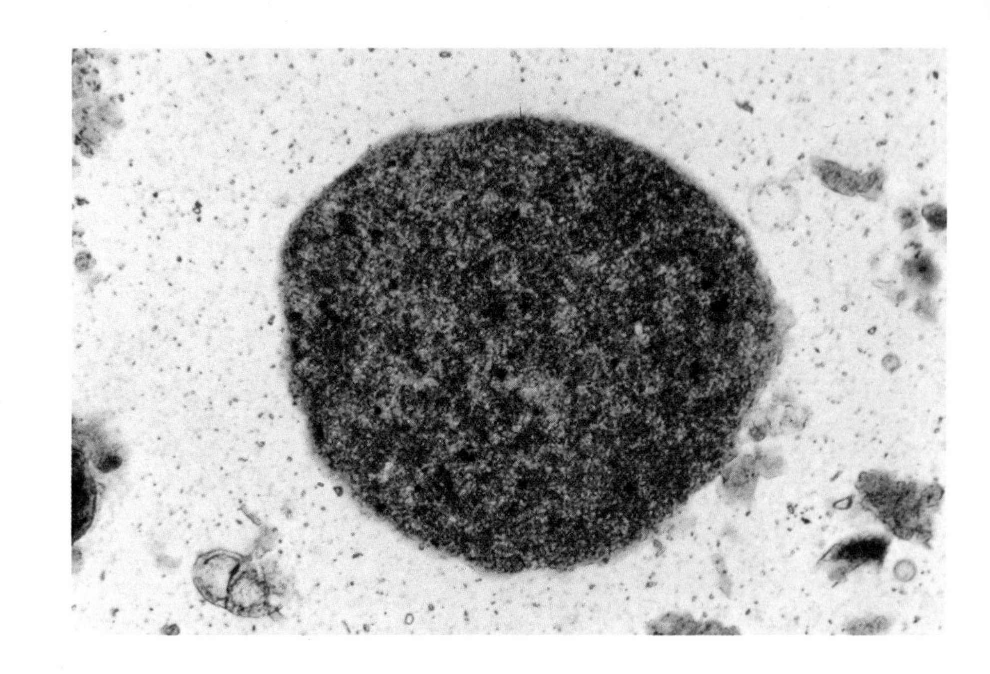

Up. Oligocene Suwannee Ls.
Florida

This sediment has a high percentage of micritized grains. Although a few are still recognizable as miliolid Foraminifera, grapestone, or shell fragments, many have lost all recognizable features and would have to be grouped as peloids.

G.P. 0.38 mm

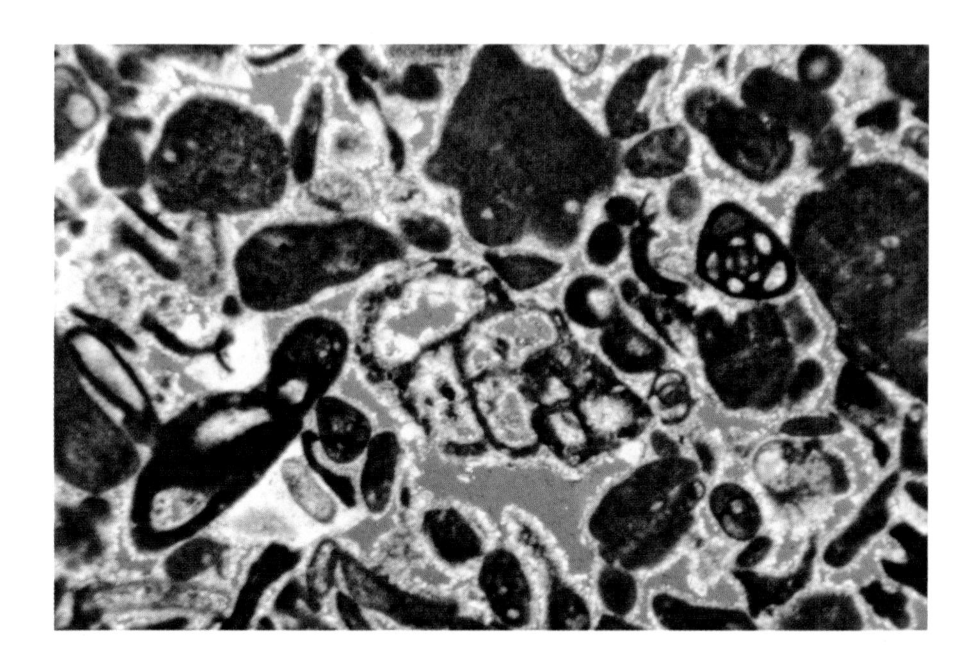

Up. Jurassic Solnhofen Ls.
Germany

A pelletal sediment which shows considerable irregularity of "pellet" size. Are some of these grains peloids produced by the micritization of other grains; are they intraclasts; or are they pellets of a number of different organisms?

0.25 mm

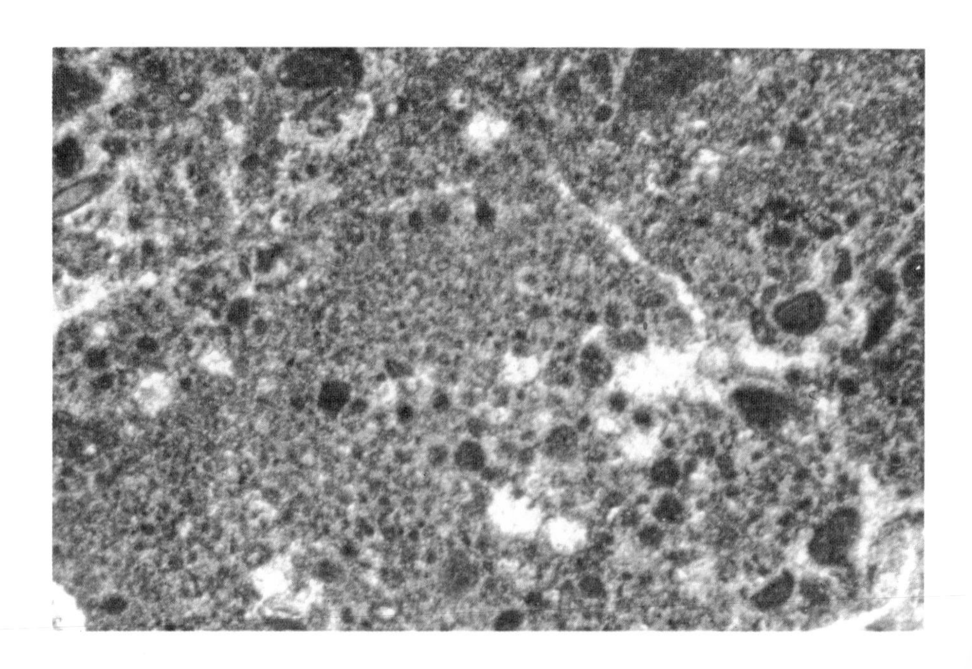

Other Minerals

Dolomites and Evaporites

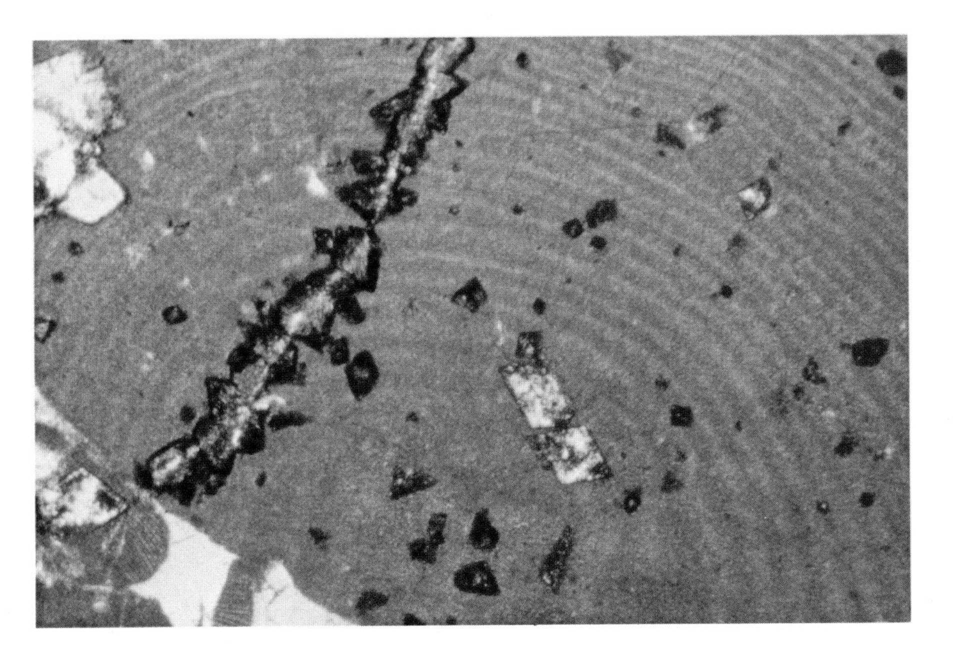

Mid. Eocene Coamo Springs Ls.
Puerto Rico

Large ankeritic (iron-rich) dolomite rhombs clearly associated with a fracture in a red algal grain. The association with both fractures and red algae is a common one for dolomite.

X.N. 0.38 mm

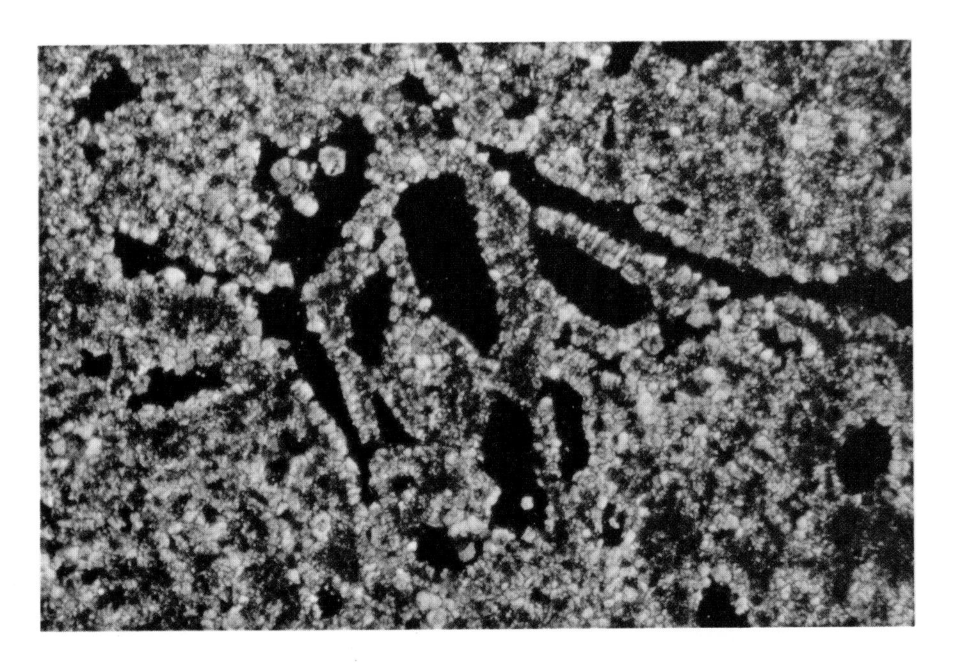

Mid. Eocene Avon Park Ls.
Florida

Finely crystalline dolomite replacement of the micritic matrix of a former biomicrite. Fossil fragments were not dolomitized and have been subsequently dissolved, leaving moldic porosity.

X.N. 0.38 mm

Lo. Ordovician Stonehenge Ls.
Pennsylvania

Coarse replacement dolomite (note euhedral rhombic outline of crystals). Intercrystalline porosity in such rocks can be very high.

X.N. 0.38 mm

Up Permian Tansill Fm.
New Mexico

The dark areas in this section are micritic and pisolitic sediments which have been replaced by aphanocrystalline dolomite considered to have been virtually penecontemporaneous. Light-pink stained areas are later sparry calcite. Such dolomite is very difficult to recognize without staining or chemical analysis.

0.64 mm

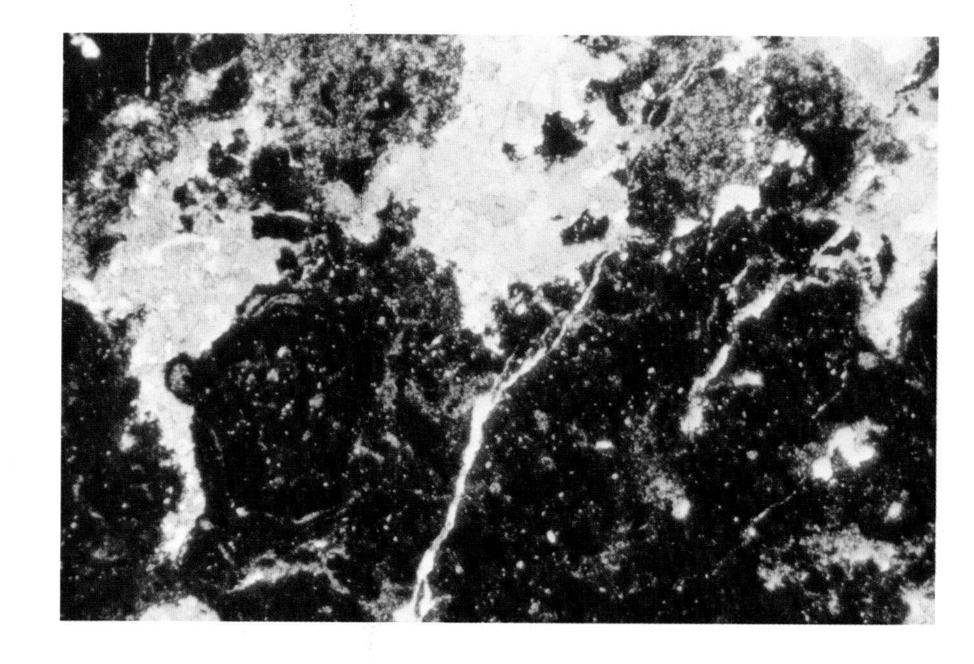

Lo. Ordovician Ellenburger Ls.
Texas

A medium crystalline replacement dolomite showing considerable iron zoning. Note the consistency of zonation from crystal to crystal, indicating that crystals formed simultaneously and during a period of uniformly fluctuating conditions.

0.11 mm

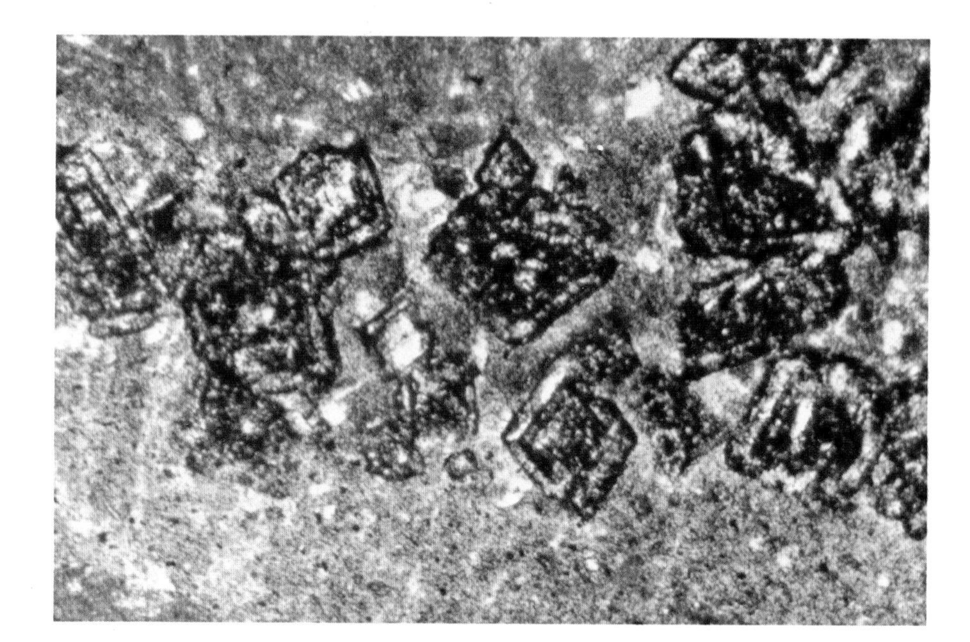

Up. Cretaceous Chalk
France

A tightly interlocked, medium-crystalline, complete replacement dolomite with considerable zonation. This dolomite probably formed rather late in the history of the sediment and is recognizable as dolomite on the basis of both the euhedral crystal shapes and the internal zonation.

0.11 mm

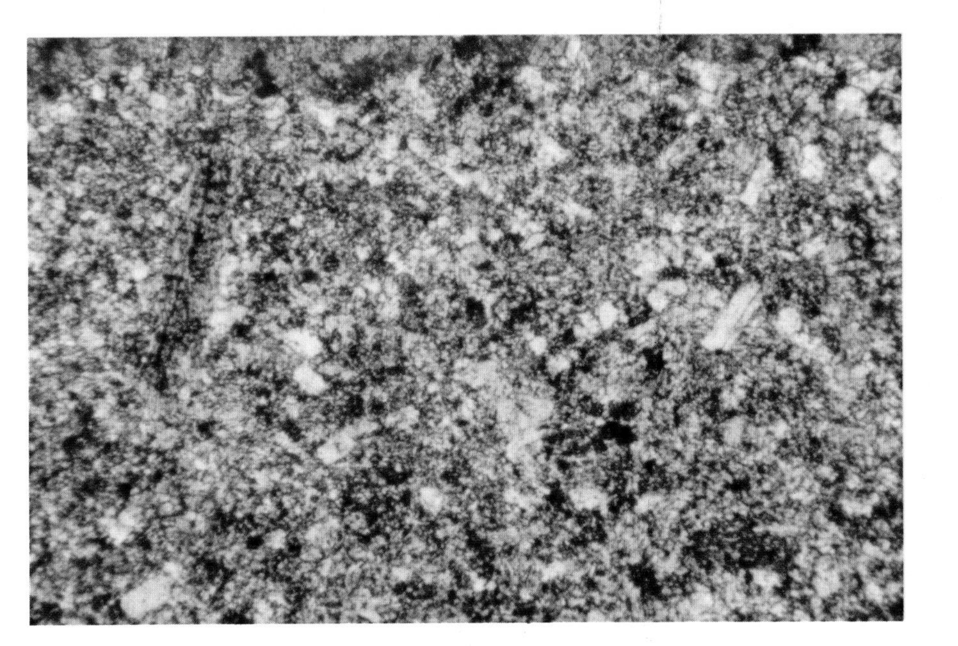

Up. Permian Castile Fm.
New Mexico

A bedded and varved gypsum deposit. At the top of the photo is a lamina of carbonate while the rest of the slide shows interlocking crystals and crystal fragments of gypsum. In the deeper subsurface this rock is anhydrite and probably has inverted to gypsum in near-surface areas. Note characteristic low birefringence

X.N. 0.38 mm

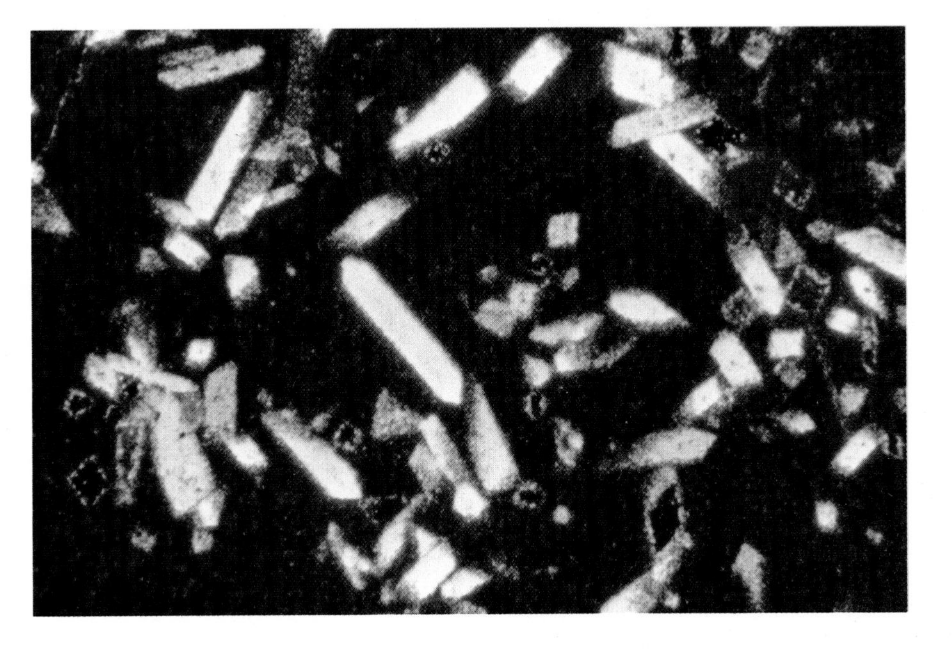

Mississippian upper Debolt Fm.
Canada

Gypsum crystals as replacement of or displacement of micritic sediment. Note the low birefringence and characteristic crystal shapes of these gypsum grains. Several have been replaced by calcite spar while still retaining their crystal outline.

X.N. 0.25 mm

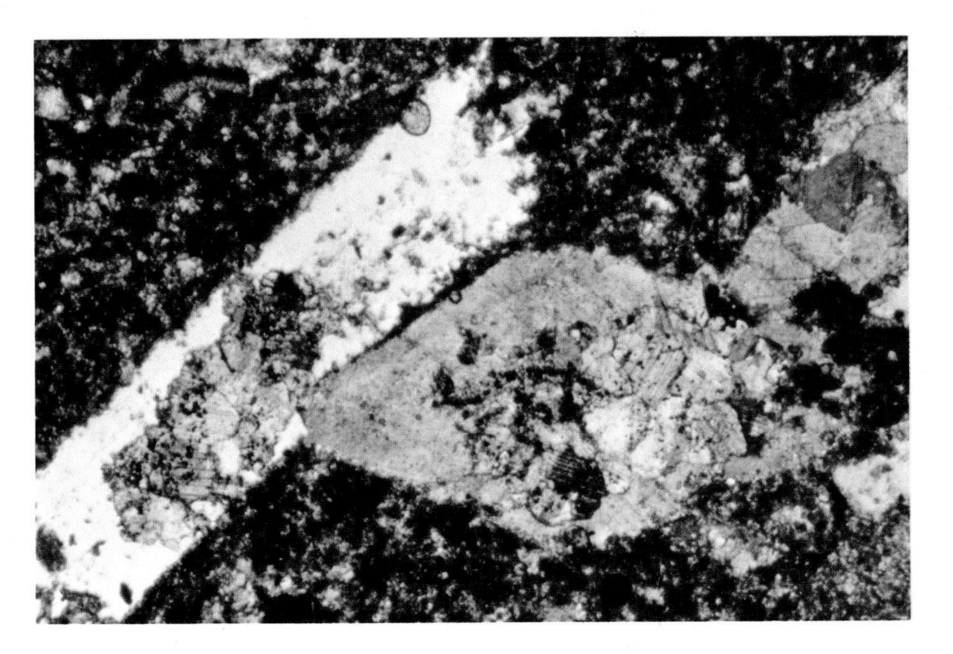

Mid. Pennsylvanian Paradox Fm.
Utah

Several large replacement crystals of gypsum. Low birefringence and crystal shape are diagnostic, while presence of abundant carbonate inclusions is common in the case of replacement grains. Gypsum must be carefully distinguished from authigenic quartz, and can be lost through prolonged or improper thin-section grinding.

X.N. 0.25 mm

Lo. Cretaceous Ferry Lake(?)
Anhydrite
Texas

Fine-grained "chicken-wire" anhydrite replacement of micritic sediment. Shows "auto fractured" crystal fragments in a nodular texture with interspersed, displaced original micritic sediment. Anhydrite can be lost from thin sections unless they are carefully (and rapidly) prepared.

X.N. 0.38 mm

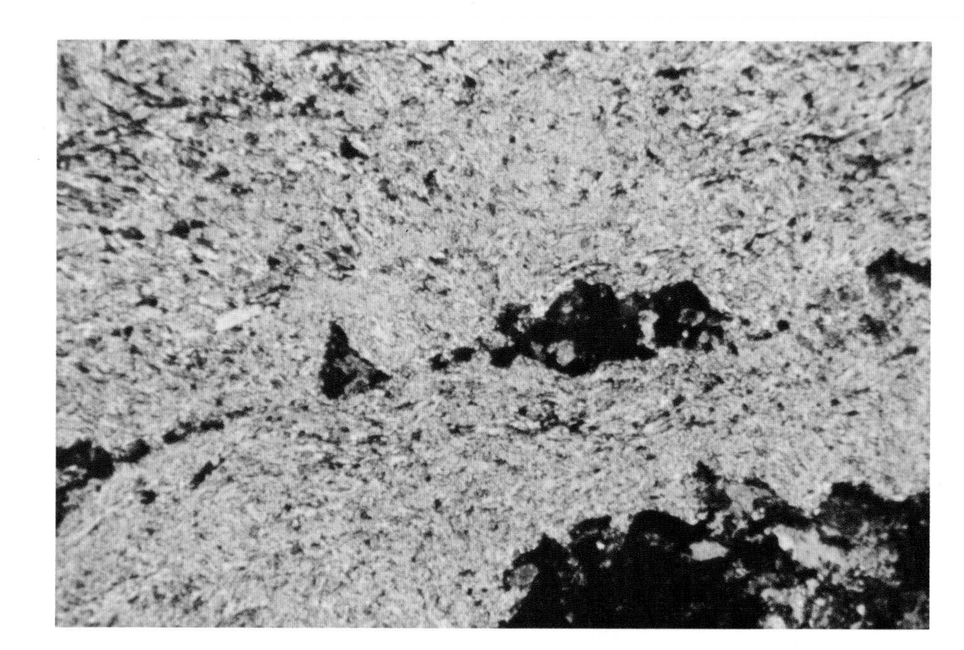

Lo. Cretaceous Ferry Lake
Anhydrite
Texas

Fibrous to bladed anhydrite shown filling the central cavity of a serpulid worm tube. Coarse euhedral anhydrite replaces the serpulid tube walls.

X.N. 0.38 mm

Lo. Cretaceous Glen Rose Ls.
Texas

Because of the high solubility of most evaporites they are very susceptible to postdepositional removal by leaching. Yet even in those cases, evidence can be seen for the original presence of evaporites. In this example it takes the form of clear molds with euhedral crystal outlines, probably of anhydrite. Pink areas are now voids.

G.P. 0.25 mm

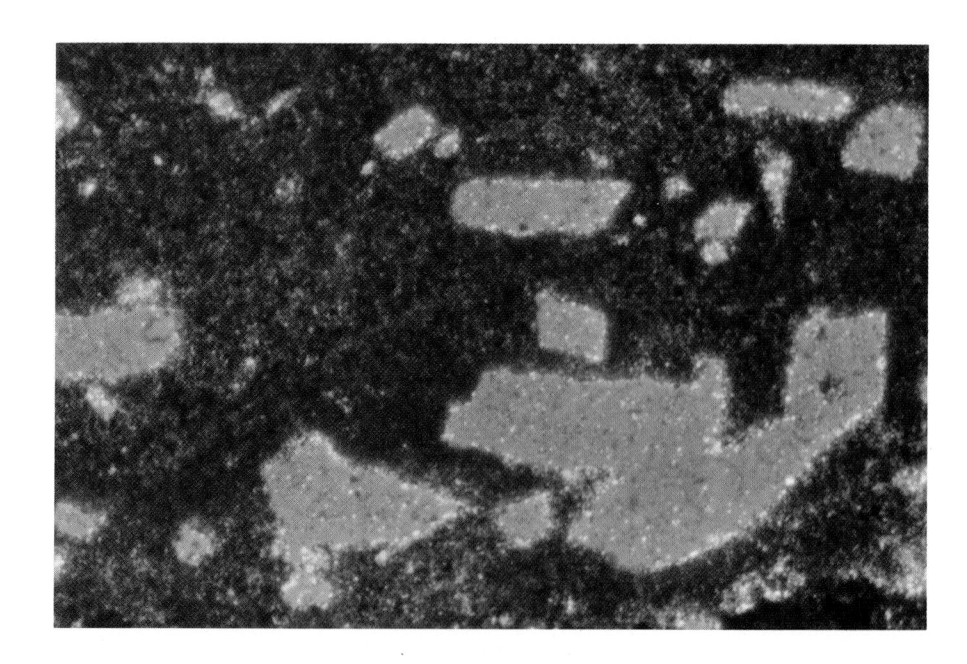

Lo. Cretaceous Glen Rose Ls.
Texas

Celestite ($SrSO_4$) filling large vugs in a sparse biomicrite. Celestite formation is often associated with early supratidal diagenesis (releasing Sr from aragonite). Celestite shows very low (first-order) colors.

X.N. 0.38 mm

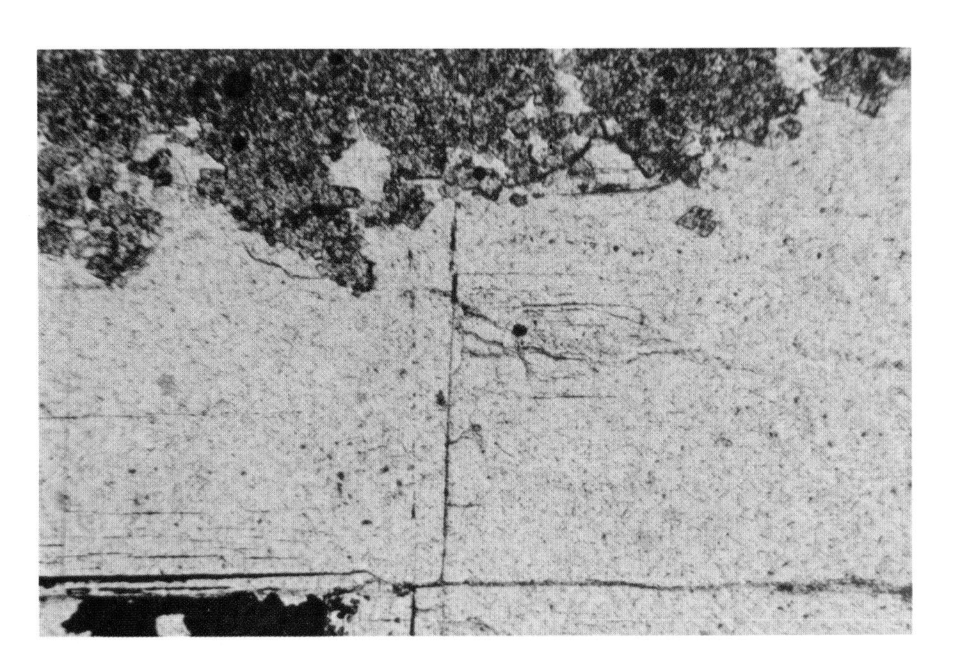

Mid. Devonian Keg River equivalent
Canada

Halite is rarely found in outcrop samples but can be present in subsurface materials. It is isotropic and thus difficult to recognize. However, the presence of intersecting cleavages at right angles and inclusions within what otherwise looks like pore space is indicative of halite. Halite is commonly lost during thin section preparation unless sections are ground in oil.

0.37 mm

Silica Minerals

Up. Cambrian Copper Ridge Dol. and Conococheague Ls.
Virginia

Well-rounded detrital quartz sand grains scattered throughout a dolomitized carbonate mudstone. Quartz grains are at various stages of extinction, but none show colors higher than first order.

X.N. 0.38 mm

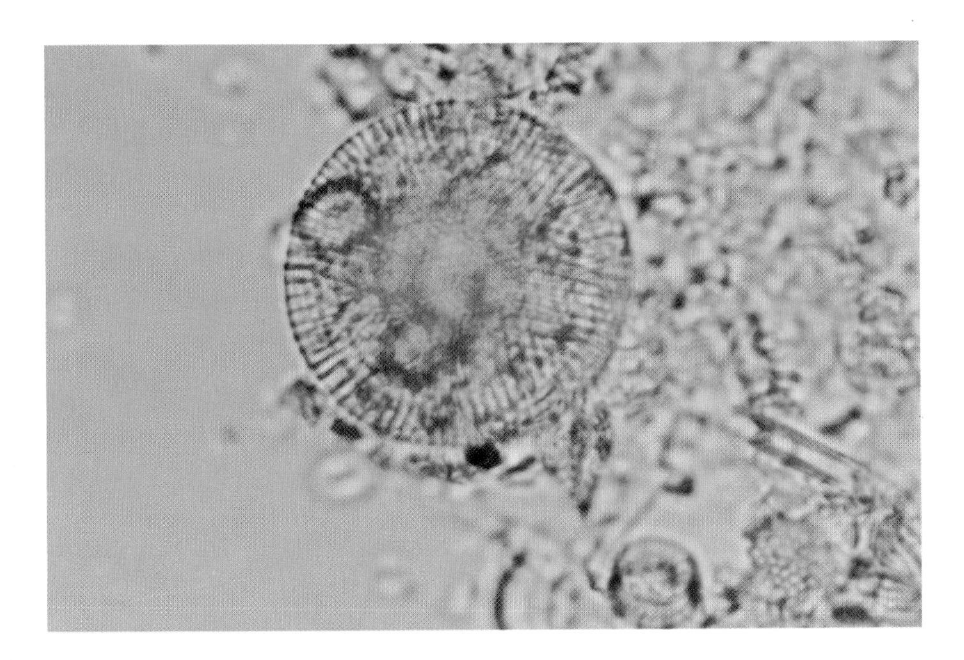

Eocene deep-water ooze
Bermuda Rise, North Atlantic Ocean

Opaline diatom (large circular grain with intricate pattern) and opaline sponge spicule fragment (lower right center). A large percentage of the chert in sedimentary rocks is derived from the dissolution and reprecipitation of biogenic opal from diatoms, radiolarians, sponges, and other organisms.

0.016 mm

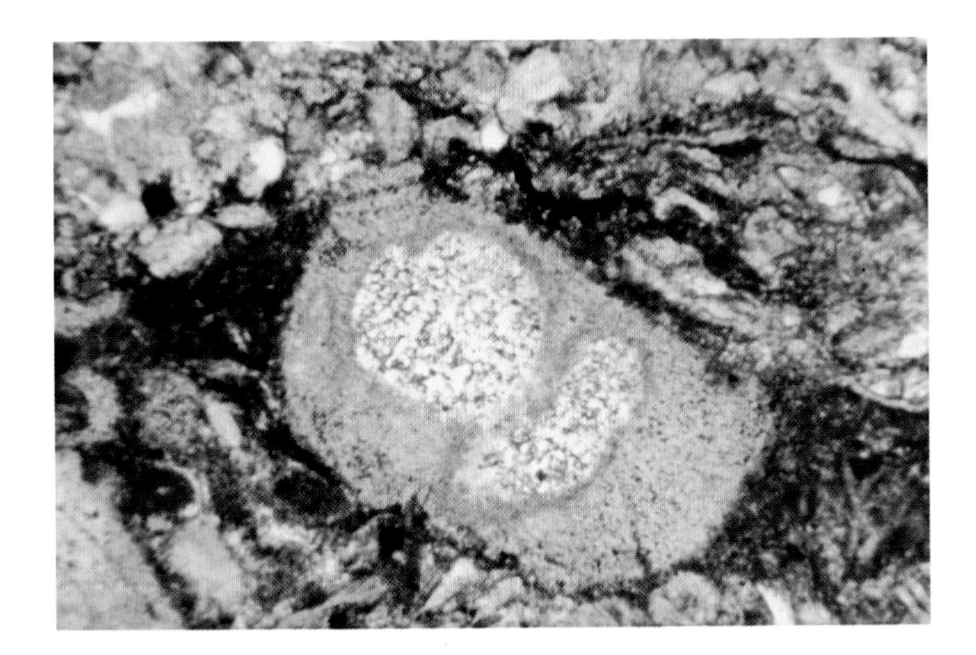

Up. Silurian part of Tonoloway-Keyser Ls.
Pennsylvania

A chert (microcrystalline quartz with included water bubbles) replacement of the central portion of a crinoid fragment. Chert is a very common replacement and cementation mineral in carbonate rocks.

X.N. 0.30 mm

Mid. Eocene Coamo Springs Ls.
Puerto Rico

Silica replacement of part of a mollusk fragment. This is a coarse fibrous chalcedony/megaquartz texture which clearly crosscuts primary skeletal fabric.

X.N. 0.38 mm

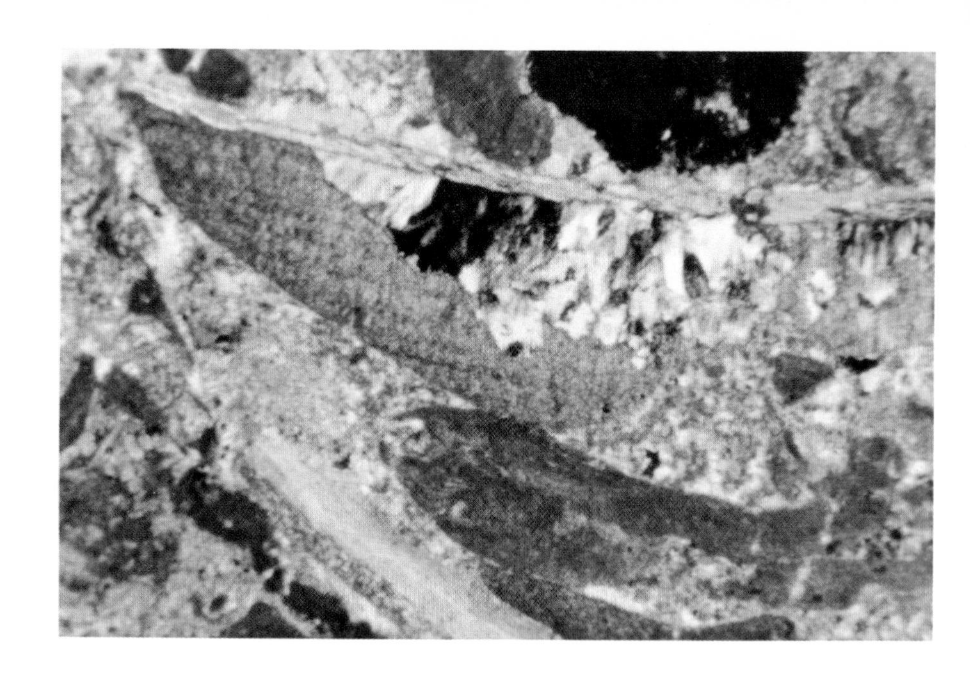

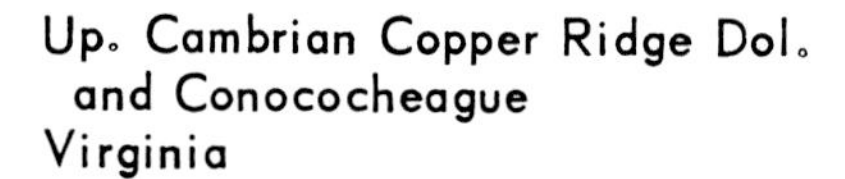

Up. Cambrian Copper Ridge Dol.
and Conococheague
Virginia

Authigenic quartz crystal in a calcite vein. Position in vein indicates certain authigenic origin, and the perfectly euhedral termination is characteristic of such authigenic minerals.

X.N. 0.08 mm

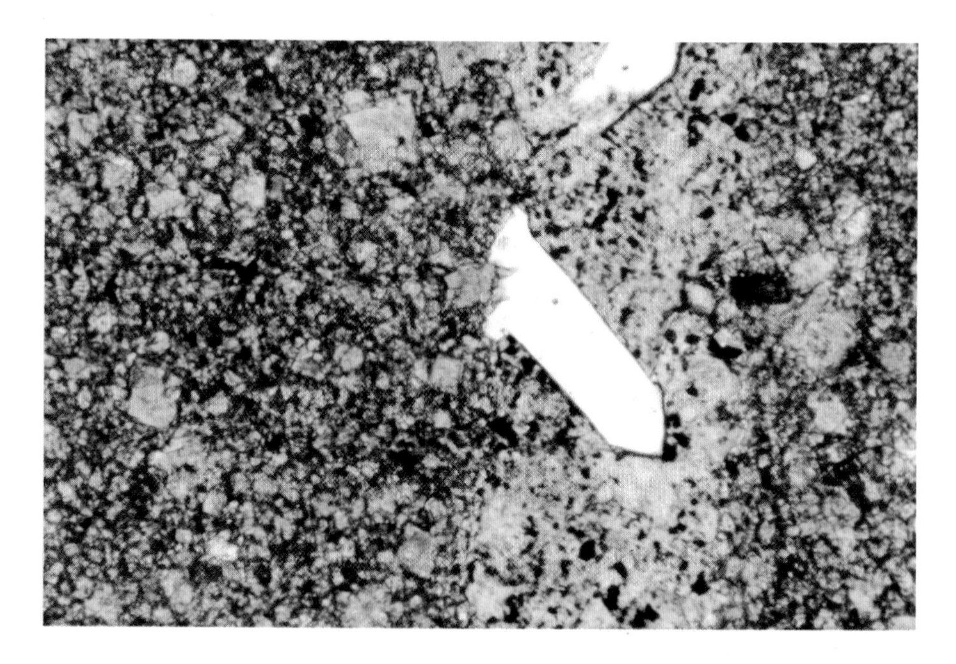

Mid. Pennsylvanian Paradox Fm.
Utah

Coarse replacement of a bivalve wall by authigenic quartz. Crystals retain many carbonate inclusions and extend from shell wall into the surrounding matrix. Silica replacement is often very selective and is commonly associated with microenvironments in and between shells or in other areas of higher organic content.

X.N. 0.25 mm

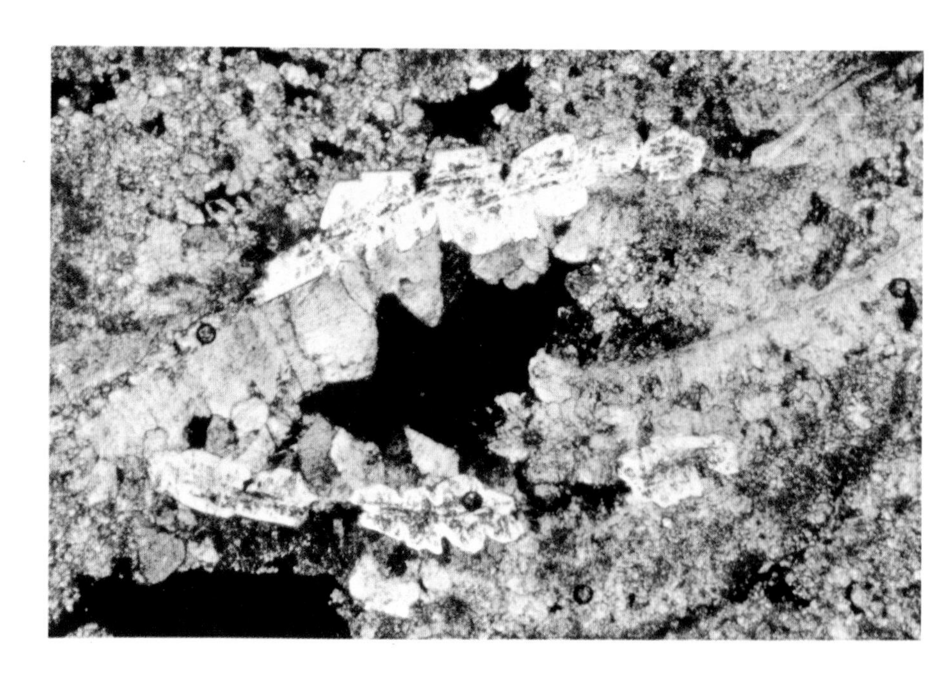

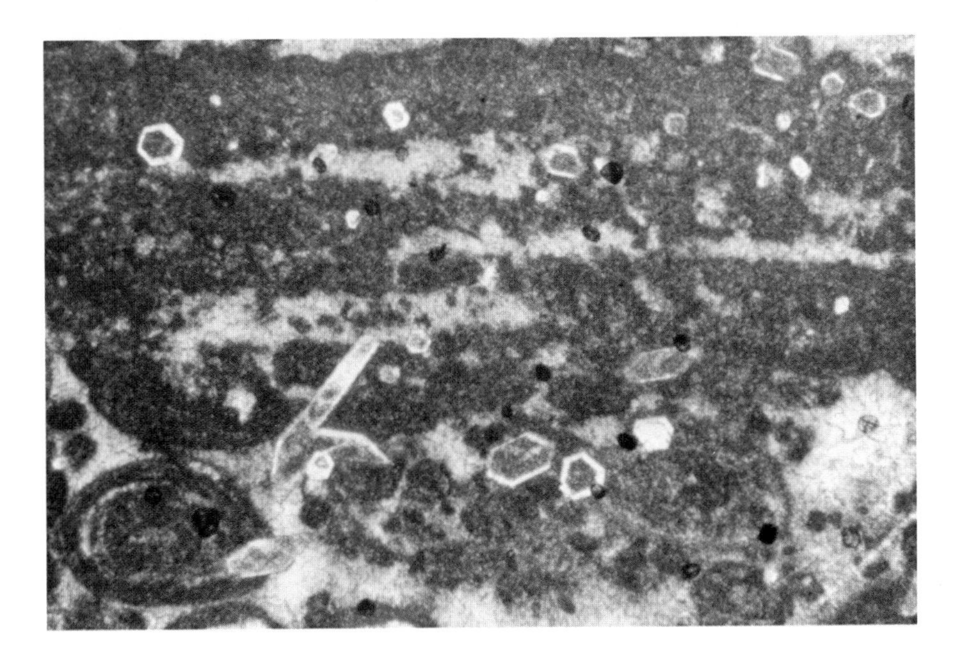

Up. Jurassic Zuloaga Fm.
Mexico

Extensive authigenic quartz replacement. Note that most of the authigenic quartz crystals display excellent crystal terminations but have enormous central cores of almost completely undigested carbonate.

0.25 mm

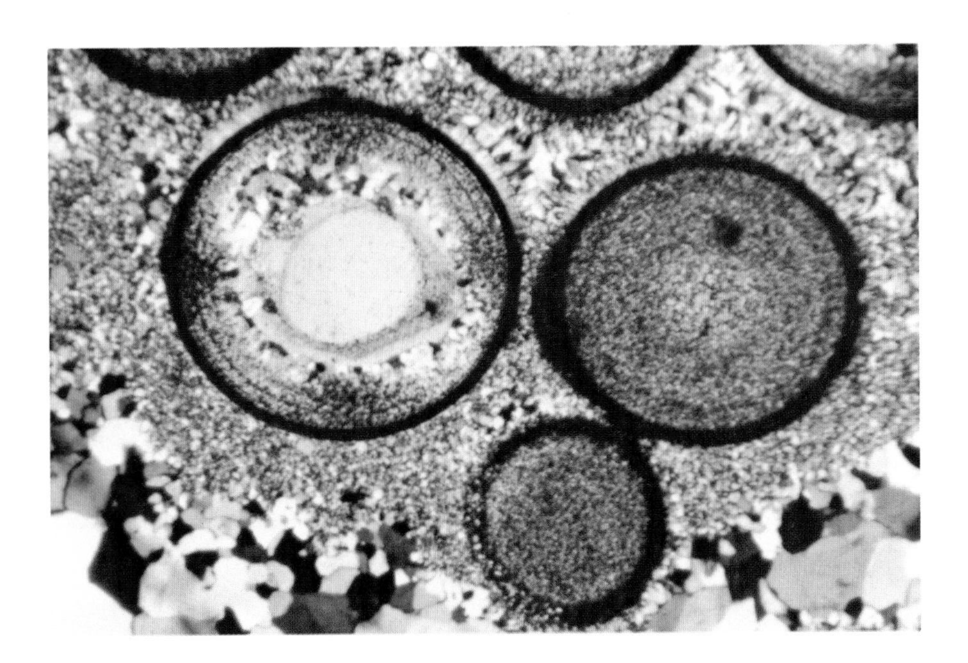

Up. Cambrian Mines Dolomite Mbr. of Gatesburg Fm.
Pennsylvania

Complete silica replacement of an originally carbonate (oolitic) sediment. Note quartz overgrowths on nuclei of ooids, chert replacement of ooids and part of matrix, chalcedonic infilling of former voids, and coarse megaquartz vein at bottom of photo. A complete range of silica fabrics thus can be found even in small areas.

X.N. 0.38 mm

Up. Cambrian Copper Ridge Dol. and Conococheague Ls.
Virginia

Multiple diagenetic generations. Chert (black-speckled background in this photo) replaced original limestone. Later, euhedral dolomite rhombs grew in chert. Finally, dolomite and chert are cut by megaquartz veins.

X.N. 0.08 mm

Eocene Green River Fm.
Wyoming

Carbonate ooids cemented (and partly disrupted) by coarse chalcedonic quartz. Chalcedonic quartz is normally a void filling fabric.

X.N. 0.38 mm

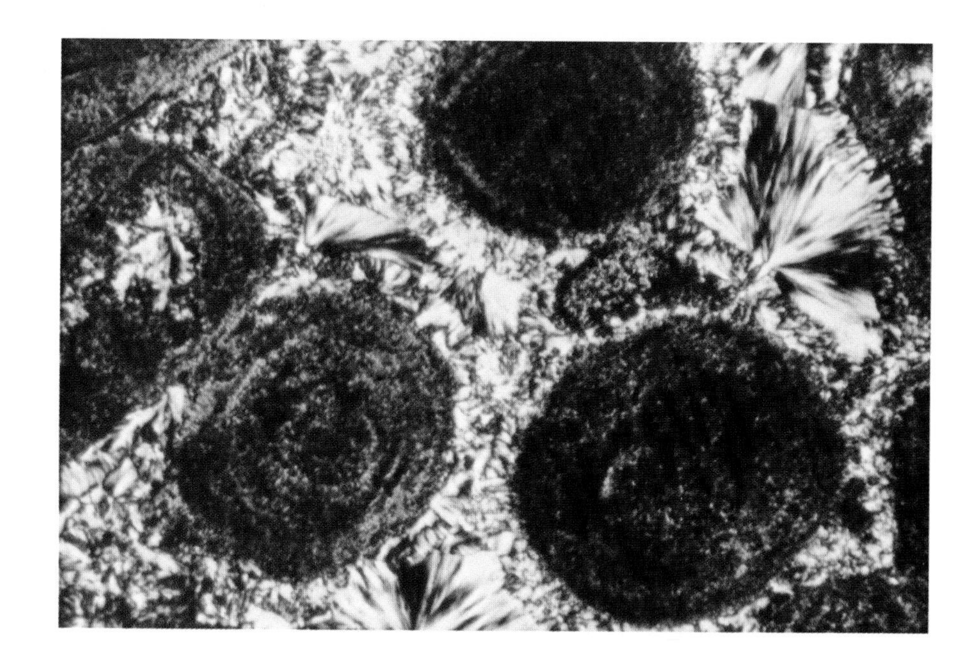

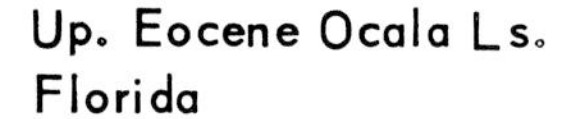

Up. Eocene Ocala Ls.
Florida

Chert cementation (partial) of a carbonate grainstone. Chert is just beginning to fringe carbonate grains (including large Foraminifera in center).

X.N. 0.30 mm

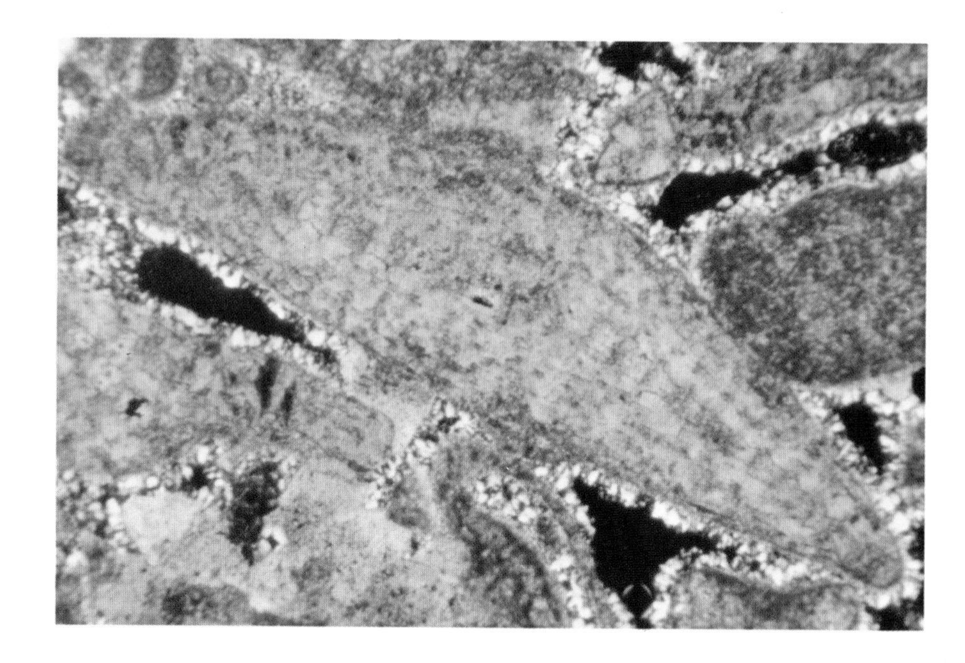

Lo. Permian Getaway Ls. Mbr. of Cherry Canyon Fm.
Texas

Chalcedonic quartz infilling of a calcareous sponge. Note the well-developed radiating bundles of chalcedony fibers cut by uniform growth banding, as well as the growth interference between adjacent chalcedony bundles.

X.N. 0.64 mm

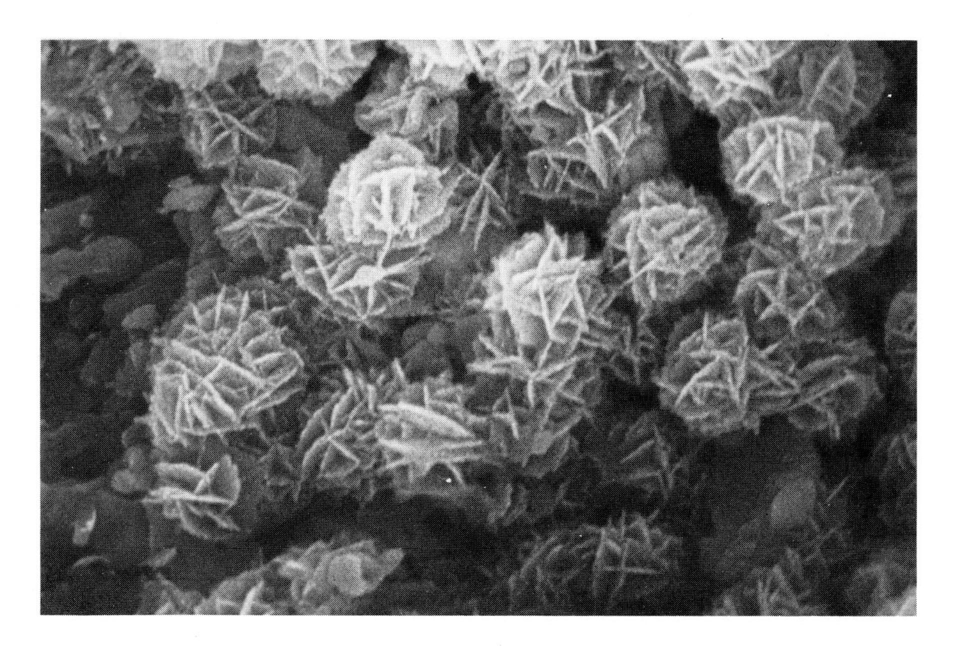

Up. Cretaceous Craie grise
Netherlands

SEM view of siliceous area within chalk. The photo shows small spherules (sometimes called lepispheres) of cristobalite. Recent work has shown that the alteration of opal to quartz chert generally goes through the intermediate stage of cristobalite, and that the transitions between these three phases are largely temperature controlled.

SEM Mag. 5000X 2.6 μm

Up. Cretaceous Monte Antola Fm.
Italy

Authigenic feldspar replacing limestone. Albite is most common replacement although other types are known. Recognized on basis of crosscutting texture and euhedral or twinned outline (not well shown here) as well as by the presence of carbonate inclusions. Rarely forms a large percentage of the sediment.

X.N. 0.04 mm

Iron Minerals

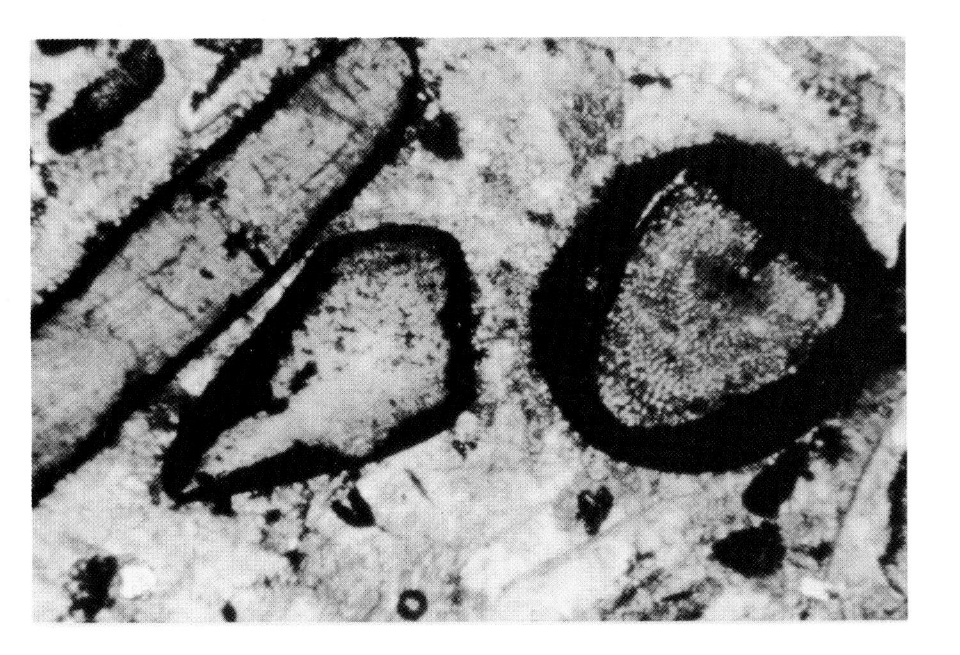

Mid. Silurian Clinton Fm.
Pennsylvania

Hematite oolitic coatings on carbonate grains (fossil fragments). These coatings are good indicators of shallow, agitated water (and in this case, form an economic iron deposit). Hematite may be after chamosite, which was a replacement of original carbonate.

X.N. 0.24 mm

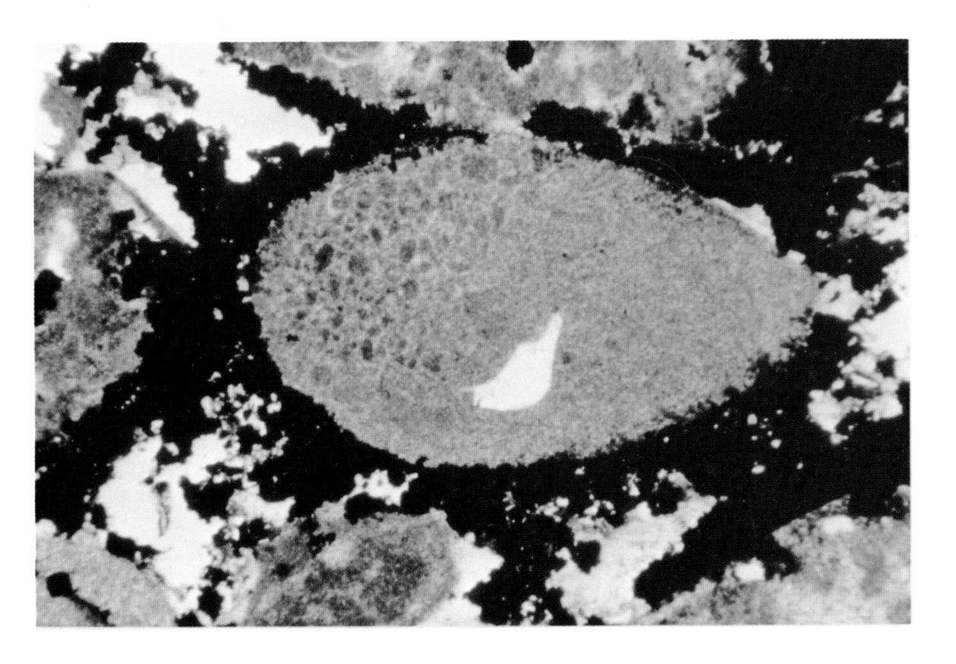

Lo. Cretaceous Glen Rose Ls.
Texas

Hematite filling pore space between internal fillings of small *Corbula* (pelecypods). Hematite is distinguishable from other opaque minerals (pyrite, magnetite, etc.) under reflected light. Under very strong transmitted light hematite may show typical blood-red color.

0.30 mm

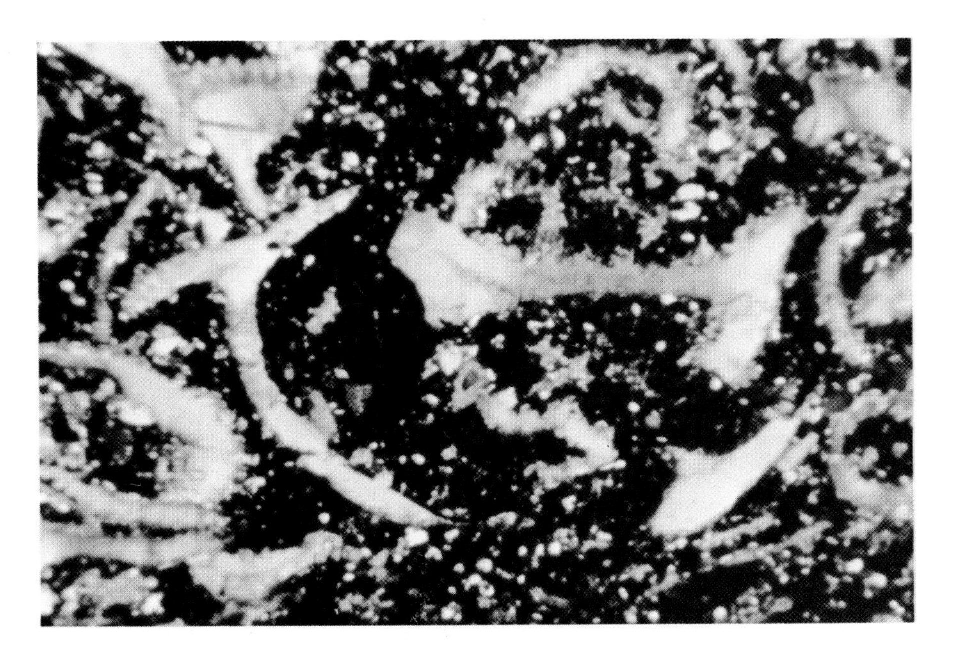

Mid. Silurian Clinton Fm.
Pennsylvania

Dense hematite cement with abundant quartz silt grains and trilobite fragments. This was an economic iron ore.

X.N. 0.38 mm

Jurassic Eisenoolith
Germany

Hematite ooids illuminated with conoscopic condenser in place--gives very intense transmitted light which shows reddish-yellow color of this slightly weathered hematite. These were originally carbonate ooids which were replaced by various iron minerals.

0.24 mm

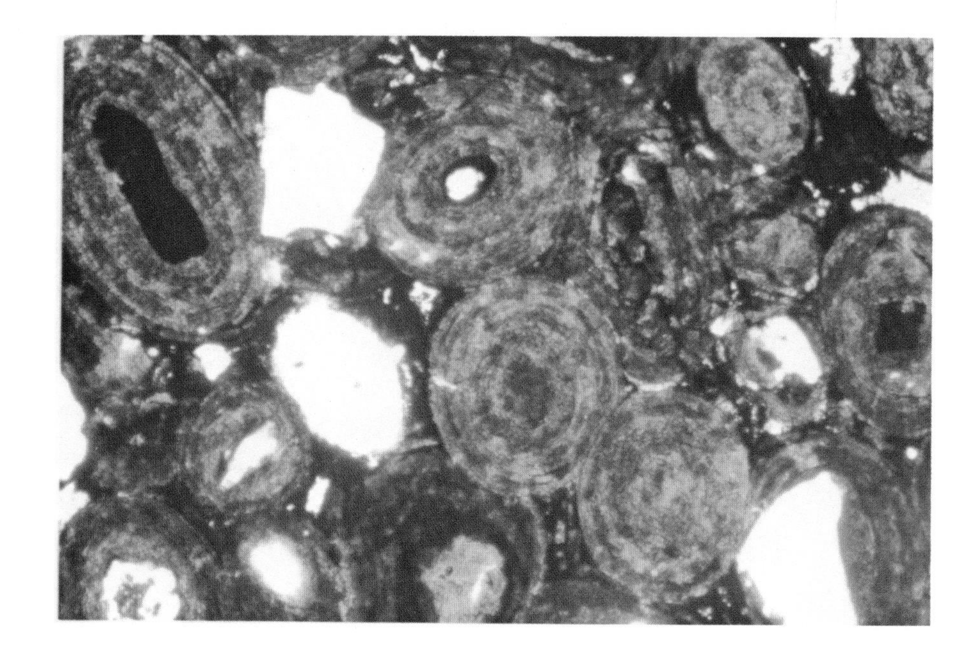

Up. Cretaceous Monte Antola Fm.
Italy

A pyrite grain in transmitted light. Although some detrital pyrite can be found in carbonates, most is authigenic. It is often found as spherules (framboids) associated with patches of organic matter or fossils, because it forms under reducing conditions where sulfur is present.

0.10 mm

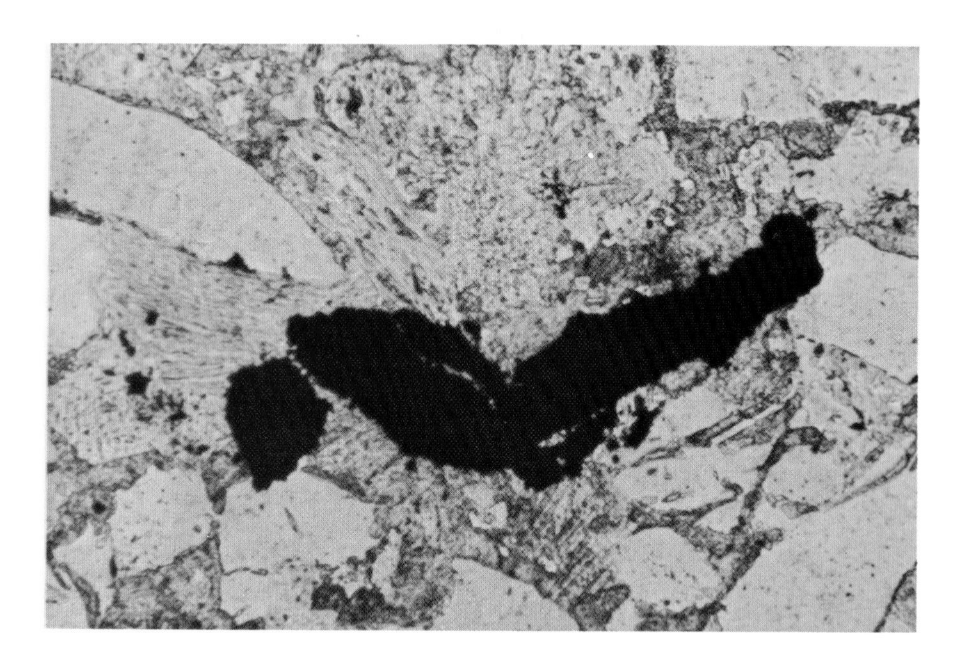

Up. Cretaceous Monte Antola Fm.
Italy

Photographed with a combination of reflected and transmitted light, this example shows a burrow filled with framboidal pyrite and surrounded by an alteration halo of limonite. Pyrite in association with burrows is common because of the high organic content of such structures.

R.L. 0.04 mm

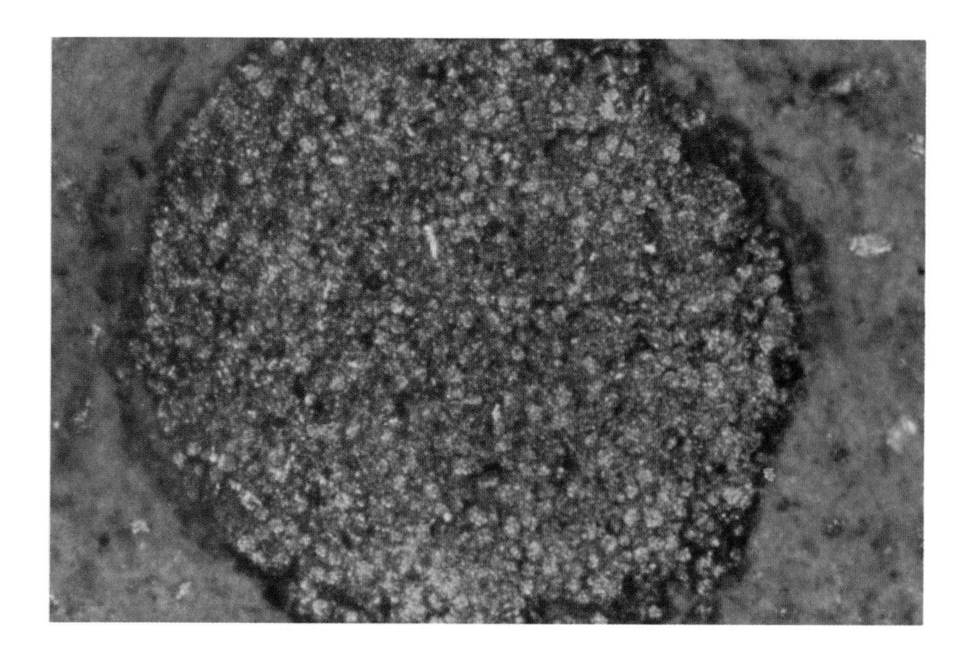

Up. Cretaceous Chalk
British North Sea

SEM view of a pyrite framboid. Note the cluster of small interlocking crystals and the smooth, almost perfectly spherical exterior. These can form singly or in groups as an authigenic product.

SEM Mag. 4300X 3 μm

Phosphate and Glauconite

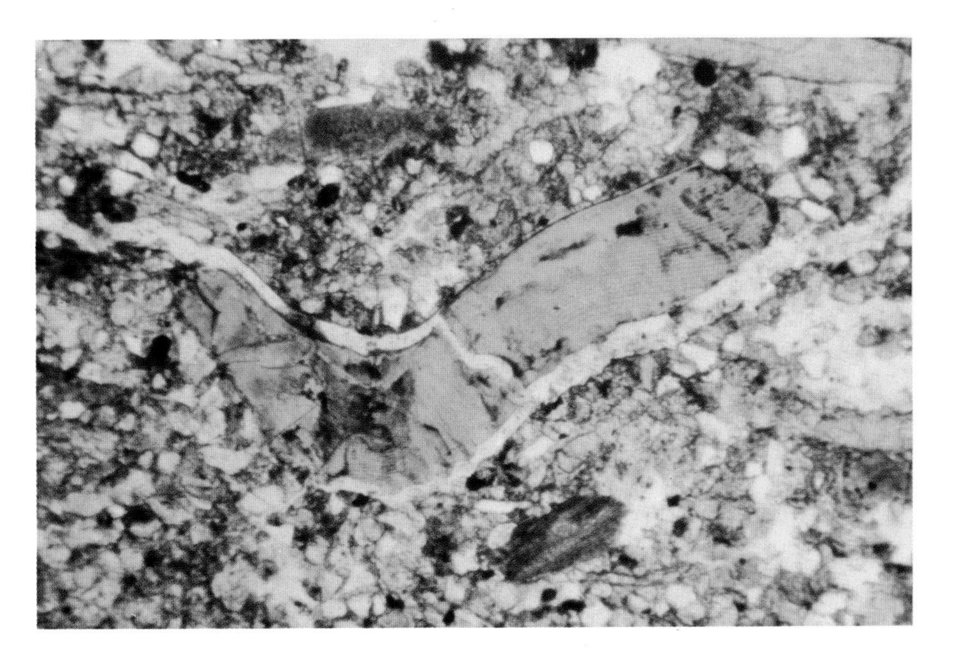

Up. Cretaceous San Carlos Fm.
Texas

A phosphatic shell fragment (made of collophane) in a silty carbonate sediment. Phosphate typically shows a brownish color in transmitted plane light.

0.24 mm

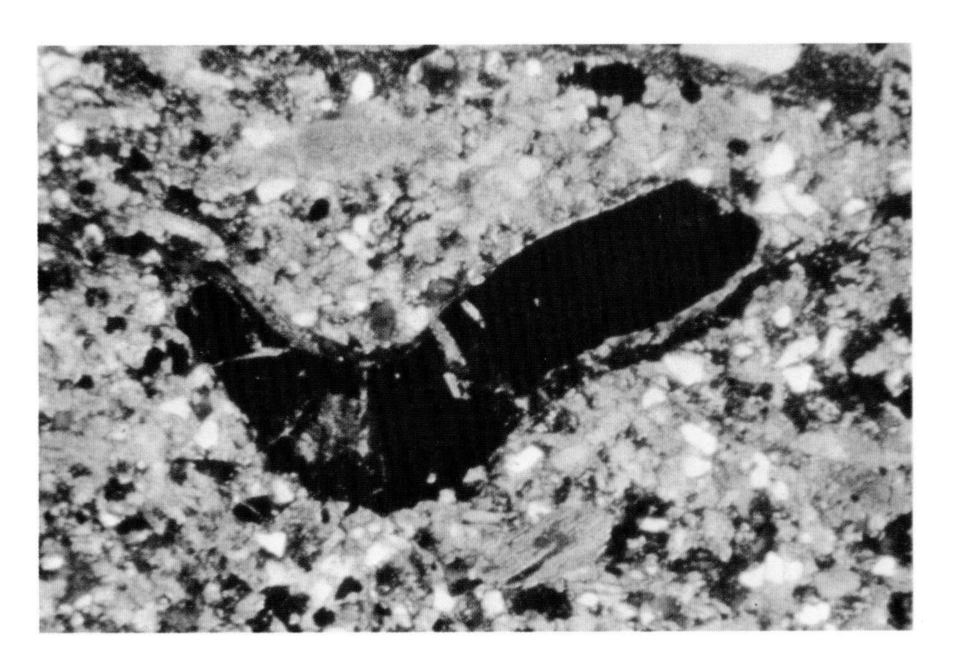

Up. Cretaceous San Carlos Fm.
Texas

Same as above photo but with polarized light to show the very low order (virtually isotropic) interference colors typical of phosphate (collophane grains.

X.N. 0.24 mm

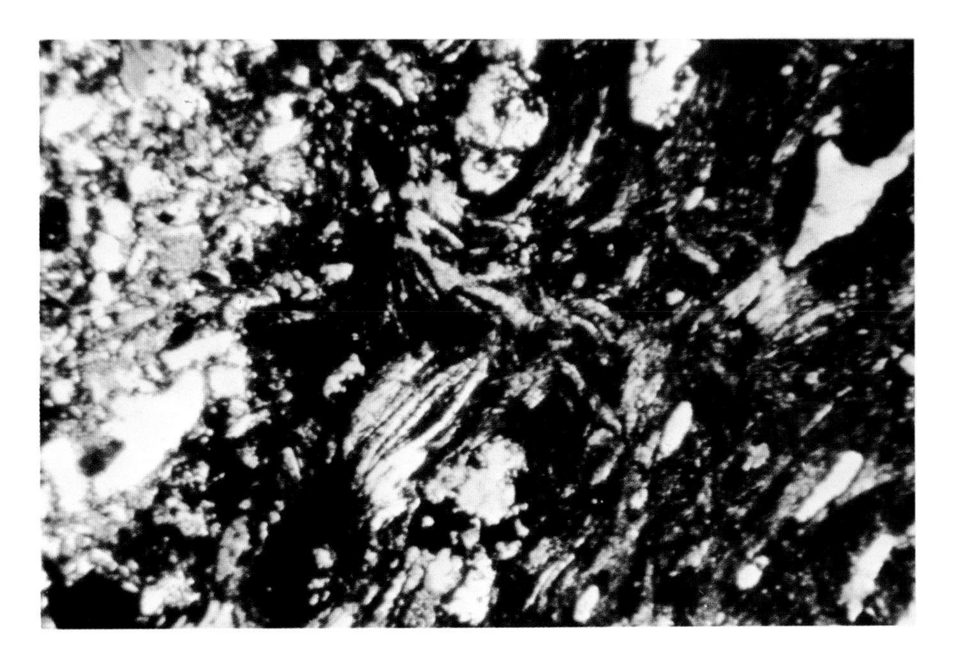

Lo. Tertiary sediment
Texas

Section of a piece of phosphatic bone material (collophane) with a very common cross-fibrous texture, and the very low order interference colors.

X.N. 0.38 mm

Up. Cretaceous Navesink Fm.
New Jersey

Large, rounded glauconite grains in a glauconitic marl. These may have a fecal pellet origin. Green color and speckled greenish birefringence are characteristic for this mineral.

X.N. 0.38 mm

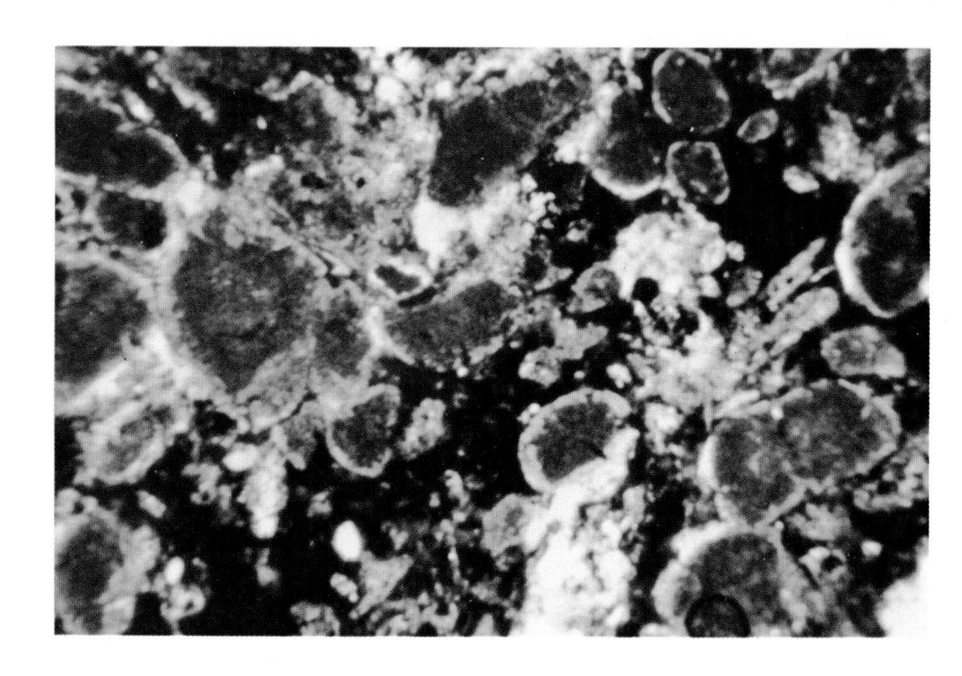

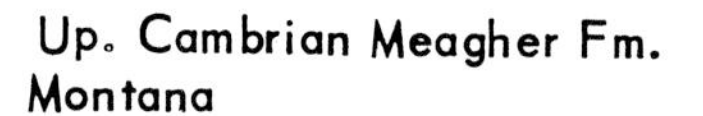
Up. Cambrian Meagher Fm.
Montana

Angular glauconite grains showing typical very fine granular texture and dark green color. The angularity indicates these glauconite grains may have been formed by alteration of detrital biotite.

X.N. 0.06 mm

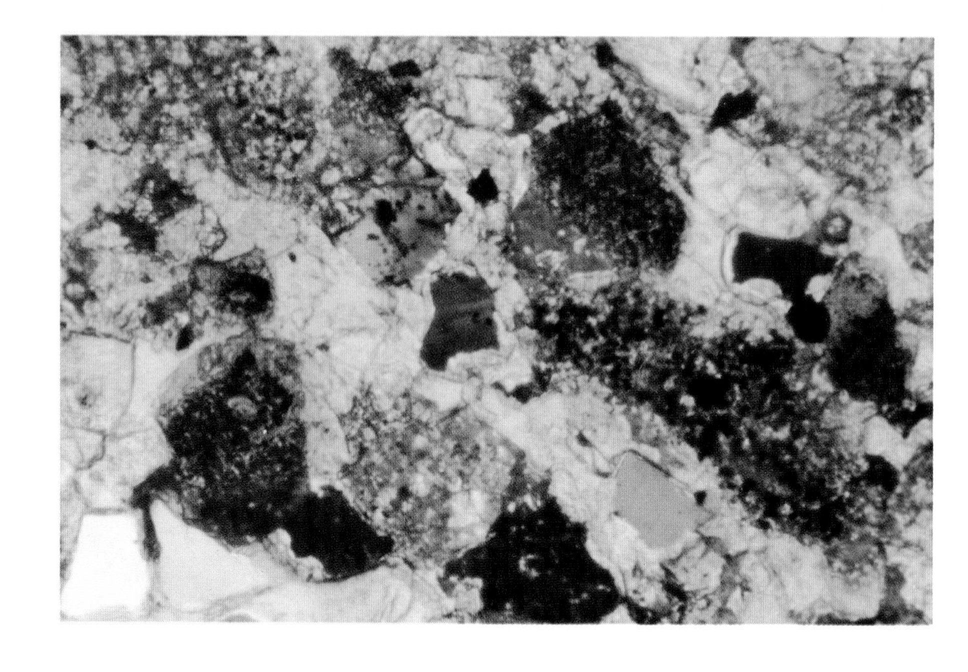

Carbonate Cements

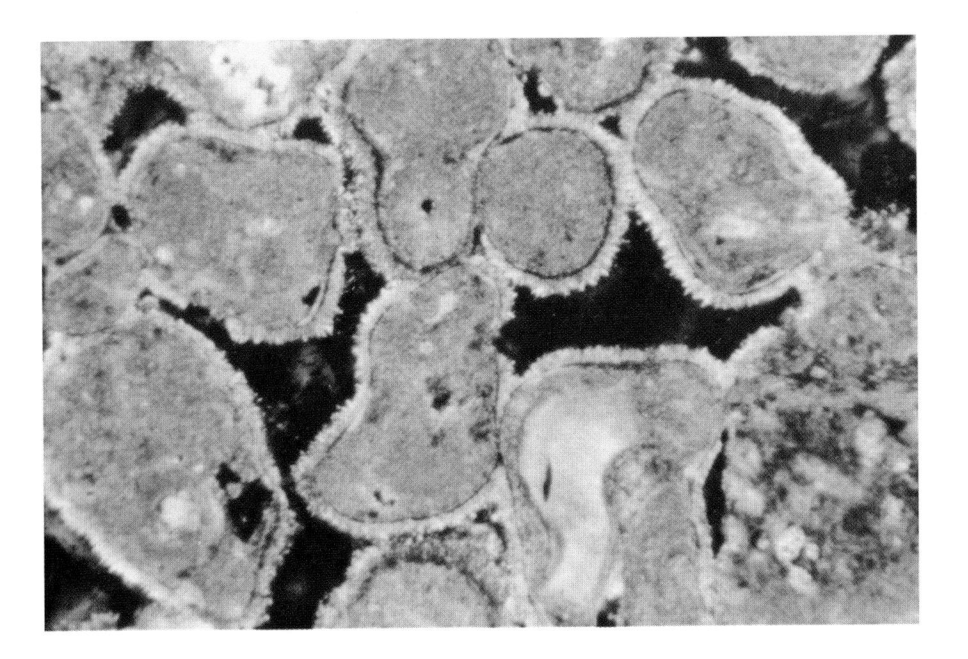

Holocene beachrock
Bahamas

Uniform fringe of fibrous aragonite cement around oolitic and micritized grains. The needle character and uniform (isopachous) fringing are characteristic of lower intertidal and submarine cement.

X.N. 0.24 mm

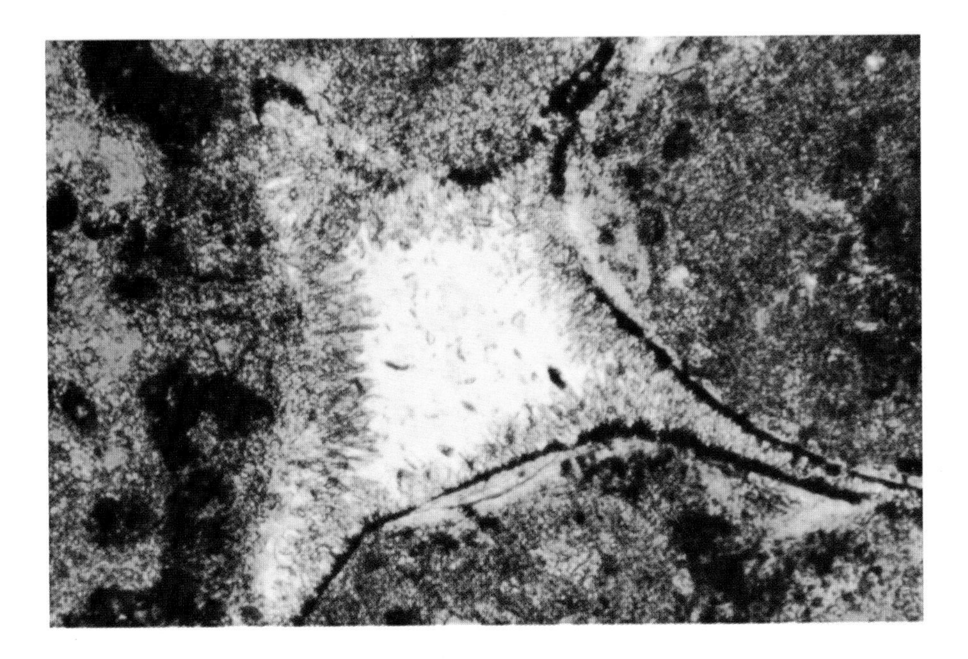

Holocene beachrock
Bahamas

Closeup of uniform isopachous fringe of fibrous aragonite-needle cement holding together oolitic grains. Fringe smoothly follows outline of voids and is not concentrated near grain junctions as in some vadose cements.

0.06 mm

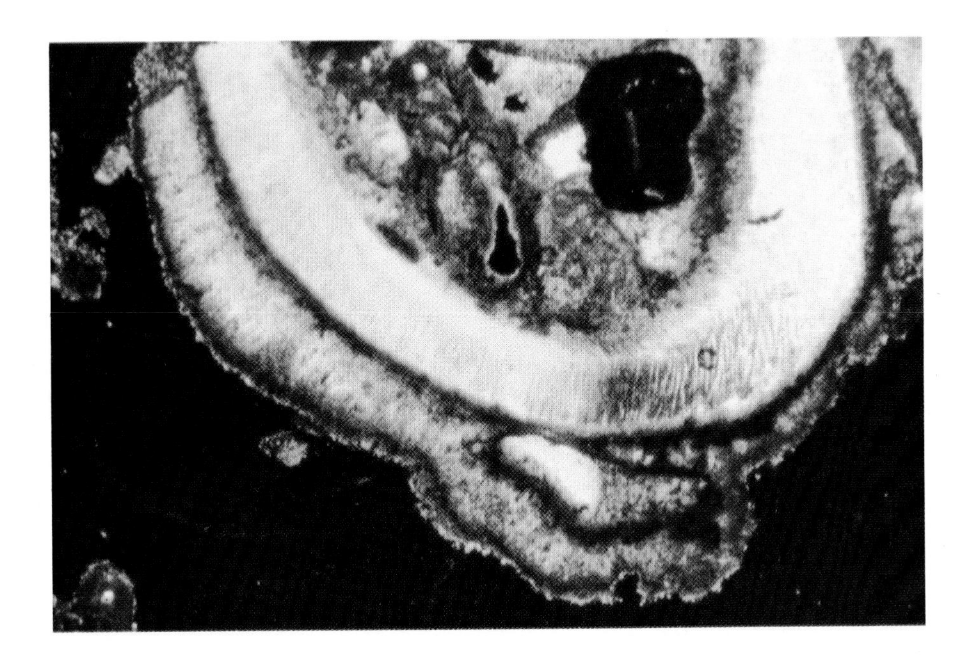

Quaternary sediment
Abu Dhabi

Fringing cements of aragonite (light-colored, radial fibrous crystals) and later high-Mg calcite (micritic) cement. The mineralogies, compositions, and crystal morphologies are those of cements produced from seawater while the depositional textures (meniscus and microstalactitic) are vadose. This results from the fact that vadose waters in Abu Dhabi sabkhas are marine to hypersaline in composition.

X.N. 0.22 mm

Quaternary sediment
Abu Dhabi

Complete cementation of pore space by aragonite needles with small amounts of high-Mg calcite. Modern submarine and intertidal sediments can be fully cemented in the depositional environment. However, both the primary grains and the cements are of unstable mineralogies and the rock is likely to undergo very significant future diagenetic modification with possible improvement of porosity.

X.N. 0.25 mm

Holocene sediment
Belize (British Honduras)

Submarine aragonite-needle cement forming isopachous linings of voids within a coral skeleton. Note irregular but basically radial arrangement of fibers.

SEM Mag. 3000X 4.1 μm

Holocene sediment, Tobacco Cay
Belize (British Honduras)

Cemented internal sediment within voids in a coral skeleton. Cement is pelloidal high-Mg calcite lime silt (stained red with Clayton Yellow). Sediment is less than 500 years old and is from the barrier reef crest. Photo courtesy of R. N. Ginsburg.

0.17 mm

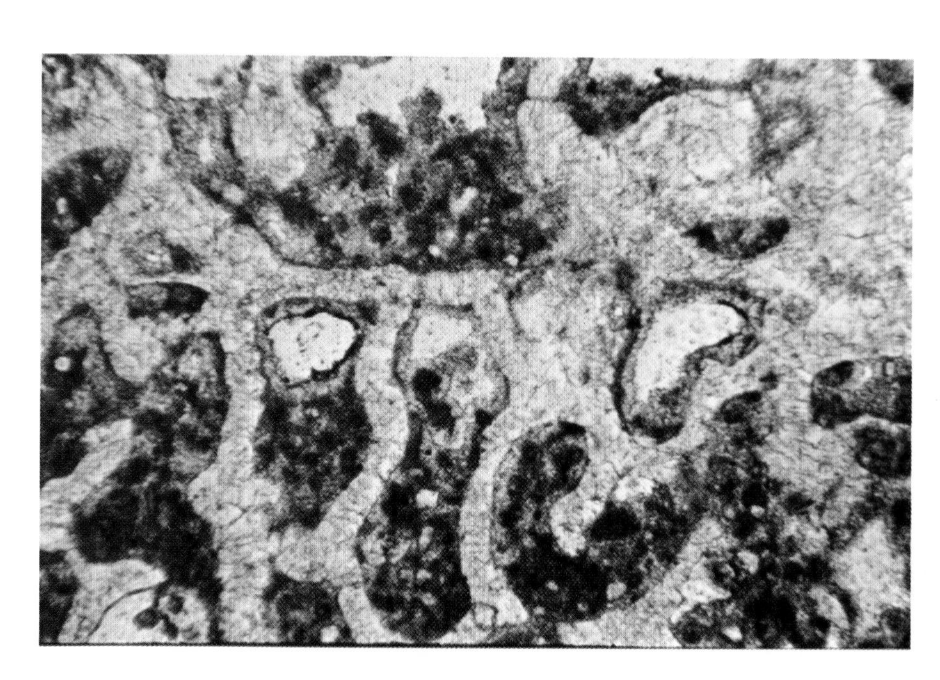

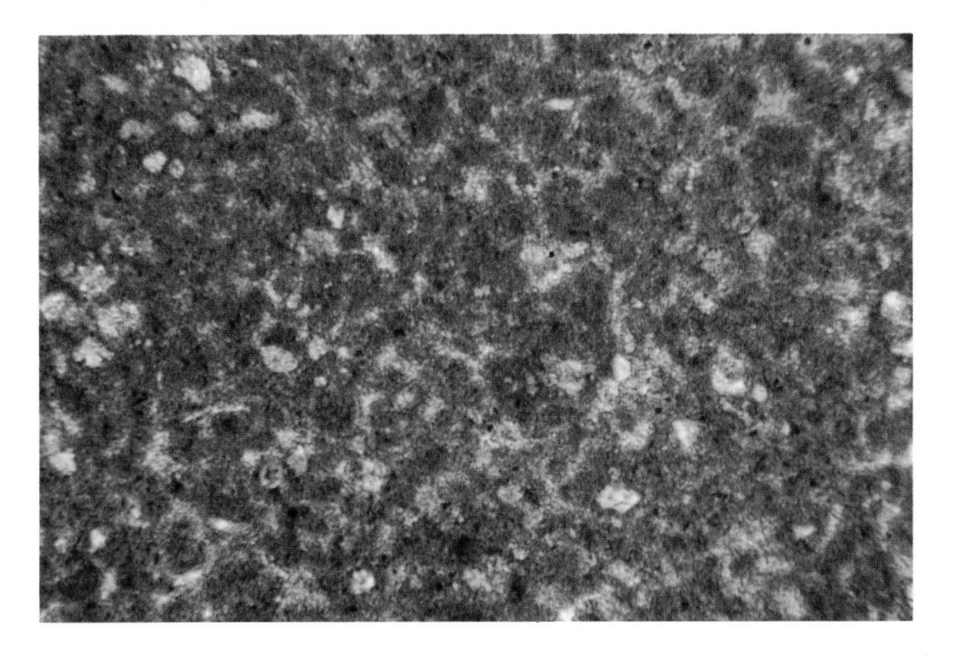

Holocene sediment, Tobacco Cay
Belize (British Honduras)

Detail of a large area of cemented internal sediment of pelloidal high-Mg calcite lime silt stained red with Clayton Yellow. Such cementation is volumetrically very important in the cementation of many modern reefs and other shallow marine sediments. Photo courtesy of R. N. Ginsburg.

0.09 mm

Holocene sediment
Virgin Islands (St. Johns)

SEM view of a high-Mg calcite rim cement. In thin section these small, equant crystals appear very similar to a micritic or pelletal coating. Only through scanning electron microscopy can one resolve the excellent crystal terminations. This cement is from a shallow marine setting.

SEM MAG. 3000X 4.1 μm

Quaternary sediment, Tobacco Cay
Belize (British Honduras)

Cementation of large voids in barrier reef wall. Fibrous aragonite is arranged in dense, botryoidal arrays interlayered with gray marine sediments and red-stained high-Mg calcite cement and sediment. Sediment is about 12,000 years old. Photo courtesy of R. N. Ginsburg.

Scale shown on photo

Quaternary sediment, Tobacco Cay
Belize (British Honduras)

Cementation of large voids in reef wall of barrier reef. Shows dense botryoidal aragonite in association with red (Clayton Yellow stained) high-Mg calcite cement. Note dark, elliptical borings of marine organisms which cut cement. This is excellent evidence for syndepositional cementation. Sediment is less than 12,000 years old. Photo courtesy of R. N. Ginsburg.

X.N. Scale shown on photo

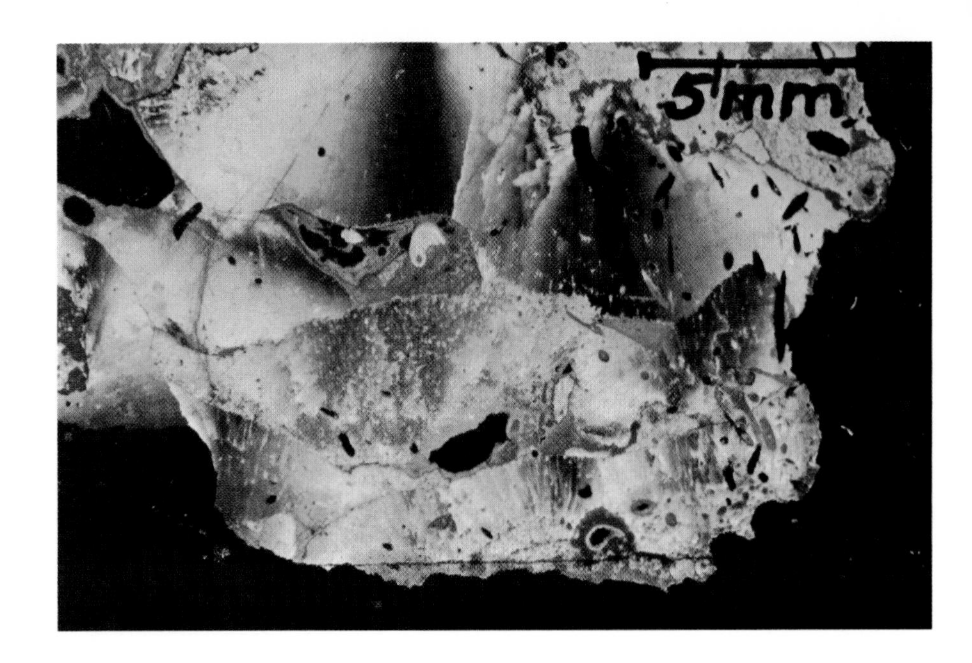

Up. Cretaceous White Ls.
Northern Ireland

This is a hardground or synsedimentary lithification surface. These form during breaks in sedimentation and reflect, in some areas, lowered sea level (but not subaerial exposure). The sediment on the right has been cemented with carbonate, phosphate, and glauconite and is bored, encrusted, and eroded into pebbles (upper left). It is important to distinguish such submarine hiatuses from unconformities produced by subaerial exposure and erosion. Encrustation and repeated boring by *marine* organisms coupled with *marine* cements are the important criteria for proving submarine origin.

0.64 mm

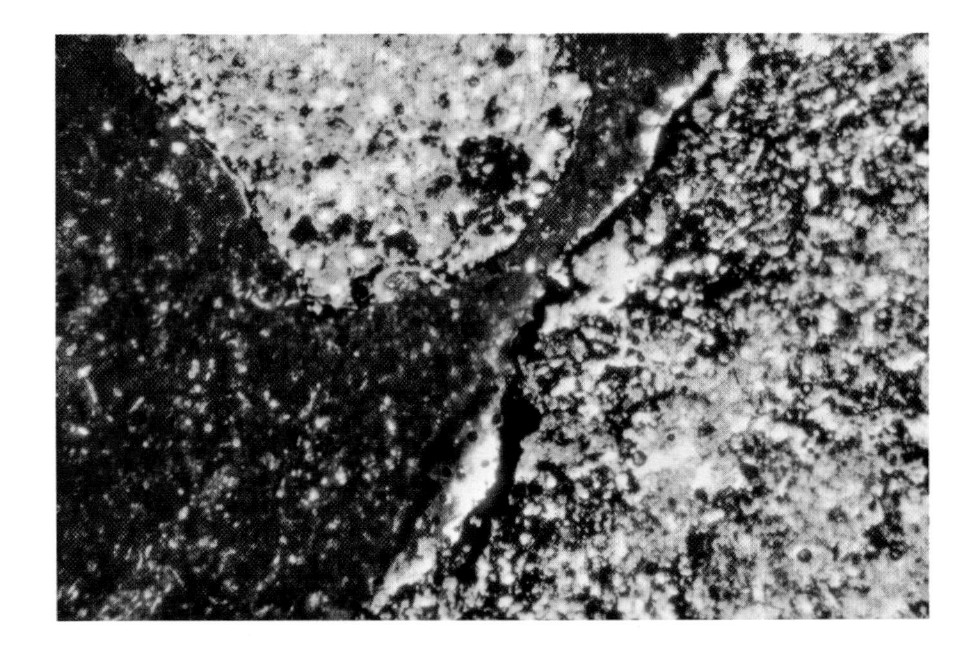

Holocene dune sands
Mexico

Meniscus cement. Note that cement is found only at grain contacts (presumably where meniscus water films were trapped)--no complete cavity linings. This type of cement is characteristic of vadose cementation. Crystals are calcite precipitated from freshwater.

X.N. 0.10 mm

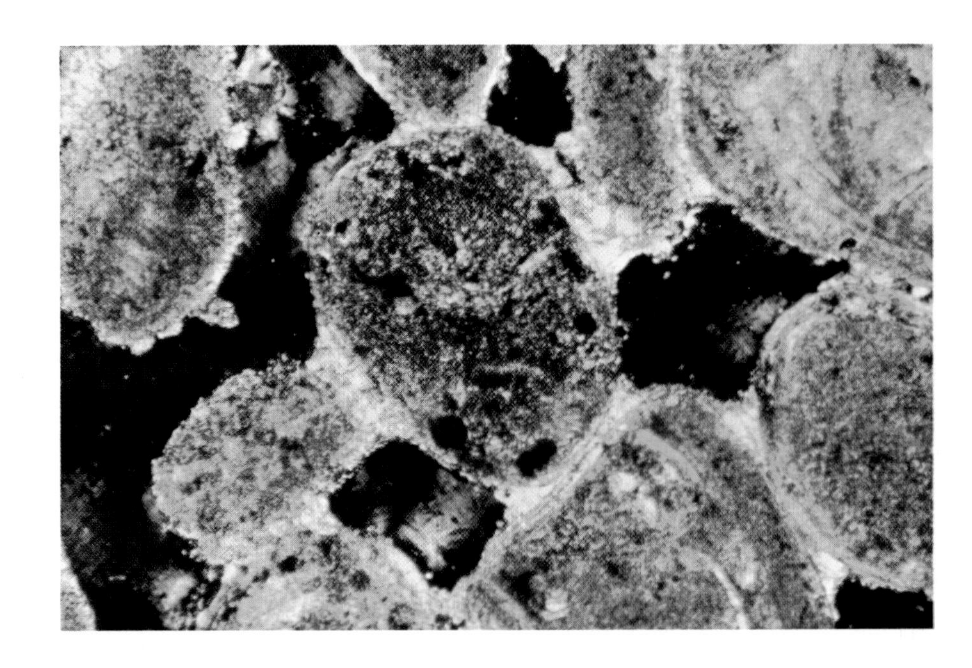

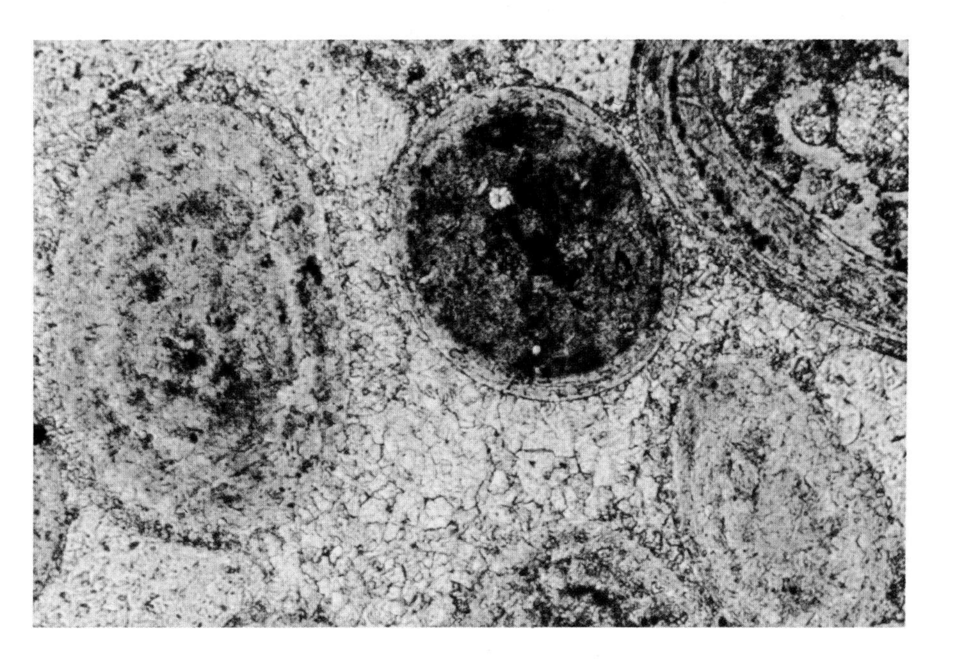

Quaternary sediment
Joulters Cay, Bahamas

An example of extensive freshwater cementation, probably within the vadose environment. Note the formation of a blocky calcite mosaic with relict meniscus texture and the lack of dissolution features on the still-aragonitic grains. Later diagenesis of the rock may yield moldic porosity.

0.11 mm

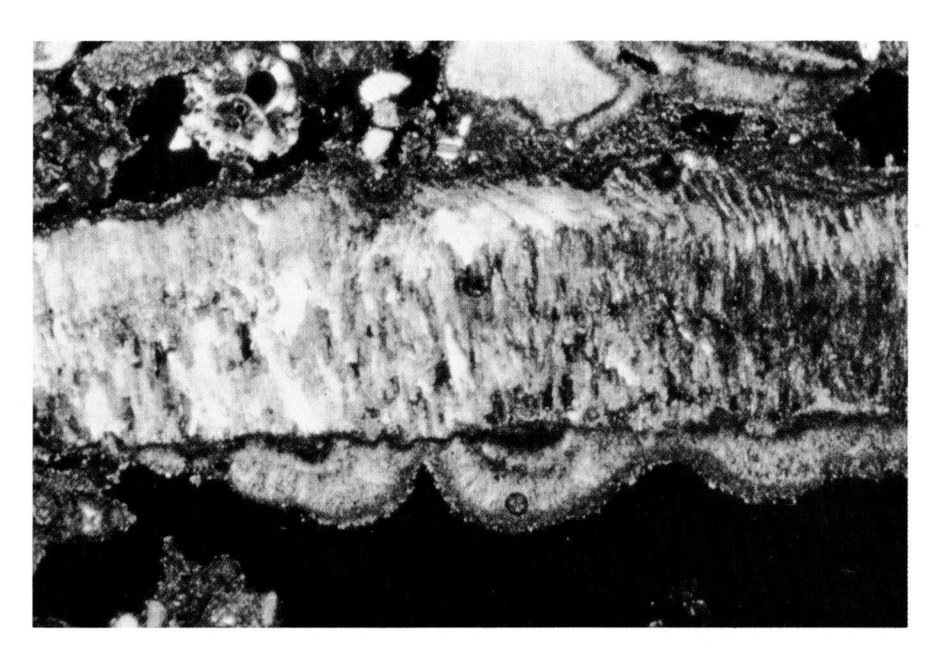

Quaternary crust
Abu Dhabi

A modern example of microstalactitic cement on a mollusk. The cement (here aragonite and high-Mg calcite, but more commonly low-Mg calcite) hangs as pendulous bulges from the undersides of grains. This is clearly the product of the concentration of water drops on the bottoms of grains. Such cement, occasionally termed pendulous or gravitational cement, is characteristic of vadose exposure.

X.N. 0.25 mm

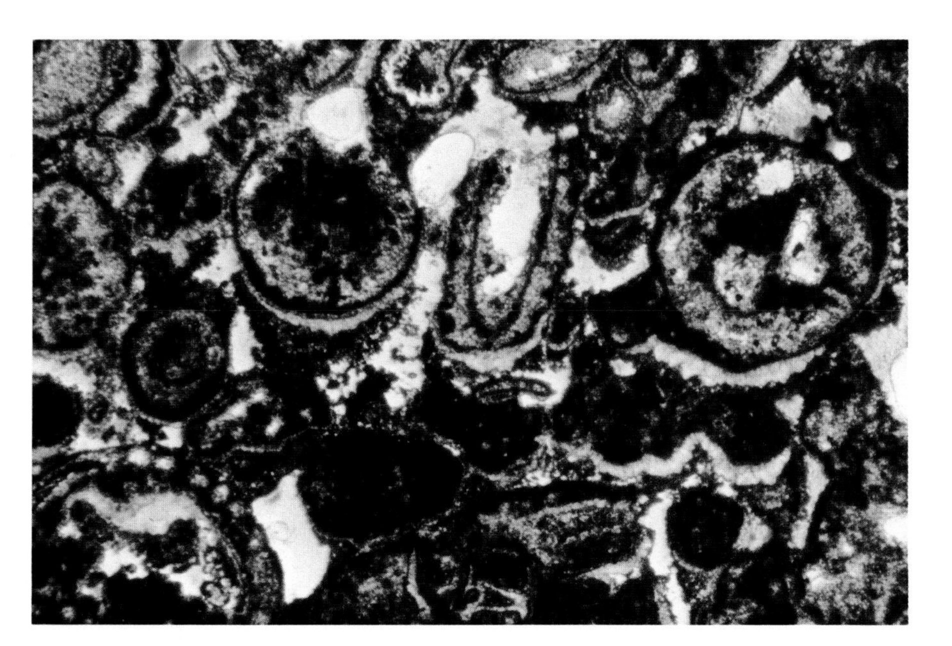

Up. Permian Tansill Fm.
New Mexico

An ancient example of preserved microstalactitic cement. Note the slight corrosion of the upper surfaces of grains and the thick cement crusts on the undersides of grains or internal voids in the *Mizzia* grainstone. The green areas are voids filled with stained plastic.

0.64 mm

Cretaceous El Abra Fm.
Mexico

Small leached voids of this dismicrite fabric are filled with a geopetal crystal silt (vadose silt) which commonly is found in sediments which have been exposed to vadose diagenesis. The filling of leached voids and the crystal size and shape of the silt are the diagnostic features in this key to subaerial exposure.

0.64 mm

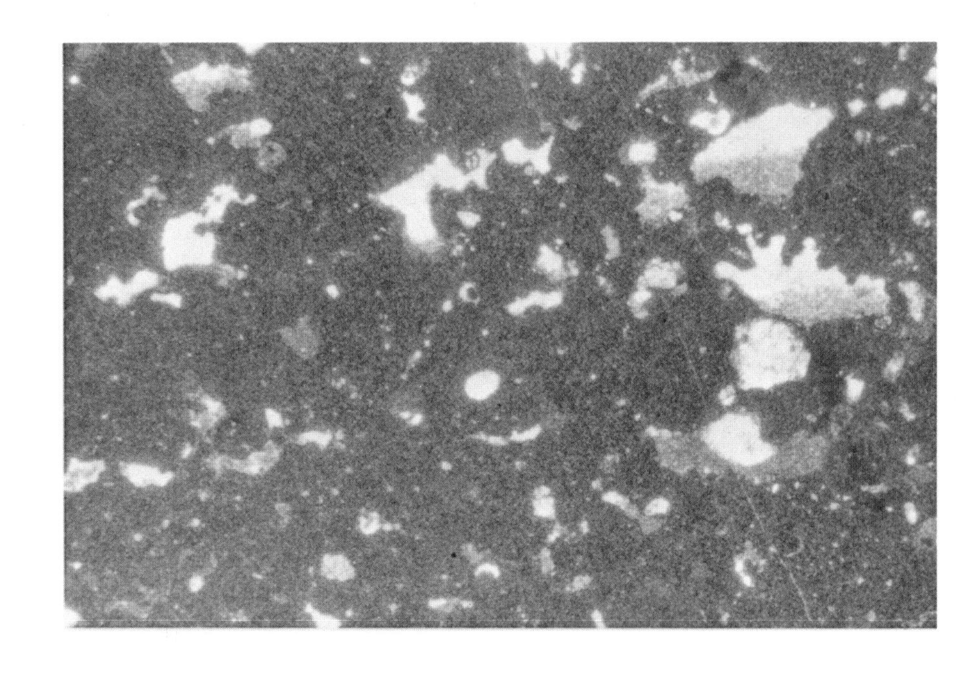

Pleistocene Key Largo Ls.
Florida

A partially leached void filled with low-Mg calcite whisker crystal cement. These are thin, randomly oriented calcite crystals which have been found in a number of modern localities. They are produced in meteoric waters, probably within the vadose zone, but are not commonly recognized in ancient environments.

X.N. 0.38 mm

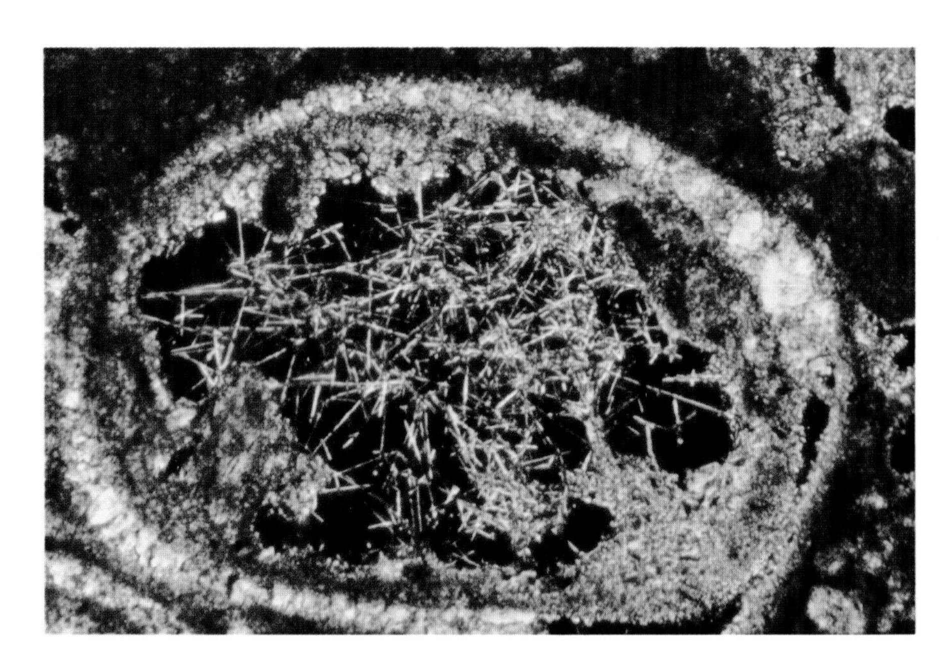

Pleistocene dunes
Isla Mujeres, Mexico

Whisker crystal cement. This closeup view shows details of orientation and size of these low-Mg calcite crystals.

X.N. 0.09 mm

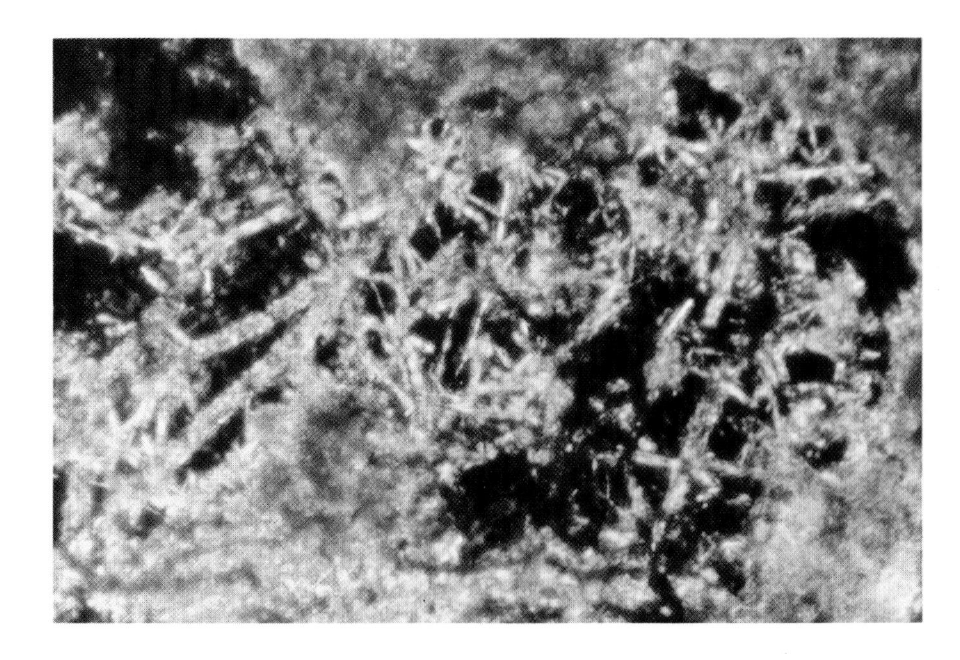

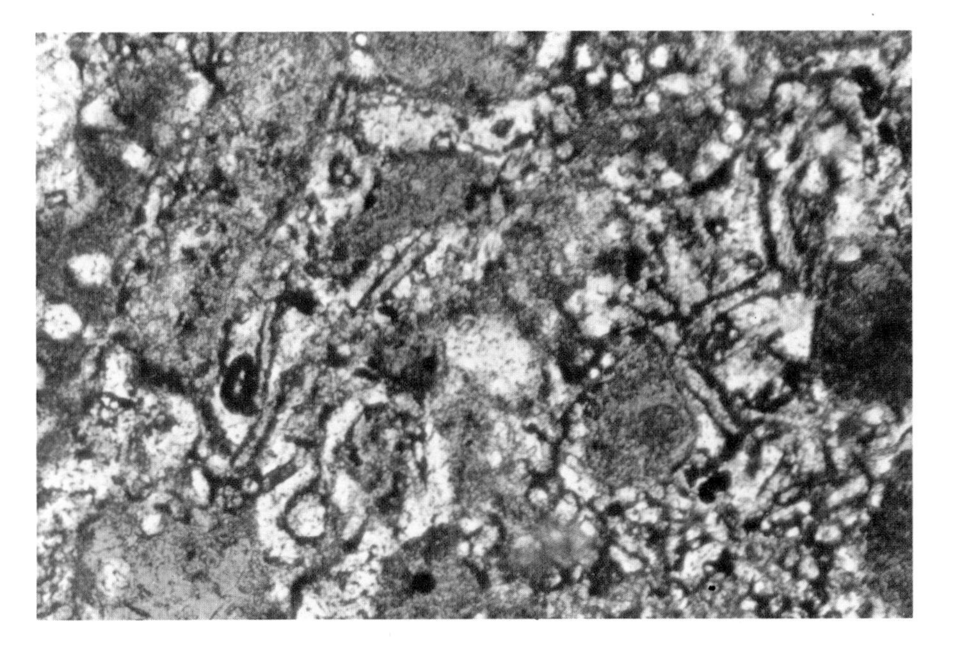

Pleistocene dunes
Isla Mujeres, Mexico

Root-hair cement. Interstitial areas are crisscrossed by a network of randomly oriented tubes of calcite which have been precipitated around the root hairs of salt bushes and other plants which colonize these dunes. These reflect meteoric vadose conditions and could be preserved in the geologic record.

0.09 mm

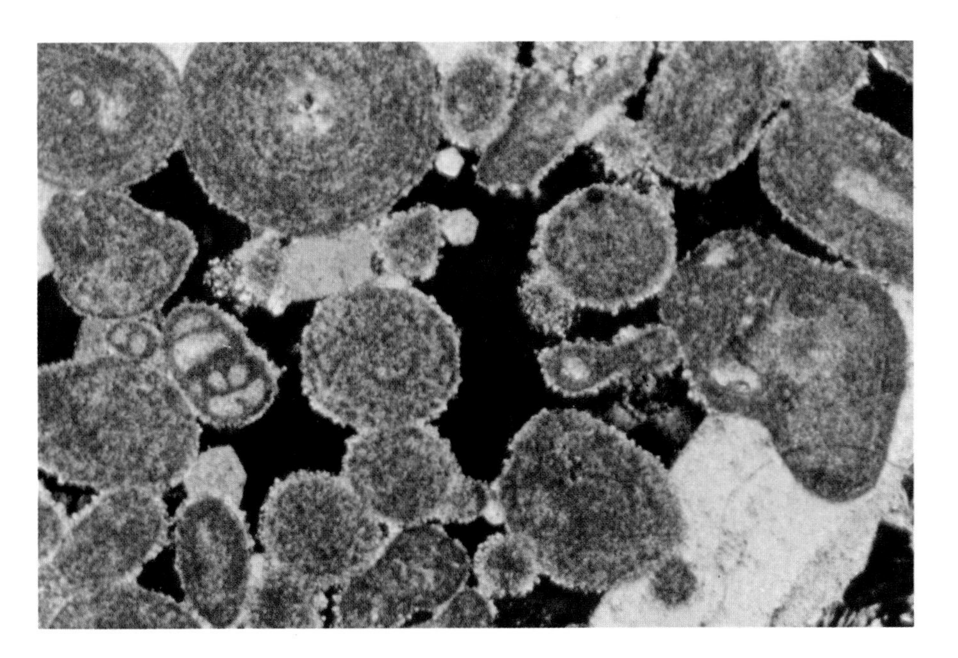

Up. Mississippian Ste. Genevieve Ls.
Indiana

An ancient example of an oolitic limestone which, despite probable alteration of the ooids from original aragonite to calcite, has undergone very little cementation. Intergranular pore space is completely preserved except near echinoderm fragments where syntaxial rims have formed. Does this reflect preserved intergranular porosity or has an unstable cement been leached out?

X.N. 0.25 mm

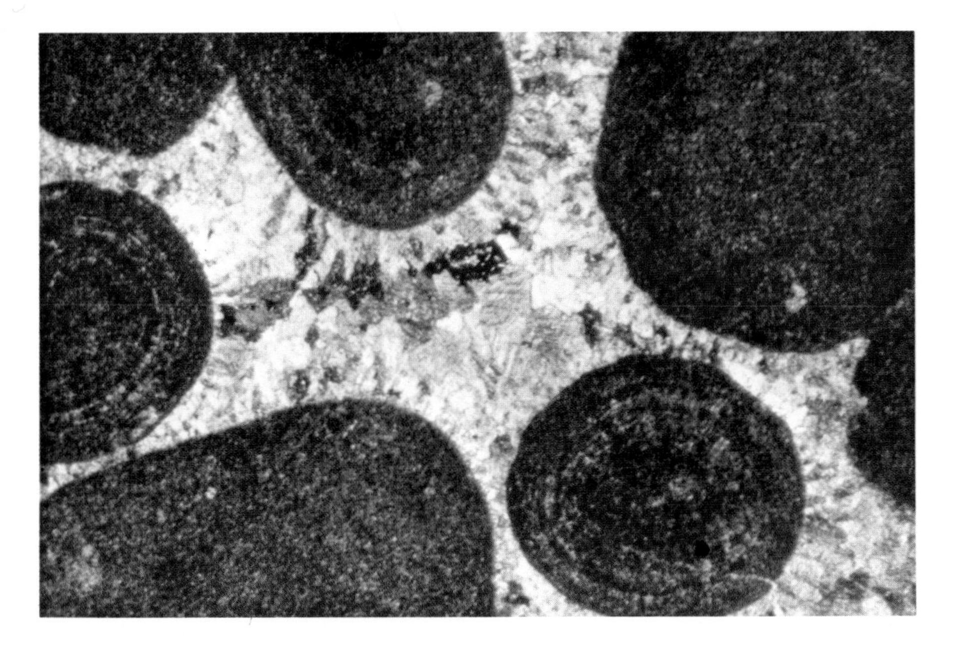

Jurassic-Cretaceous pisolitic ls. series
Italy

Ooids (micritized and radially recrystallized) with a coarse fibrous cement fringe (perhaps indicating intertidal or subtidal cementation).

X.N. 0.38 mm

Mississippian Coral Ls.
England

Small clam or ostracode showing coarse sparry calcite cavity filling. Fine-grained fringe on inside of shell is followed by very coarse calcite spar. This is a typical void-filling fabric.

X.N. 0.38 mm

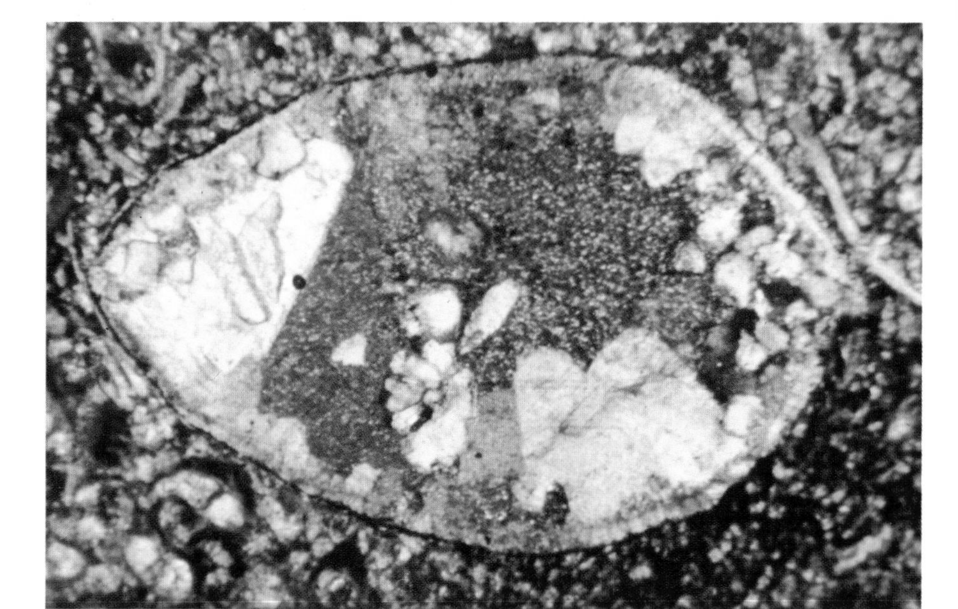

Mississippian part of Lodgepole Ls.
Montana

Cloudy, radiaxial fibrous spar filling of a mollusk-bounded void. The reasons for the unusual optical properties and inclusions in this type of spar are not clear but this type of cement is apparently very early and is common in Mississippian bioherm mounds. It may represent alteration of original submarine aragonite cement.

X.N. 0.38 mm

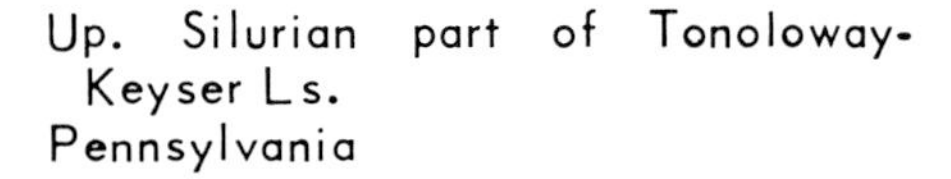

Up. Silurian part of Tonoloway-Keyser Ls.
Pennsylvania

Crinoid biosparite with extensive rim cementation of crinoid fragments. Each crinoid grain is surrounded by an optically continuous cementing rim of calcite--such syntaxial overgrowths are generally early diagenetic and often completely destroy porosity in echinoderm-rich sediments.

X.N. 0.38 mm

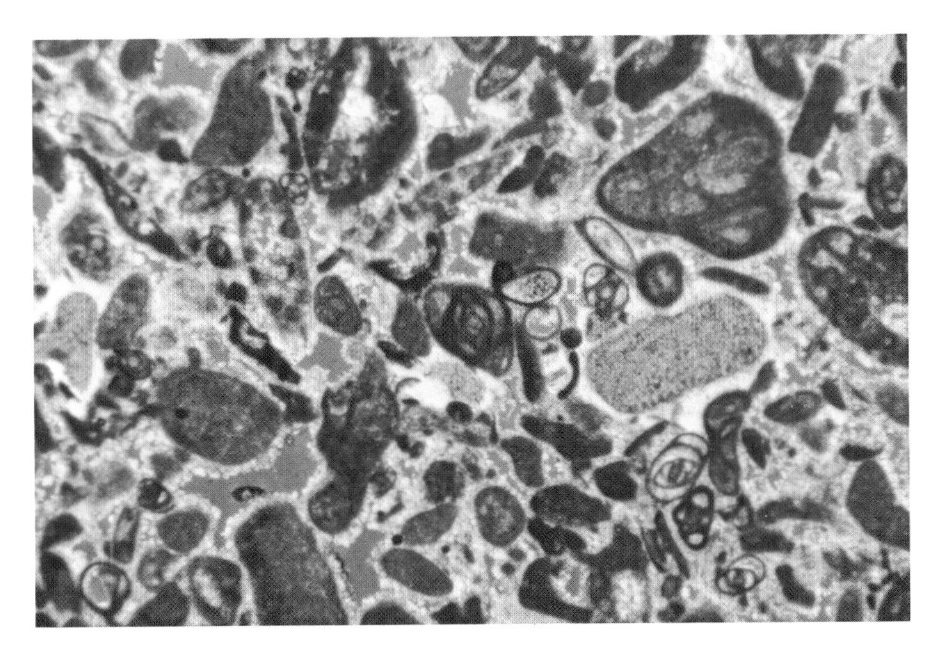

Up. Oligocene Suwannee Ls.
Florida

Highly porous biosparite. Most grains have a thin fringe of equant calcite. The echinoderm fragments, however, have large, optically continuous overgrowths around them, completely filing the surrounding porosity.

G.P. 0.38 mm

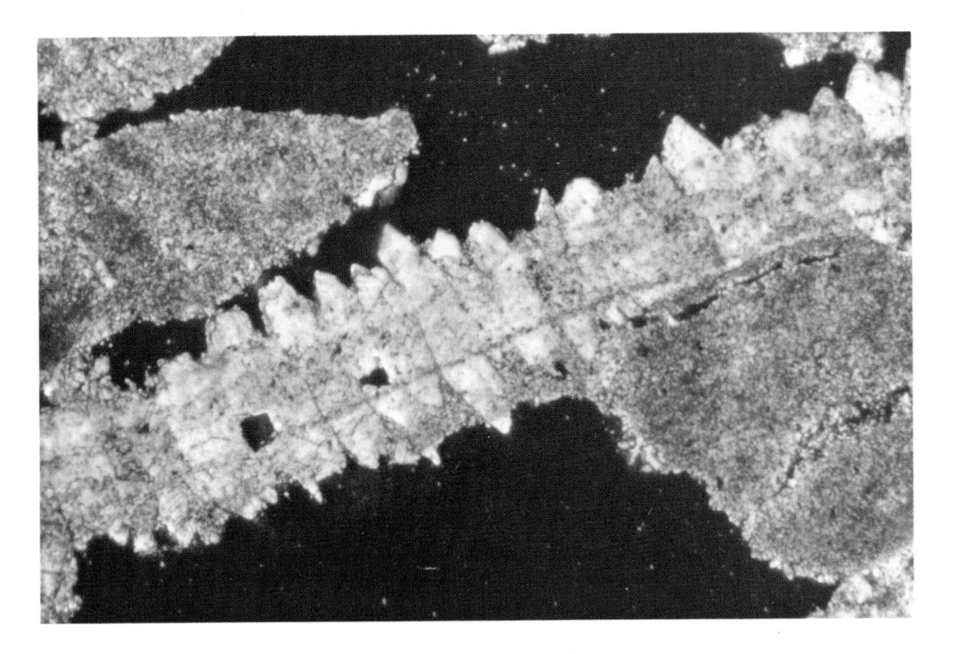

Up. Eocene Ocala Fm.
Florida

Calcite spar overgrowths on a prismatic-walled mollusk shell. Although overgrowth cementation is most common and most extensive on echinoderm fragments, it also occurs on other grains. It is most clearly recognizable when the substrate crystal structure is coarse and uniform as in this molluscan grain. Outline of original shell wall is visible due to inclusions within the calcite of the shell and traces of original micrite envelope.

X.N. 0.38 mm

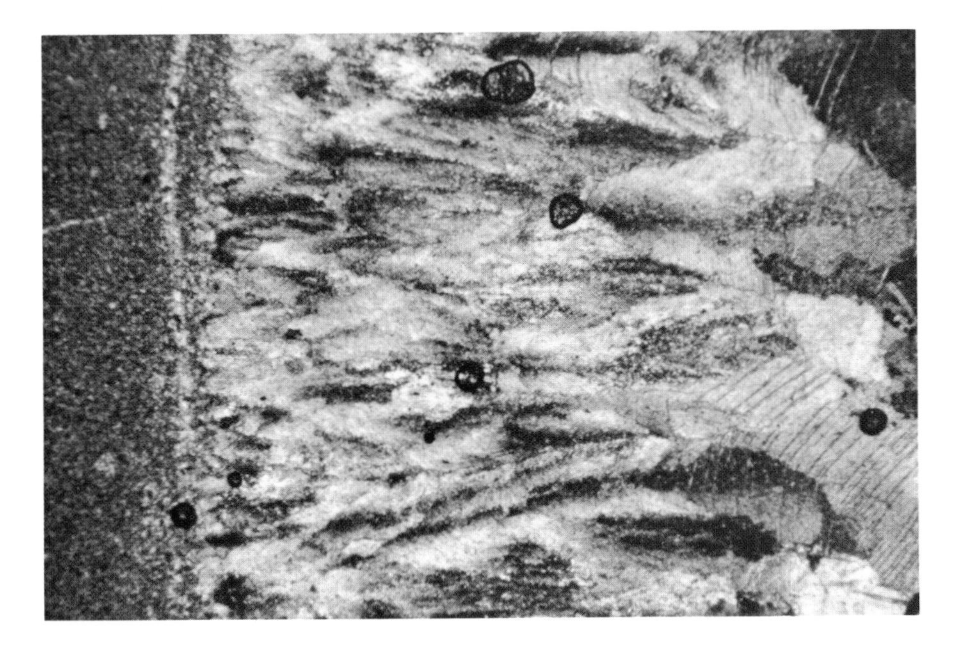

Up. Triassic Dachstein Ls.
Austria

Fibrous calcite vein with long axes of crystals oriented perpendicular to vein walls. Such textures are often (but not invariably) indicative of void filling.

X.N. 0.38 mm

Lo. Cretaceous Folkestone beds
England

Poikilotopic cement. A single calcite crystal cements a large number of detrital clastic grains. Such fabrics are most common in clastic sediments where a few carbonate grains may act as nuclei for very large cement crystals.

X.N. 0.38 mm

Lo. Pliocene Tamiami Fm.
Florida

Small clam with coarse calcite infilling of central void. Note increasing crystal size going from shell wall to center of cavity--this is common in void fill cementation (but not restricted to it).

X.N. 0.38 mm

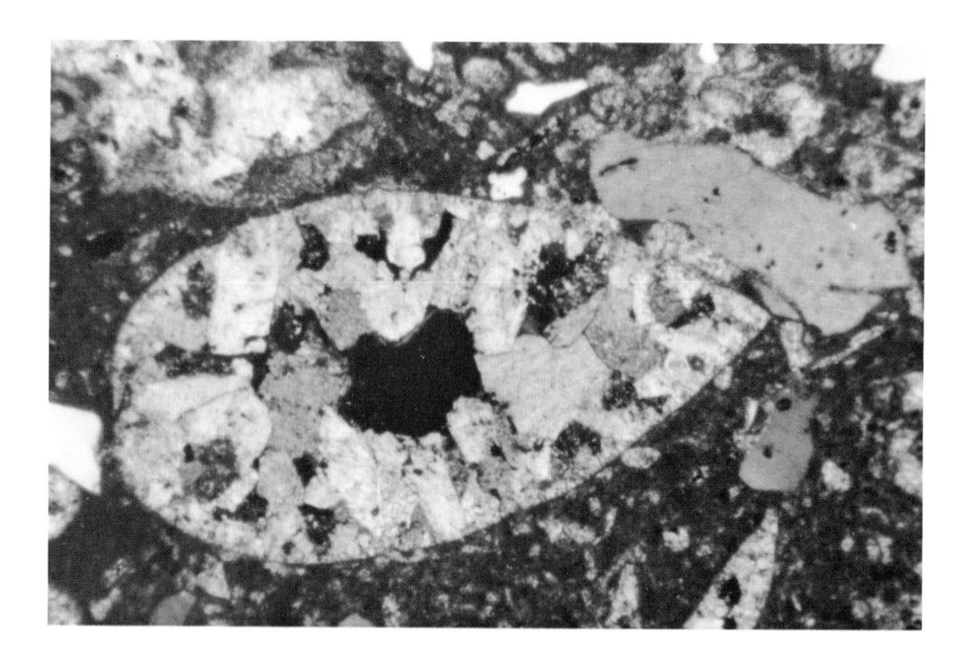

Lo. Pennsylvanian Hale Fm.
Oklahoma

An example of multiple generations of cementation in a carbonate. Bryozoan fragments are encrusted with sparry calcite void filling. The rest of the void space was filled by ankeritic dolomite which is associated with an unconformity which directly overlies this unit. Thus, the timing of cement formation can sometimes be tied into datable geological events.

0.25 mm

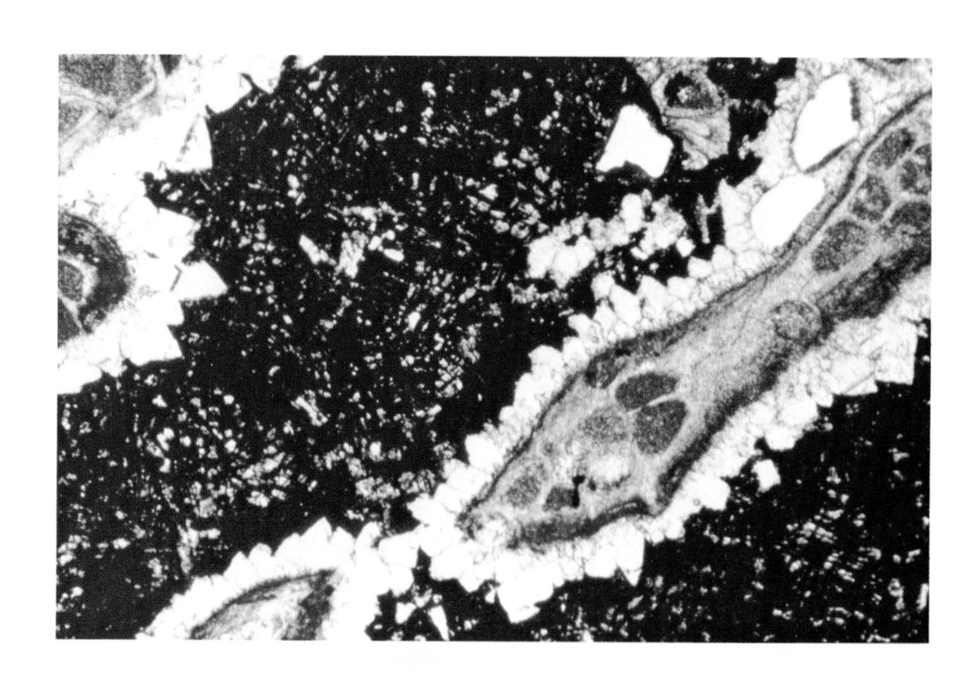

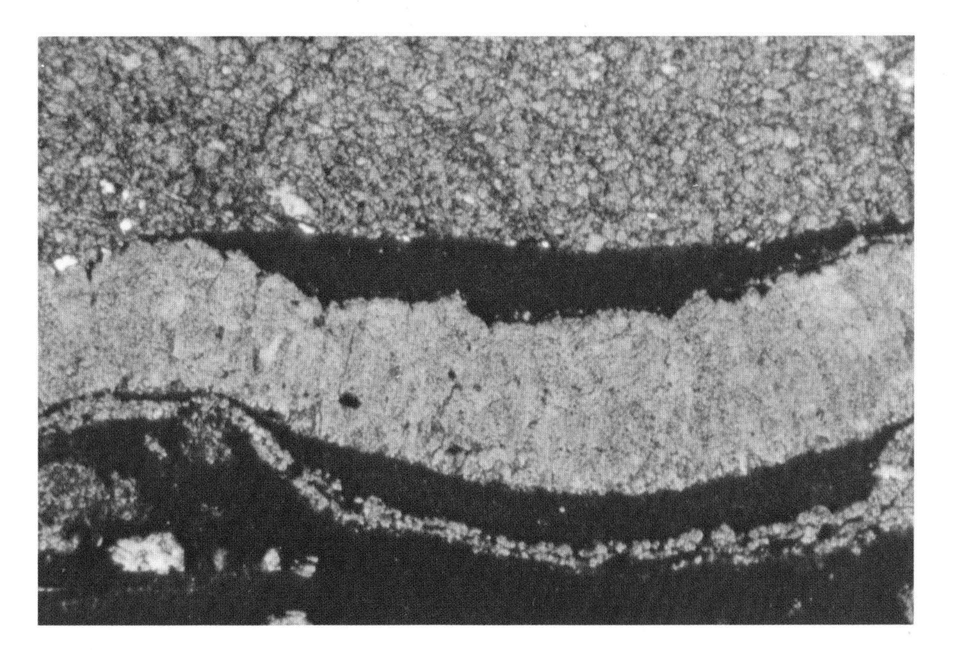

Up. Cretaceous El Abra Ls.
Mexico

A complex infilling of a large cavity. There is geopetal infiltration of silt which floors part of the cavity (avoiding highs) and alternates with thick and thin layers of sparry calcite (bladed) cement. Final filling is by uniform, equant calcite.

0.64 mm

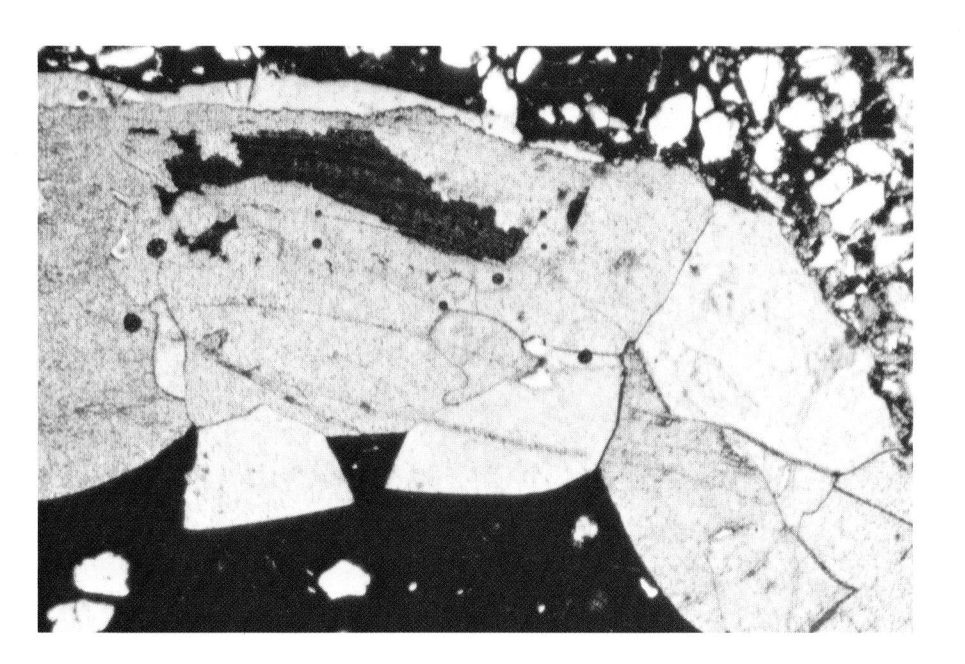

Pliocene and Pleistocene Caloosahatchee Fm.
Florida

Partial inversion of a vermetid gastropod shell. Small brown inclusion is unaltered remnant of aragonitic shell while rest has gone to a coarsely crystalline sparry calcite. Virtually no relict texture seen in inverted areas. Also note the extension of calcite crystals into the cavity-filling micrite through displacement or replacement.

0.64 mm

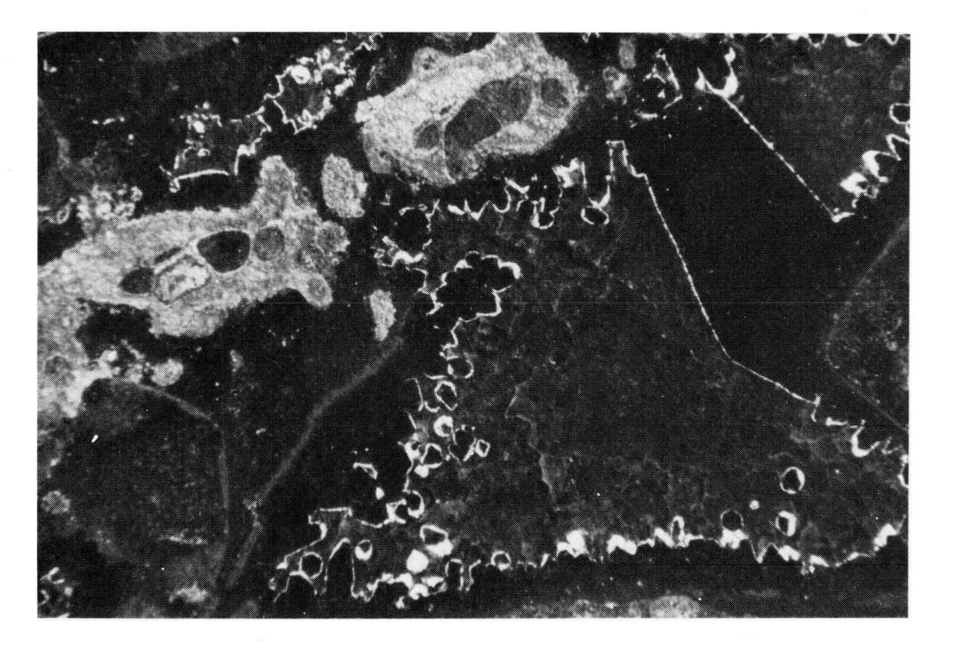

Lo. and Mid. Pennsylvanian Bloyd(?) Fm.
Oklahoma

A bryozoan biosparite seen under cathodoluminescence, and showing much more complex cementation than would be seen with normal light microscopy. This sediment would show only one generation of cement in polarized light but shows at least 2 generations here. The color differences reflect small variations in trace element composition of the calcite.

Cathodoluminescence 0.30 mm

Up. Cretaceous Chalk
British North Sea well 49/21-3

SEM view of a chalk which has been buried less than 500 meters deep. Note abundant coccoliths and small crystals derived from coccolith breakdown. Porosity is over 40 percent and crystals are rounded and non-interlocking in this sediment.

SEM Mag. 5000X 2.5 μm

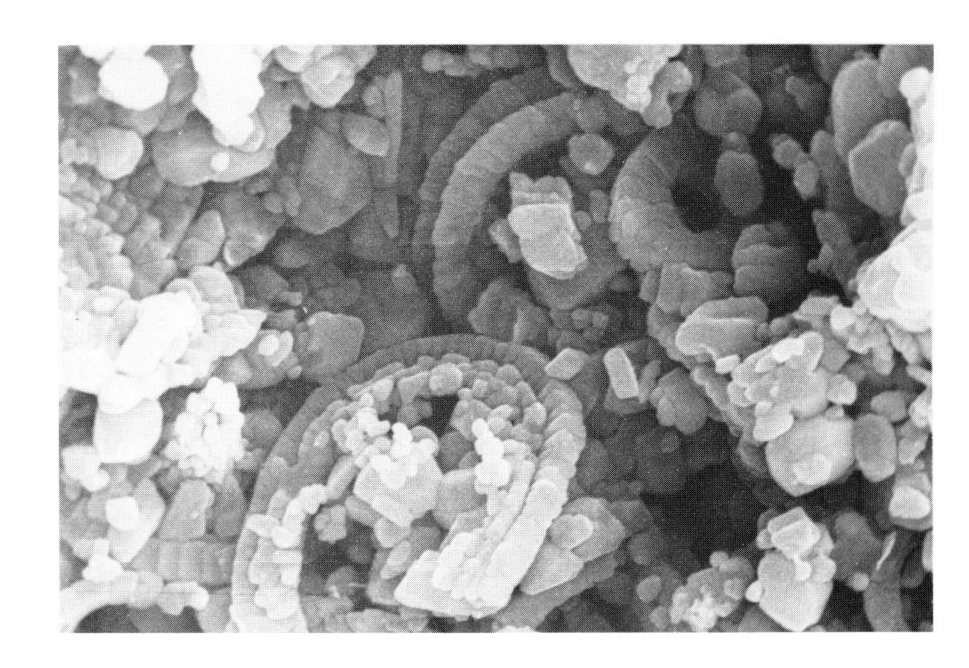

Up. Cretaceous Chalk
British North Sea well 30/16-3

SEM view of a chalk which has been buried in excess of 2000 meters deep. Although originally the same type of sediment as the previous chalk, this sediment has undergone extensive pressure solution and reprecipitation. Porosity has been reduced to 15-20 percent, crystals are angular, equant, and interlocking, yet coccolith remains can still be seen.

SEM Mag. 10,000X 1.3 μm

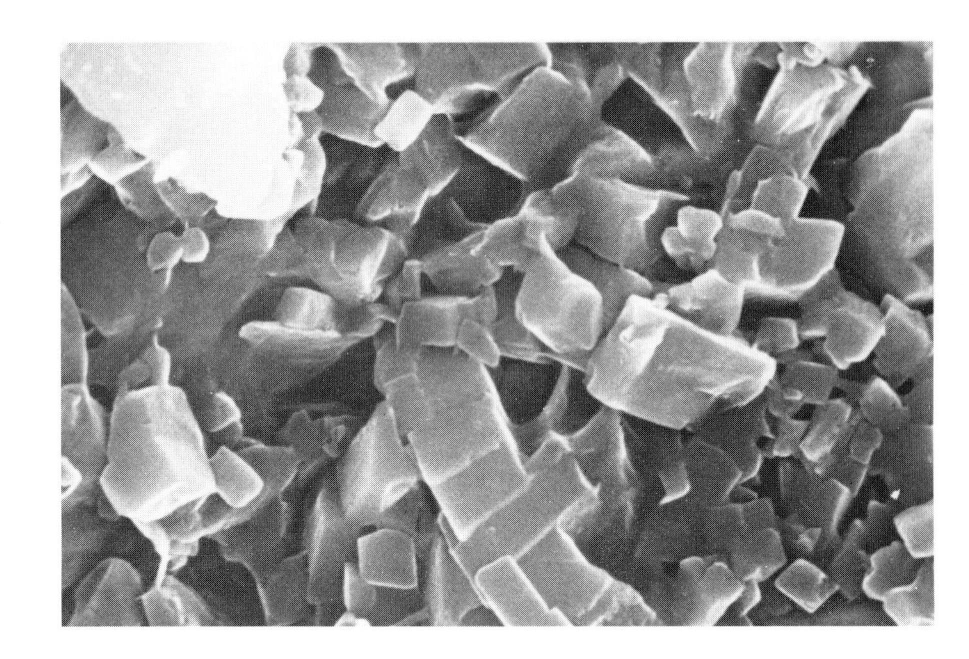

Up. Cretaceous Chalk
Norwegian North Sea well 2/4-2X

SEM view of a chalk which has been buried more than 3000 meters deep. This shows the end member of the cycle of cementation in chalks. Porosity has been reduced to less than 10 percent, crystals are large, euhedral, and closely interlocking. Although such cementation is not visible in light microscopy, it is clearly visible with SEM, and strongly affects the reservoir properties of fine-grained limestones.

SEM Mag. 10,000X 1.3 μm

Carbonate Textures

Folk Classification Types

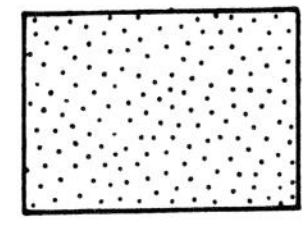

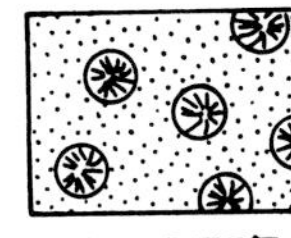
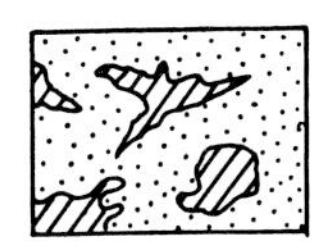

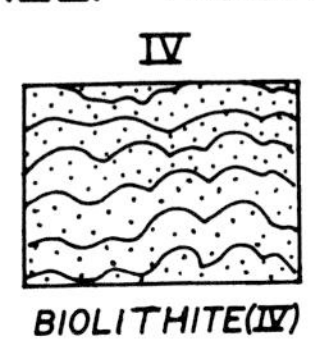
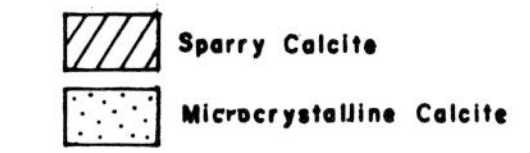

Folk's (1962) classification of carbonate rocks

	OVER 2/3 LIME MUD MATRIX				SUBEQUAL SPAR & LIME MUD	OVER 2/3 SPAR CEMENT		
Percent Allochems	0-1 %	1-10 %	10-50%	OVER 50%		SORTING POOR	SORTING GOOD	ROUNDED & ABRADED
Representative Rock Terms	MICRITE & DISMICRITE	FOSSILIFEROUS MICRITE	SPARSE BIOMICRITE	PACKED BIOMICRITE	POORLY WASHED BIOSPARITE	UNSORTED BIOSPARITE	SORTED BIOSPARITE	ROUNDED BIOSPARITE
1959 Terminology	Micrite & Dismicrite	Fossiliferous Micrite	Biomicrite			Biosparite		
Terrigenous Analogues	Claystone		Sandy Claystone	Clayey or Immature Sandstone		Submature Sandstone	Mature Sandstone	Supermature Sandstone

LIME MUD MATRIX
SPARRY CALCITE CEMENT

Folk's (1962) classification of carbonate textures

Dunham's (1962) classification of carbonate rocks

DEPOSITIONAL TEXTURE RECOGNIZABLE					DEPOSITIONAL TEXTURE NOT RECOGNIZABLE
Original Components Not Bound Together During Deposition				Original components were bound together during deposition... as shown by intergrown skeletal matter, lamination contrary to gravity, or sediment-floored cavities that are roofed over by organic or questionably organic matter and are too large to be interstices.	Crystalline Carbonate
Contains mud (particles of clay and fine silt size)			Lacks mud and is grain-supported		(Subdivide according to classifications designed to bear on physical texture or diagenesis.)
Mud-supported		Grain-supported			
Less than 10 percent grains	More than 10 percent grains				
Mudstone	Wackestone	Packstone	Grainstone	Boundstone	

Grain-Size Scale for Carbonate Rocks

Size	Transported Constituents	Authigenic Constituents	Size
	Very coarse calcirudite	Extremely coarsely crystalline	
64 mm			
	Coarse calcirudite		
16 mm			
	Medium calcirudite		
4 mm			4 mm
	Fine calcirudite	Very coarsely crystalline	
1 mm			1 mm
	Coarse calcarenite	Coarsely crystalline	
0.5 mm			
	Medium calcarenite		
0.25 mm			0.25 mm
	Fine calcarenite	Medium crystalline	
0.125 mm			
	Very fine calcarenite		
0.062 mm			0.062 mm
	Coarse calcilutite	Finely crystalline	
0.031 mm			
	Medium calcilutite		
0.016 mm			0.016 mm
	Fine calcilutite	Very finely crystalline	
0.008 mm			
0.004 mm	Very fine calcilutite		0.004 mm
		Aphanocrystalline	

Carbonate rocks contain both physically transported particles (oolites, intraclasts, fossils, and pellets) and chemically precipitated minerals (either as pore-filling cement, primary ooze, or as products of recrystallization and replacement). Therefore the size scale must be a double one, so that one can distinguish which constituent is being considered (e.g. calcirudites may be cemented with very finely crystalline dolomite, and finer calcarenites may be cemented with coarsely crystalline calcite).

The size scale for transported constituents uses the terms of Grabau but retains the finer divisions of Wentworth except in the calcirudite range; for dolomites of obviously allochemical origin, the terms *dolorudite, doloarenite,* and *dololutite* are substituted for those shown. The most common crystal size for dolomite appears to be between .062 and .25 mm, and for this reason that interval was chosen as the *medium crystalline* class (from Folk, 1962).

Ion Strength, Environment, and Carbonate Morphology

(from Folk, 1974)

Chemistry		Environment	Crystal habit
Mg High	*Na High*	Hypersaline to Normal Marine, Beachrock, Sabkha, Submerged Reefs, etc.	Steep rhombs of Mg-Calcite with vertically-oriented flutings; Fibers of Mg-Calcite and Aragonite; growth rapid in c-direction; very slow laterally because of selective Mg-poisoning. Crystals limited in width to a few microns.
(Mg Low)	*Na High*	Mainly connate subsurface waters	Complex polyhedra and anhedra of calcite; lack of Mg allows unhampered growth and equant habit.
(Mg Low)	Na moderate to low	Meteoric phreatic, to deep subsurface mingling between meteoric and connate water	Complex polyhedra and anhedra of calcite; lack of Mg and slow crystallization allows equant crystals, often coarse.
(Mg Low)	(Na Low)	Meteoric vadose; caliche; streams and lakes	Simple unit rhombohedra of calcite
(Mg Low)	(Na Low)	Streams, Lakes, Caliche	Calcite micrite. Also, calcite sheets or hexagonal crystals with basal pinacoids; sheet-structure on edges visible due to very rapid lateral growth in the absence of Mg-poisoning.

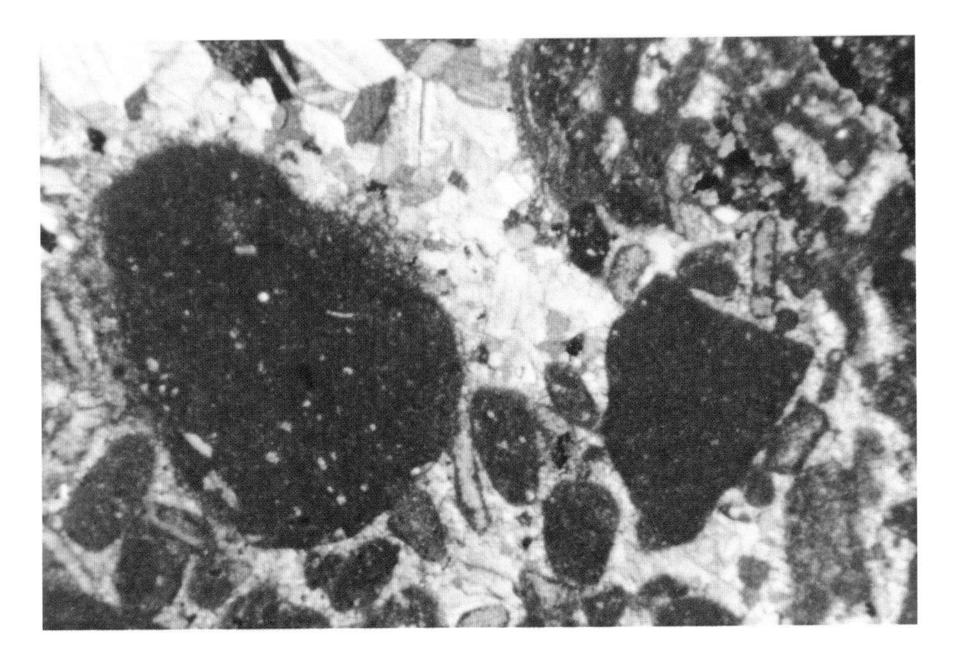

Mid. Ordovician Chazy and Black River Ls.
Pennsylvania

Intrasparite with sparsely fossiliferous micrite clasts. (Grainstone).

X.N. 0.38 mm

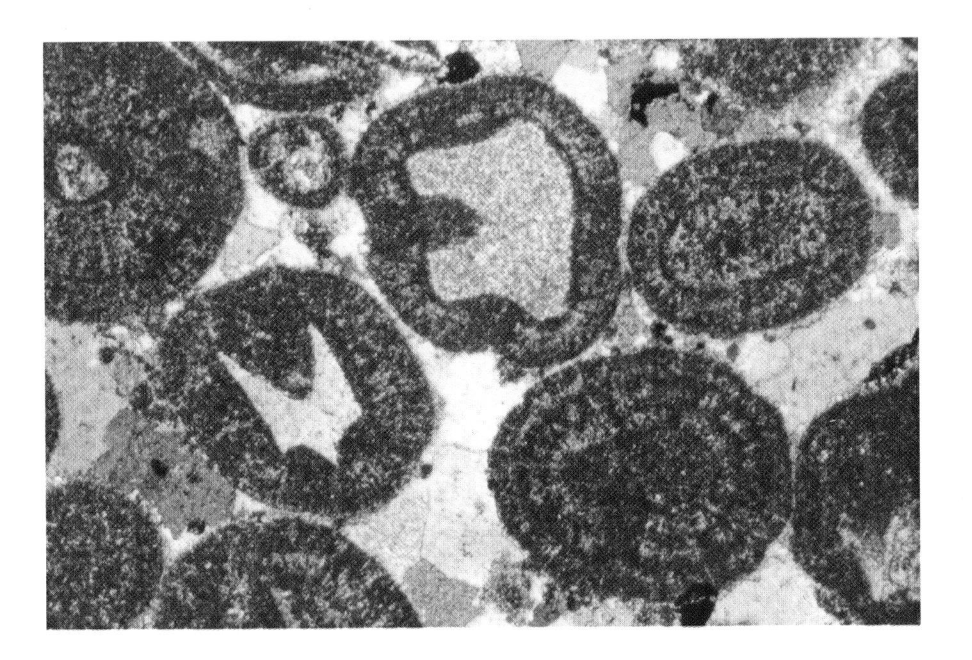

Paleozoic ls.
U.S. Midcontinent

Oosparite with coated fossil fragments. (Grainstone).

X.N. 0.30 mm

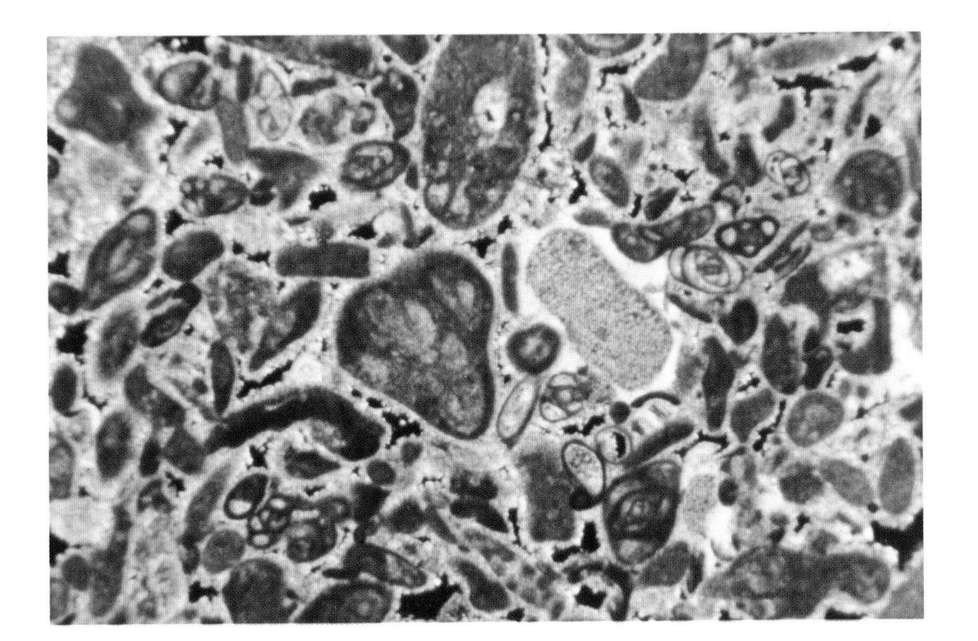

Up. Oligocene Suwannee Ls.
Florida

Biosparite with echinoderms, miliolid Foraminifera, mollusks, etc. (Grainstone).

X.N. 0.38 mm

Paleozoic (fm?)
Alabama

Pelsparite. (Grainstone).

0.38 mm

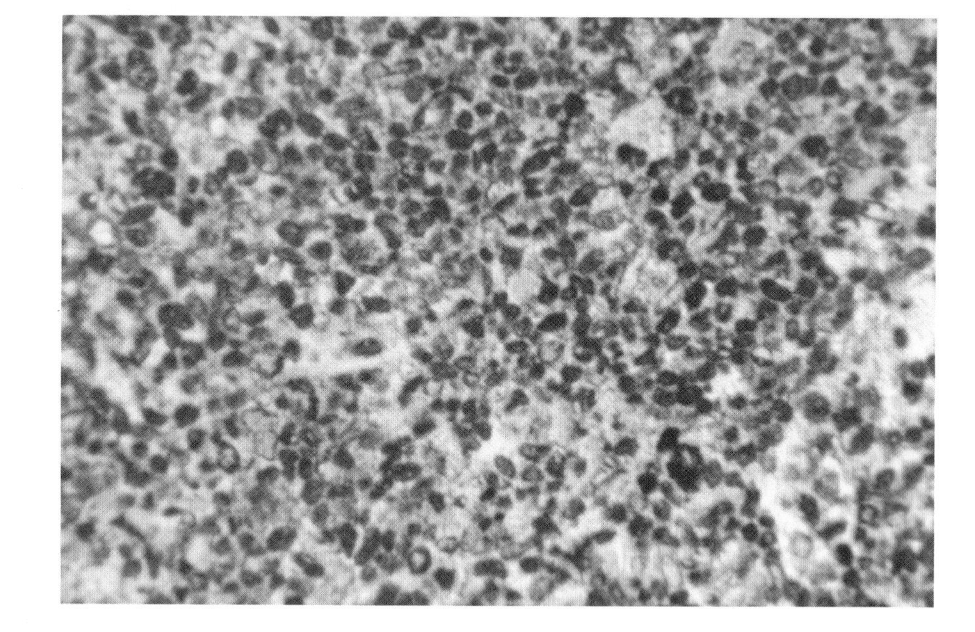

Up. Permian Capitan Ls.
New Mexico

Intramicrite with pelsparite clasts in a largely micritic matrix. Parts of this rock could be classified as a poorly-washed intrasparite. (Packstone).

0.38 mm

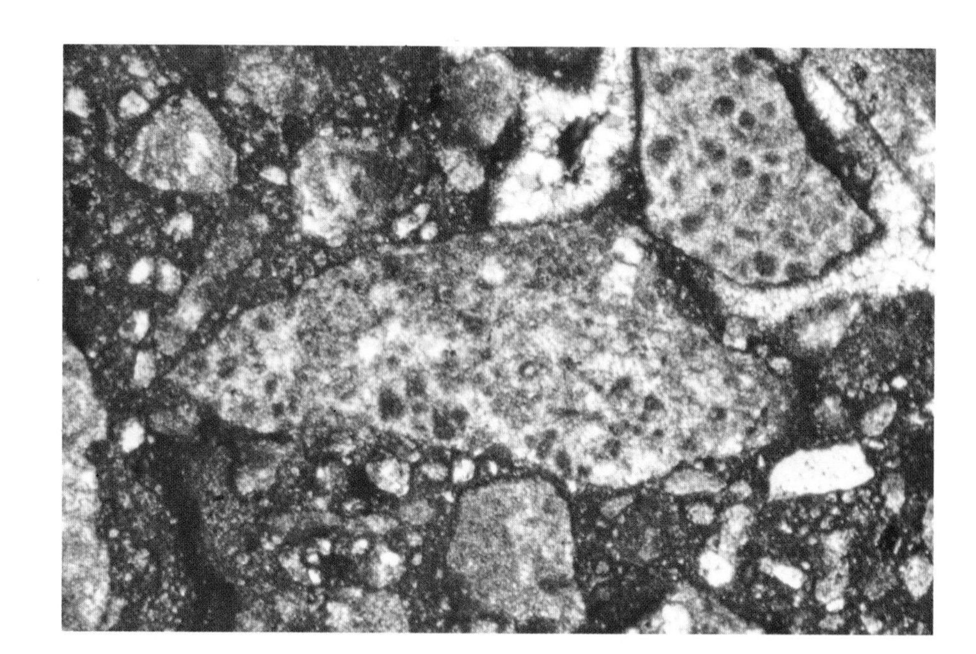

Lo. Ordovician Beekmantown Ls.
Maryland

Oomicrite (oolites have been strongly deformed). (Packstone).

0.38 mm

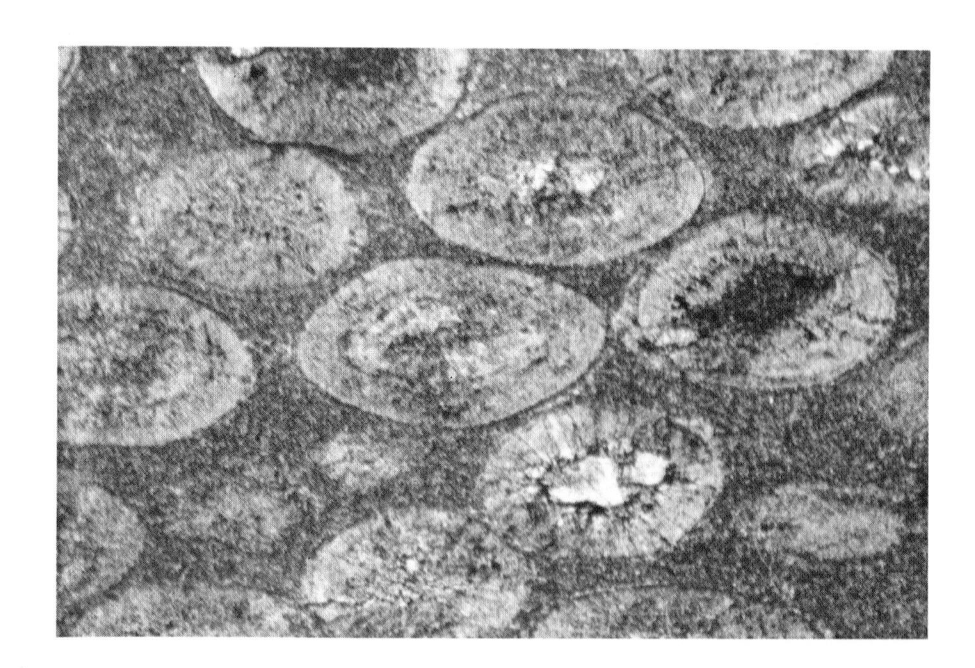

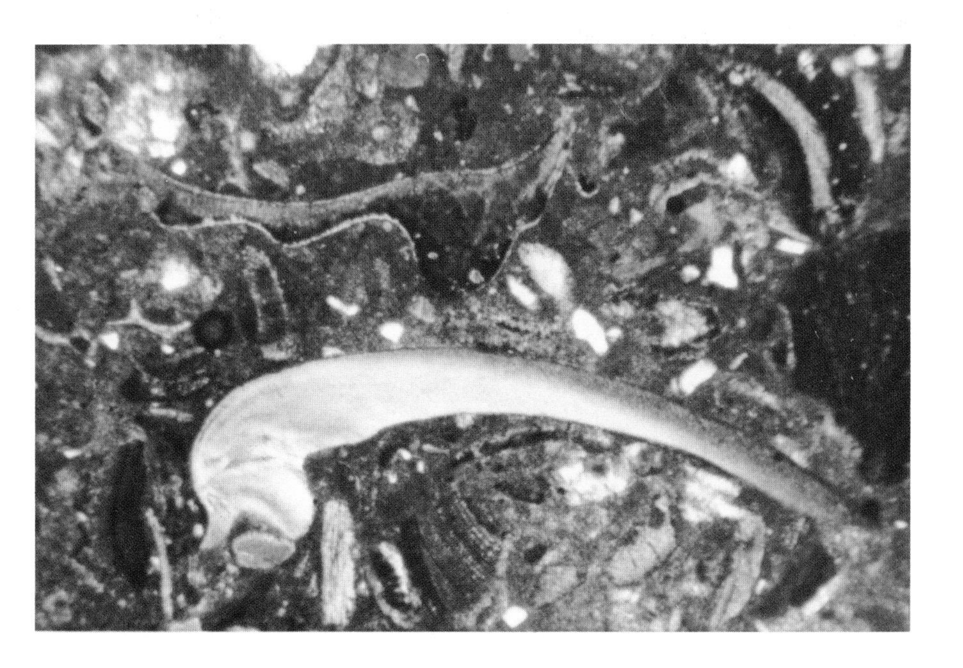

Pliocene and Pleistocene Caloosahatchee Marl
Florida

Biomicrite with mainly pelecypod fragments. (Wackestone).

P.X.N. 0.38 mm

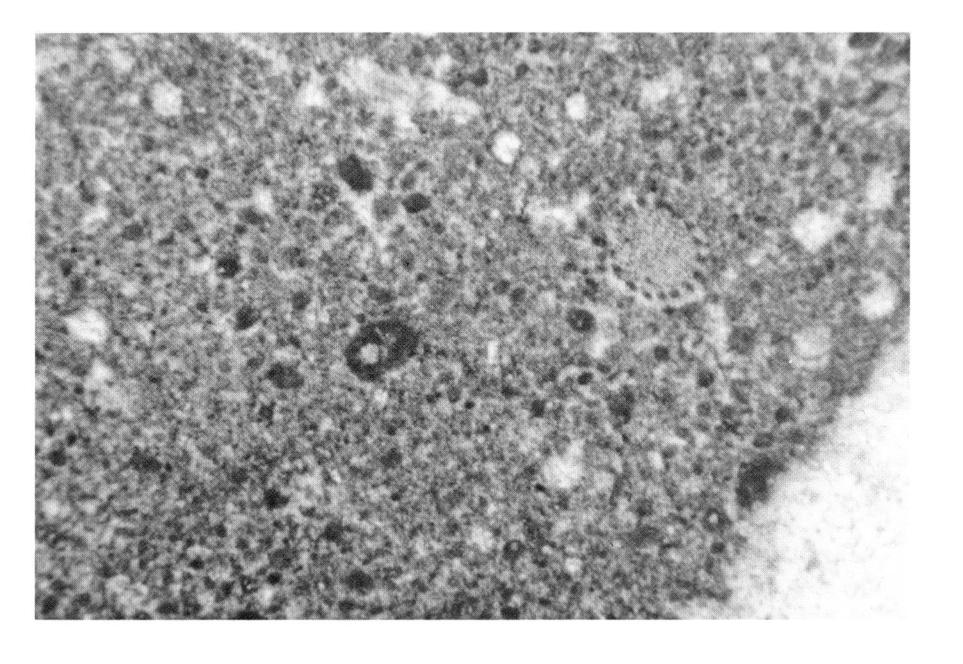

Jurassic Solnhofen Ls. (Kelheim facies)
Germany

Pelmicrite (really a biopelmicrite because of abundant fossil fragments mixed with the pellet grains). (Wackestone or packstone).

0.38 mm

Jurassic Solnhofen Ls. (Kelheim facies)
Germany

Micrite (rather pure lithified carbonate mud). (Mudstone).

X.N. 0.38 mm

Pleistocene Key Largo Ls.
Florida

Red algal biolithite. *In situ* encrustation of red algae on corals and other grains. (Boundstone).

X.N. 0.38 mm

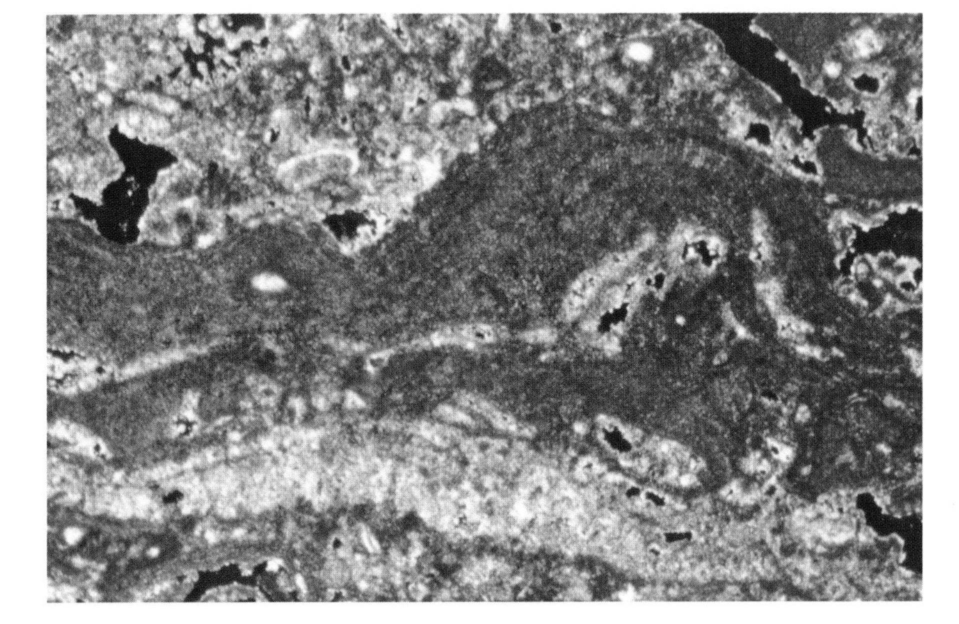

Mid. Ordovician Bays and Athens Ls. equivalent
Virginia

Dismicrite (really a dispelmicrite). Large void (not a solution vug) disrupting normal sediment texture--origin unknown.

X.N. 0.38 mm

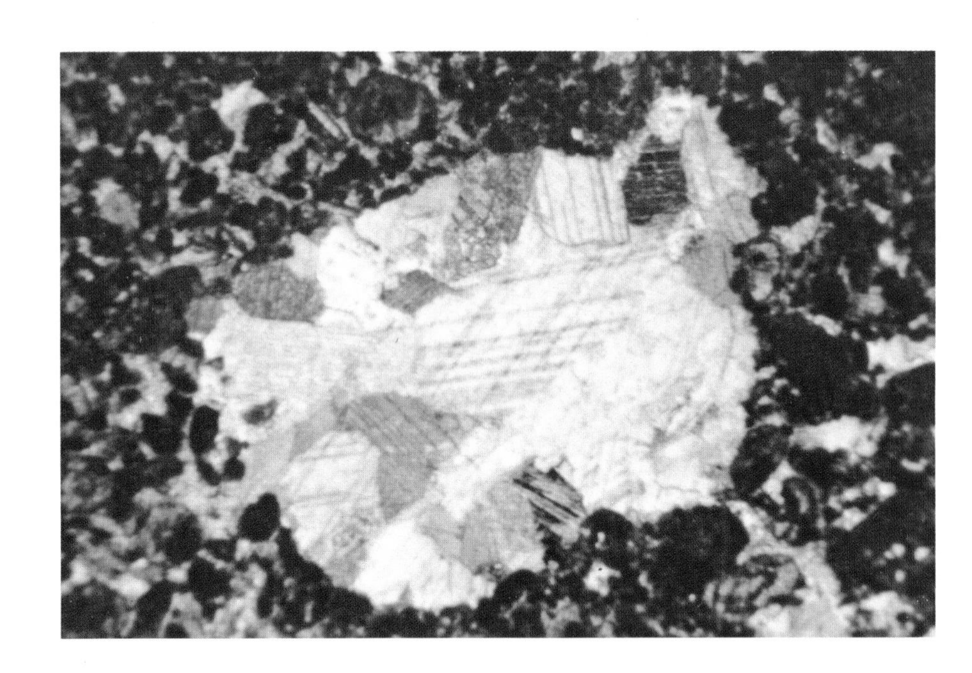

Lo. Ordovician Stonehenge Ls.
Pennsylvania

Coarse replacement dolomite.

X.N. 0.38 mm

Eocene Oberaudorf Schichten
Austria

A calclithite. More than 50 percent composed of detrital carbonate clasts (reworked, older-cycle limestones and dolomites). Often polymict, these deposits are generally important only on the downthrown sides of major faults or in extremely arid areas.

X.N. 0.64 mm

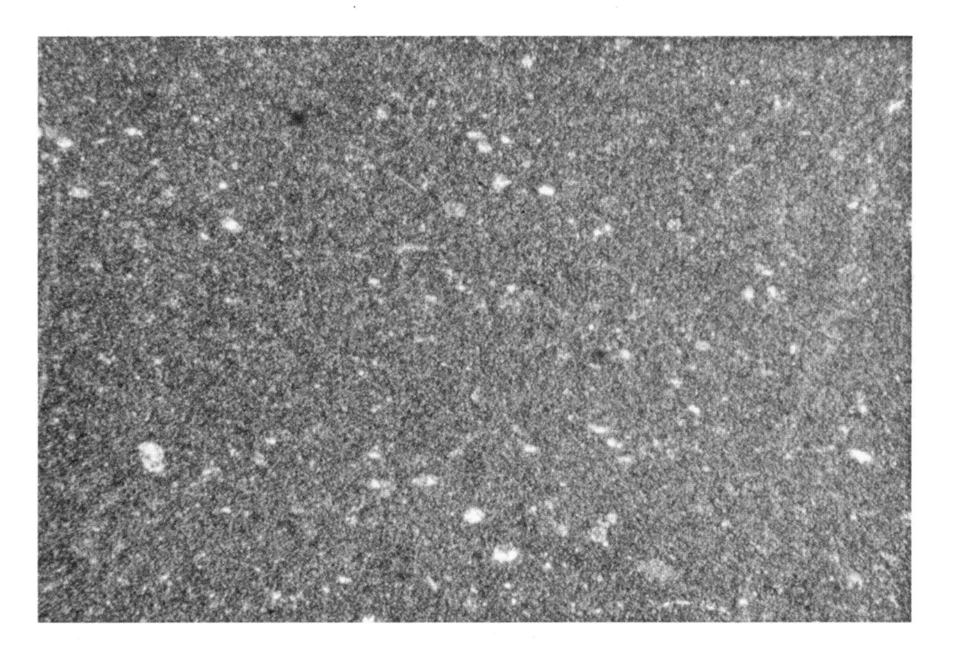

Up. Jurassic Calpionellid Ls.
Italy

This photograph, and the six that follow, illustrate the spectrum of carbonate rock textures as classified by Folk (1962). Dunham (1962) classifications given in parenthesis.

Fossiliferous micrite. Fossils are very fine-grained and include pelagic Foraminifera, calpionellids, and spicules. (Mudstone).

X.N. 0.38 mm

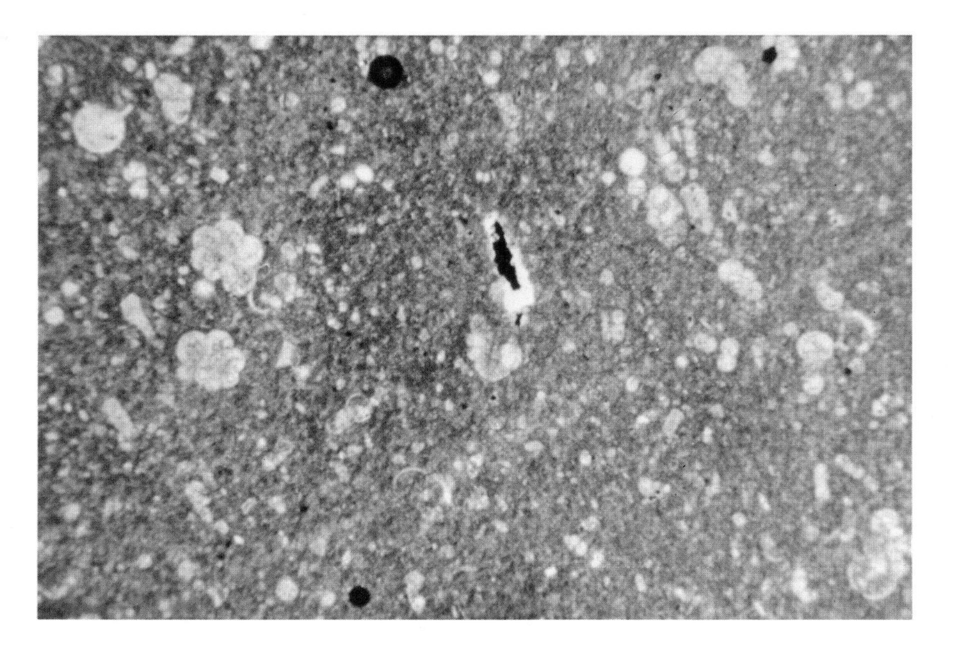

Up. Cretaceous Greenhorn Ls.
Colorado

Sparse biomicrite (mainly pelagic Foraminifera). (Wackestone).

X.N. 0.30 mm

Lo. Pliocene Tamiami Fm.
Florida

Packed biomicrite. (Peneroplid Foraminifera and quartz sand grains.) (Packstone).

X.N. 0.38 mm

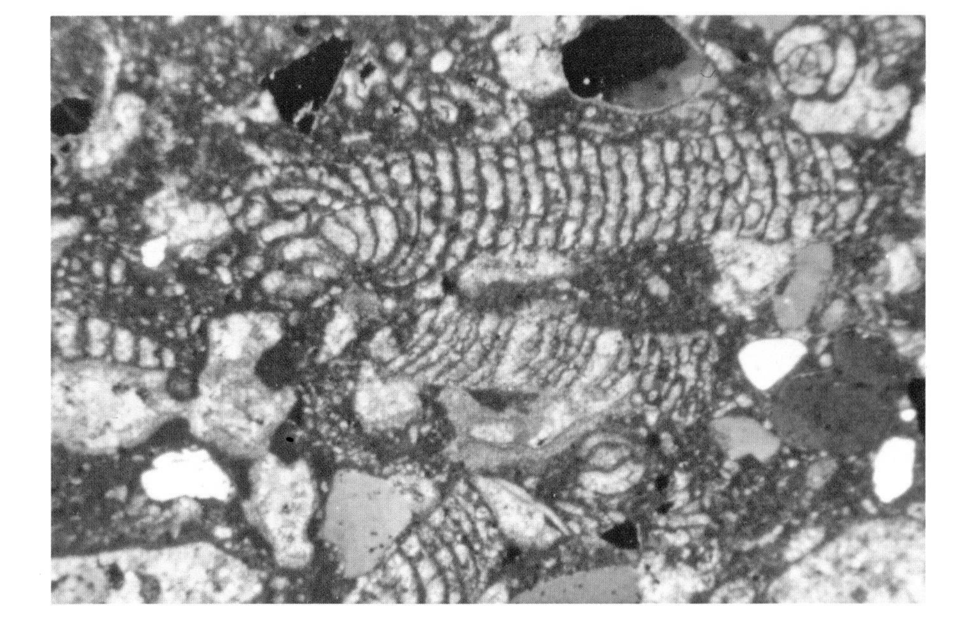

Pliocene and Pleistocene Caloosahatchee Marl
Florida

Poorly washed biosparite containing molluscan and algal fragments. (Packstone).

0.38 mm

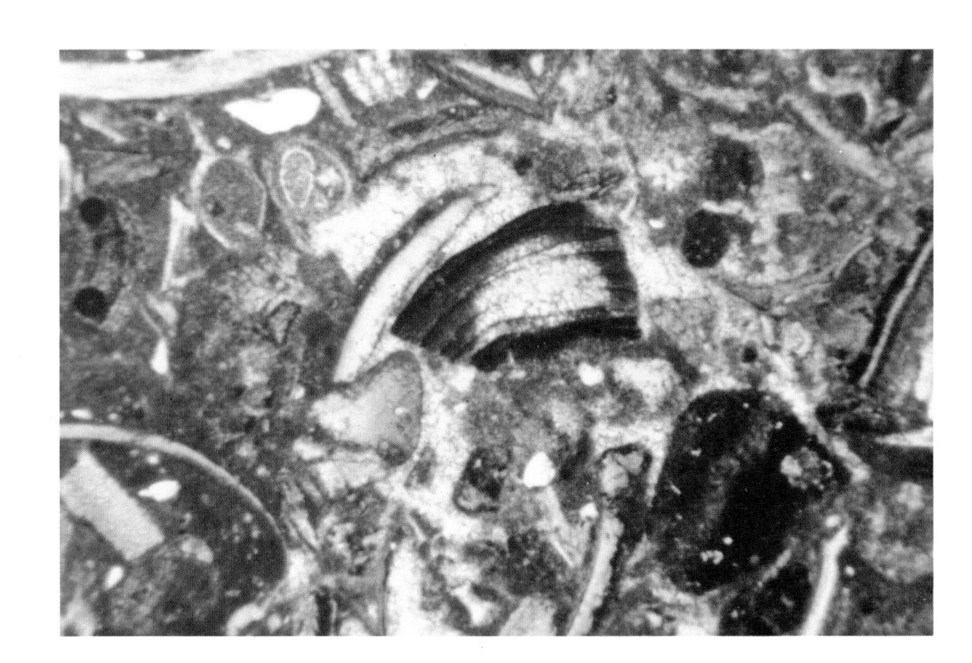

Mid. Ordovician fm.(?)
Virginia

Poorly sorted (unsorted) intrasparite. (Grainstone).

X.N. 0.24 mm

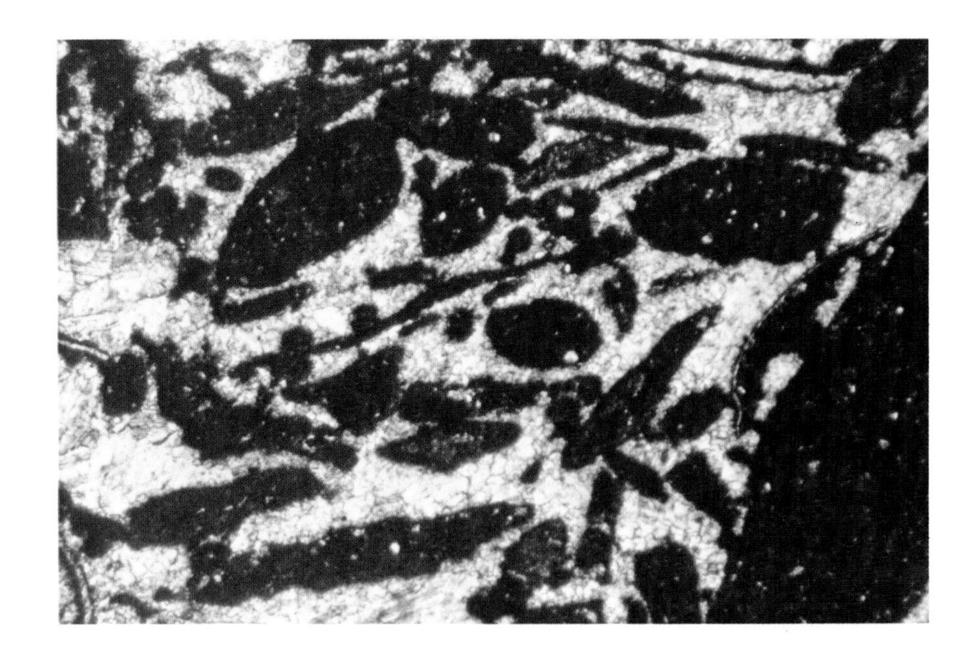

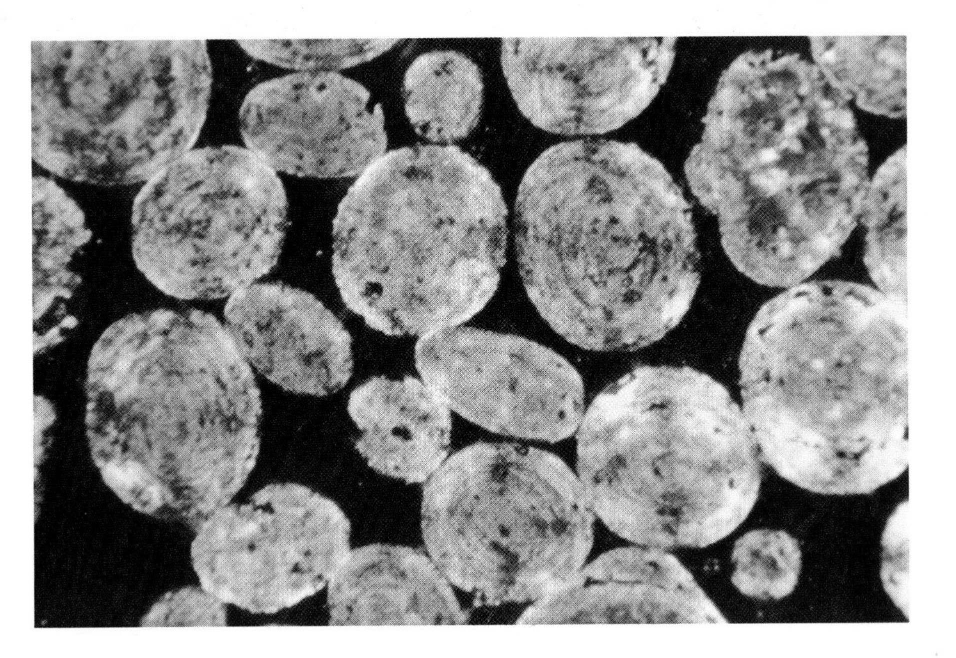

Holocene sediment
Cat Cay, Bahamas

Rounded (and sorted) oosparite. (Grainstone).

X.N. 0.38 mm

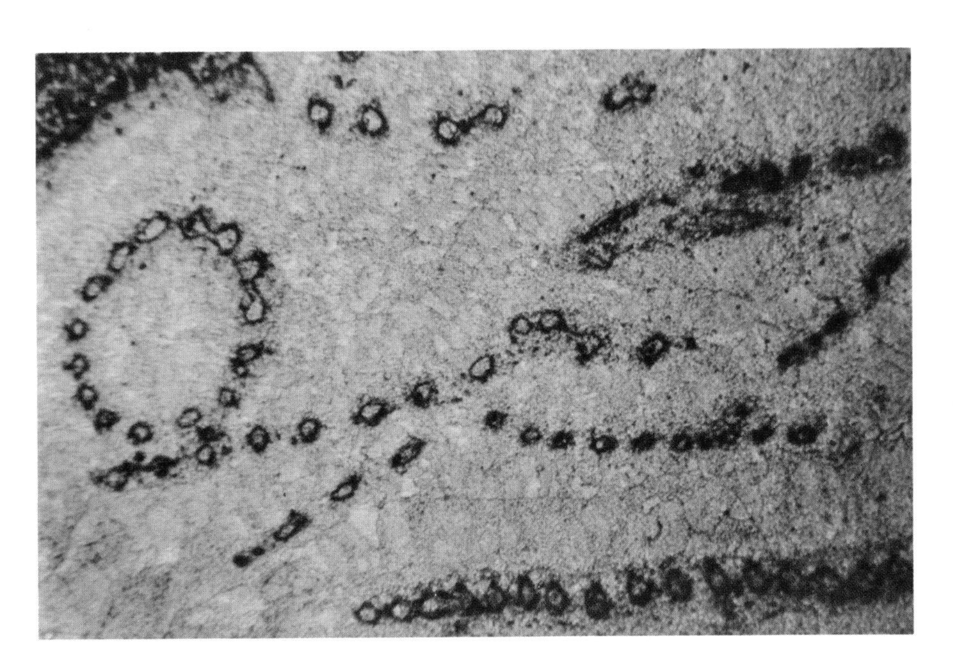

Mississippian Waulsortian Ls.
Ireland

This is a grain supported fabric which illustrates the difficulty of determining grain packing. These grains of fenestrate bryozoans are irregular in shape and, when uncemented, have as much as 90 percent porosity. If pore space were infiltrated with mud one might incorrectly class this as a mud-supported fabric.

1.24 mm

Burrows and Borings

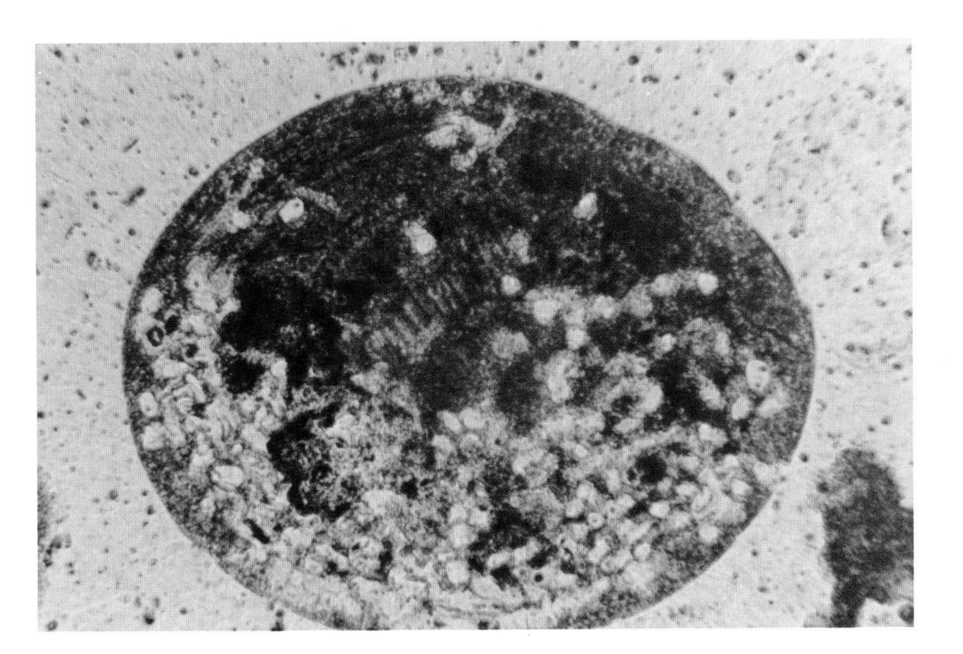

Holocene sediment
Schooner Cays, Bahamas

Example of modern subtidal boring of carbonate grains. Many organisms produce borings but sponges, algae and fungi are among the most important. Here much of the grain structure of the ooid has been destroyed by boring, possibly largely by algae.

0.05 mm

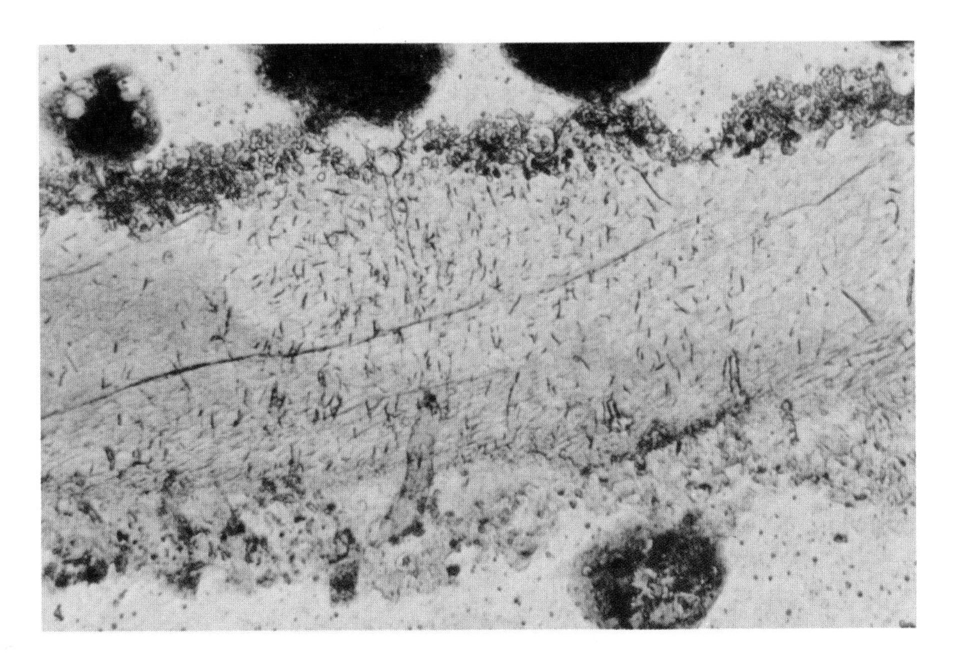

Holocene sediment
Bimini, Bahamas

A strongly bored mollusk fragment. Numerous small algal and fungal tubes have cut the crossed-lamellar structure of this grain. Recognition of algal borings is important in that it is evidence that the grain was derived from the photic zone. The borings of clionid sponges, acrothoracican barnacles, and ctenostome bryozoa could all be mistaken for algal borings unless carefully examined.

0.09 mm

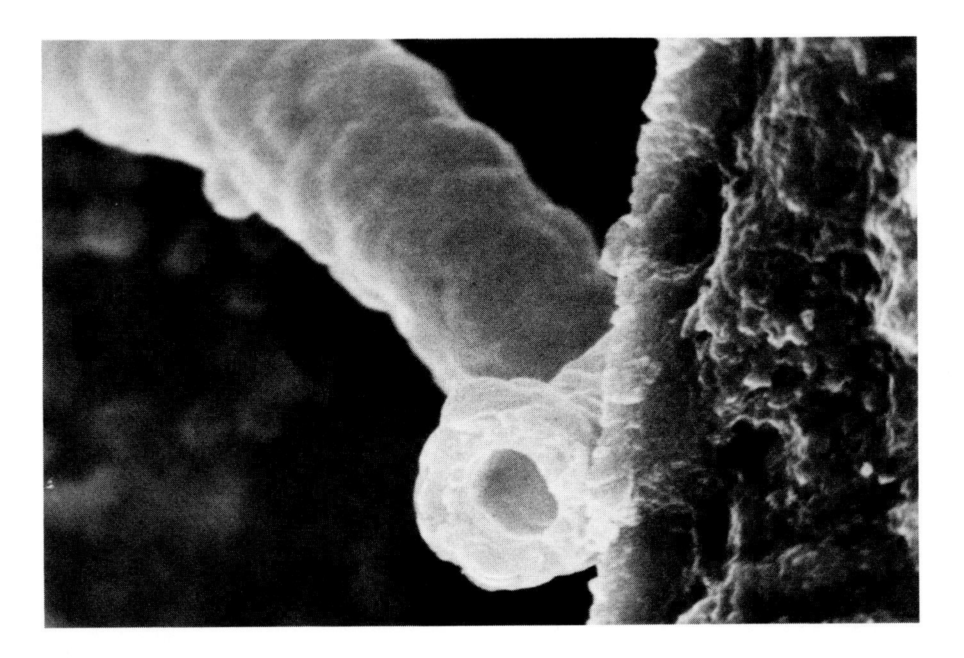

Holocene sediment
Belize (British Honduras)

SEM view of an algal tubule within a coral skeleton. Algal photosynthetic activities can lead to the precipitation of a sheath of carbonate (generally aragonite or high-Mg calcite) around the filaments; these are common in SEM views of modern sediments. Again, recognition of algal tubules allows determination of an environment within the photic zone.

SEM Mag. 3000X 4.1 μm

Up. Cretaceous White Ls.
Northern Ireland

A sponge boring within a belemnite rostrum. Sponge borings are distinguished by their relatively large size and characteristic shape. The silt-sized debris from the borings of some modern sponges adds a significant fraction to the internal sediment within reefs and fore-reefs. Sponge bioerosion is also important in the destruction of carbonate shorelines.

0.22 mm

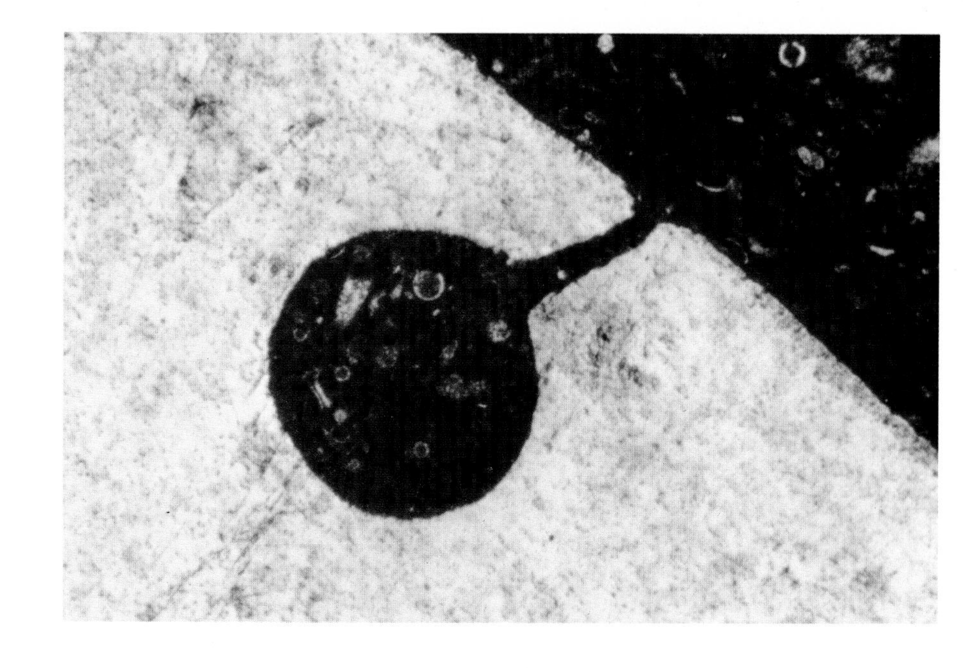

Up. Cretaceous White Ls.
Northern Ireland

Multiple sponge borings within a belemnite rostrum. Note that the area within the borings has been infilled with the same type of micritic sediment that is found in the rest of the sediment. Yet the area within the borings has been protected from compaction and thus shows the original sediment fabric while the external areas show pervasive bedding-parallel compaction and grain crushing.

0.25 mm

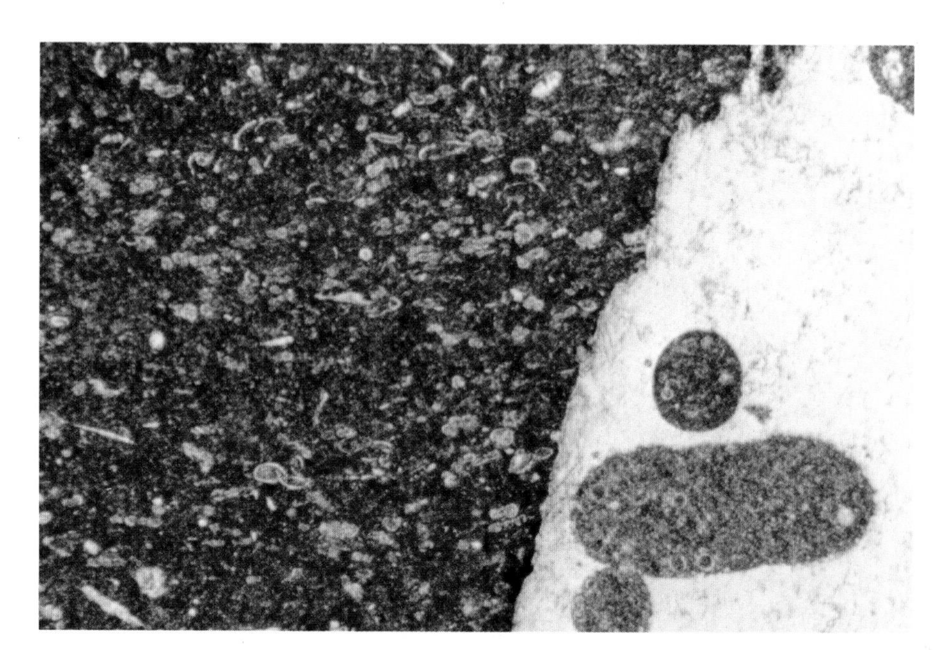

Mid. Jurassic lower Forest Marble beds
England

A boring. Borings and burrows must be carefully distinguished. Borings occur in hard substrates; burrows in soft ones. The truncation of shells by this structure indicates that the sediment must have been significantly lithified at the time of boring. Burrows normally would pass around such grains in soft sediment. Thus, one can determine presence or absence of early lithification.

0.64 mm

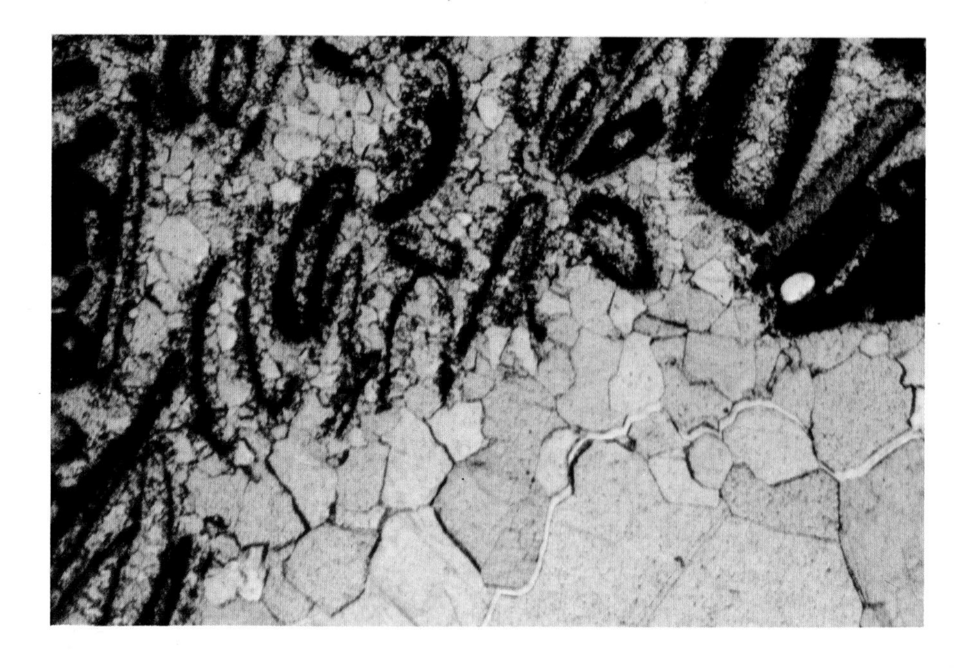

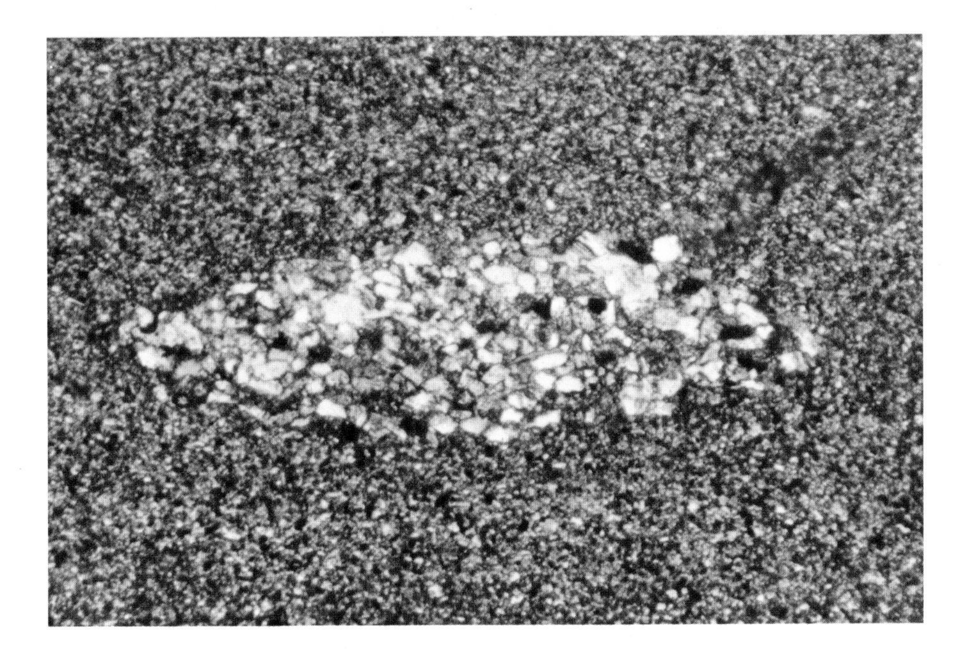

Mid. Ordovician Bays Ls.
Virginia

A sand-filled burrow in silty carbonate mudstone. Burrows are often detected by gross or subtle changes in texture as coarser or finer sediments are brought in by the burrowing organism. This burrow has been compacted during later diagenesis (although oblique sections will also give elliptical shape).

0.25 mm

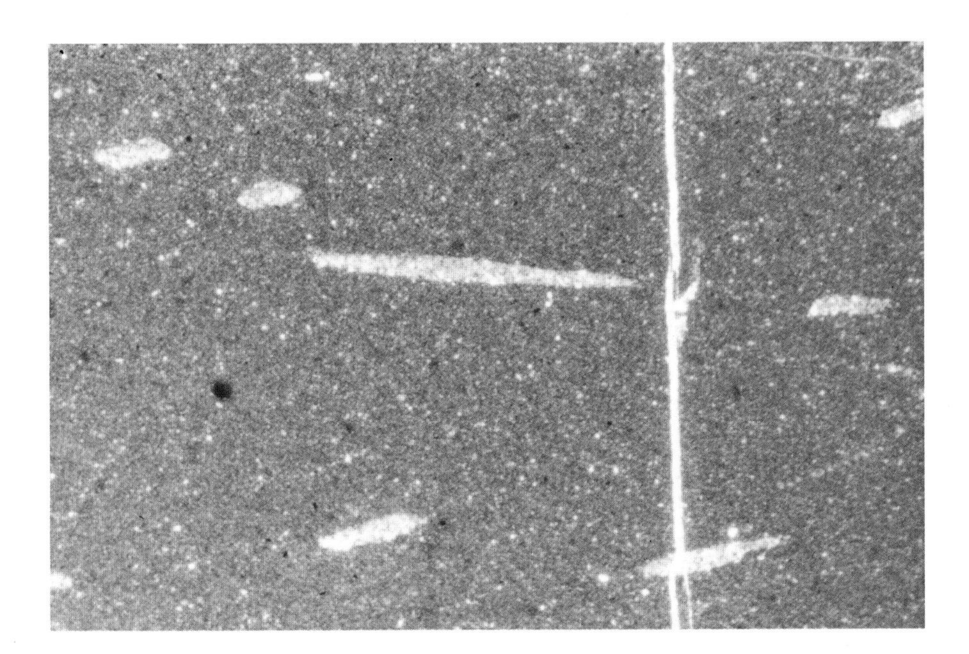

Up. Cretaceous Monte Antola Fm.
Italy

These slightly compacted burrows are part of a single burrow system probably produced by *Chondrites*. Note that they have a consistent orientation with respect to each other. Burrows can be filled with pyrite, calcite, glauconite, and other diagenetic phases, as well as with primary sediment.

0.64 mm

Crusts and Coatings

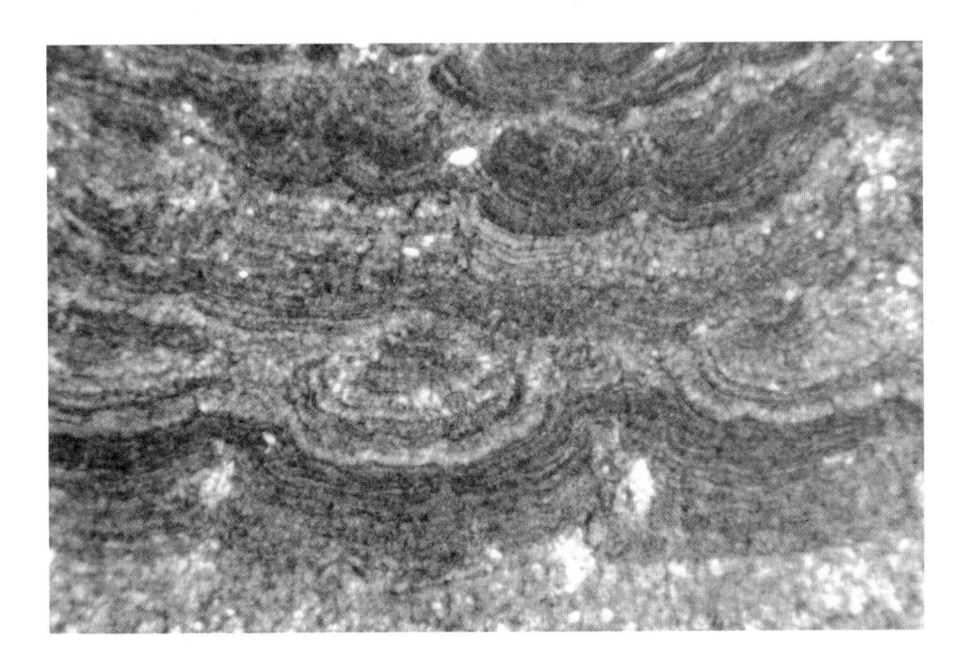

Lo. Eocene Indian Meadows Fm.
Wyoming

Part of a blue-green algal oncolite. Well, but irregularly laminated, large (about 2 cm across) grains of probable algal origin. Insoluble residues of thin sections of these grains will often still show algal tubules or filaments.

0.30 mm

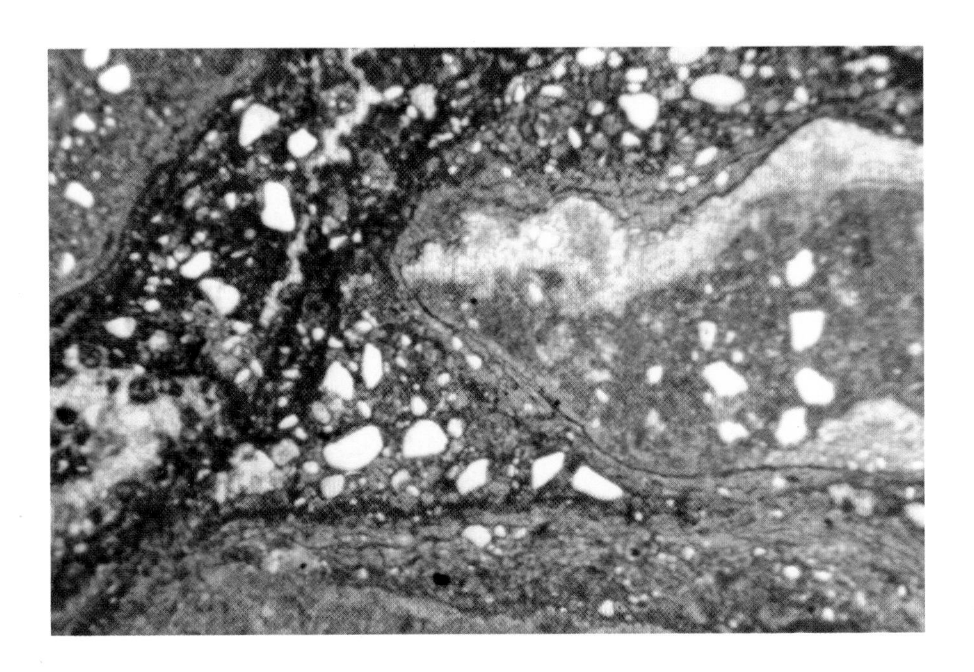

Quaternary caliche crust
Texas

Irregularly laminated caliche crust trapping carbonate and non-carbonate grains. Such crusts (which must be distinguished from algal structures) can indicate unconformities and subaerial exposure in carbonate sections. Inverse grading, autofracturing, interlocking grains, and microstalactitic textures (not shown here) aid in identification.

0.38 mm

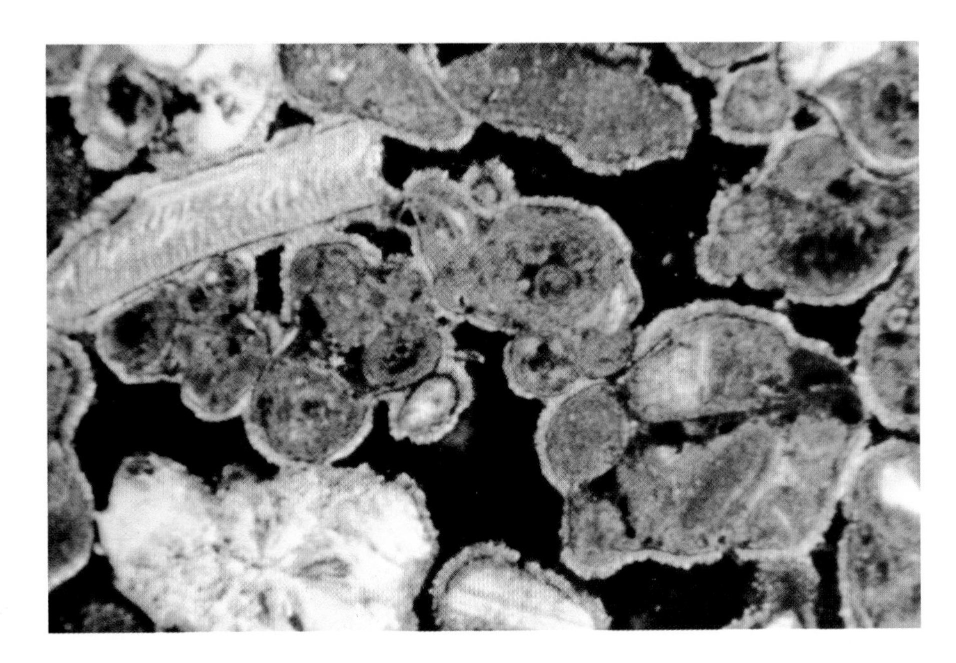

Holocene beachrock
Bahamas

Beachrock (with cementing crust of small fibrous aragonite needles) contains numerous "grapestone" grains (composite grains held together loosely and, here, secondarily oolitically coated before cementation). Illustrates encrustation (oolitic coatings and cement crusts).

X.N. 0.38 mm

Cretaceous Tamabra Ls.
Mexico

Isolated encrustations of grains can be produced by algae (as here), Foraminifera, bryozoans, and other organisms. Such encrustations can be extremely important in lithification of sediment within the environment of deposition. Such biolithites are, however, often difficult to recognize because of the vague nature of many algal products and because of later diagenesis.

0.51 mm

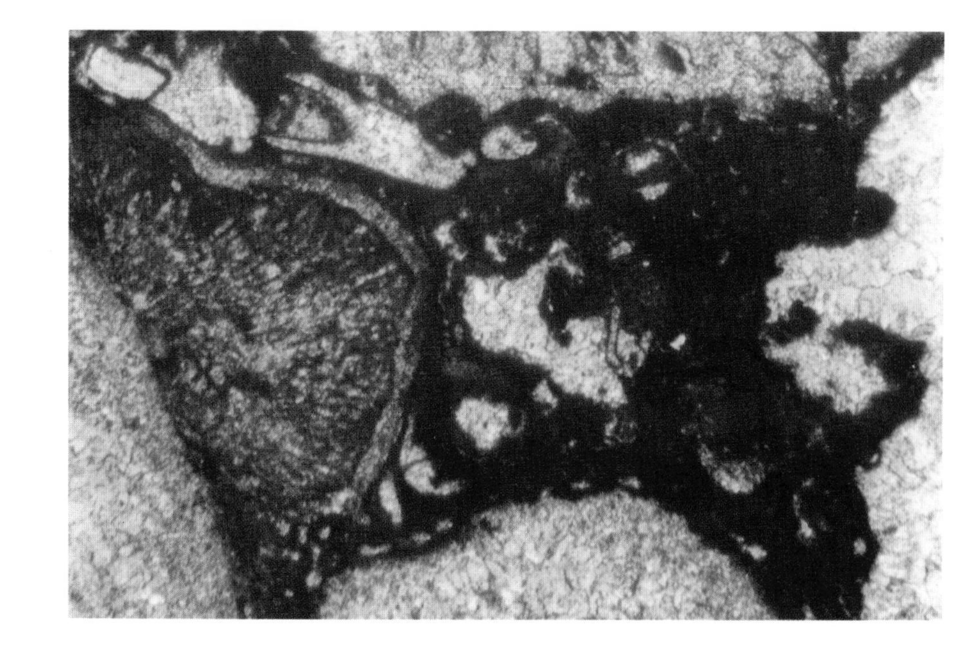

Pleistocene aeolianite
Abu Dhabi

Micrite envelopes must be distinguished from encrustations. In most cases, micrite envelopes are thin zones of surficial alteration (boring) of grains which micritizes their outer surface. Thus, there is no addition of material to the exterior of a grain. As seen here, micrite envelopes can remain even after grains are leached away.

0.09 mm

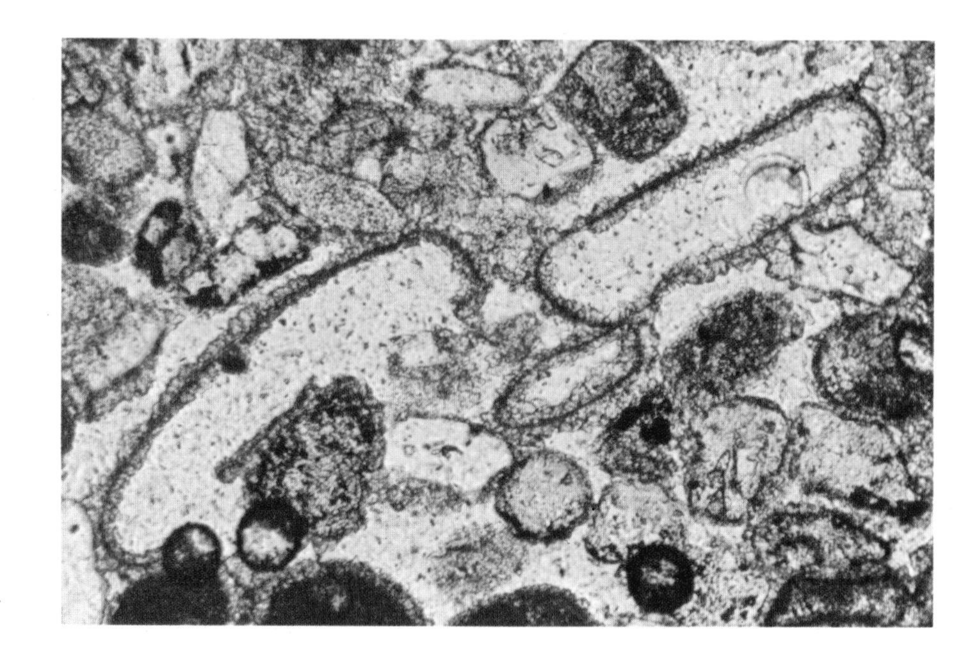

Mid. Jurassic lower Forest Marble beds
England

An example of ancient micrite envelopes and subordinate encrustations. Thin rinds of micrite can be seen with occasional irregular protrusions toward the former grain interiors--these are formed by microboring. Occasional bulges of thicker micrite are encrustations. The micrite envelopes have acted as a template for cementation of both internal (leached) and external porosity.

0.64 mm

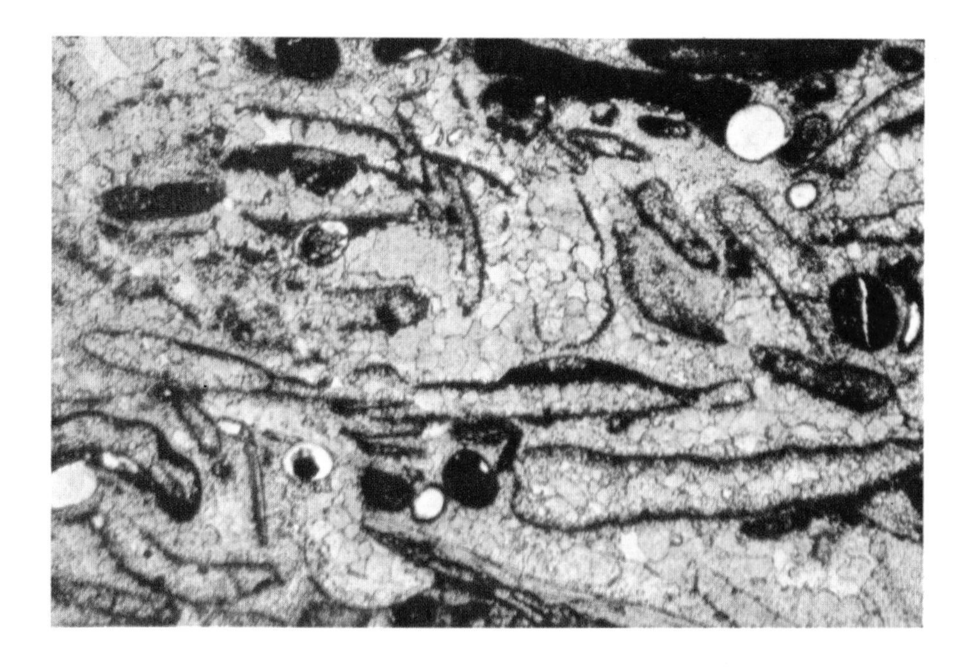

Dissolution Fabrics

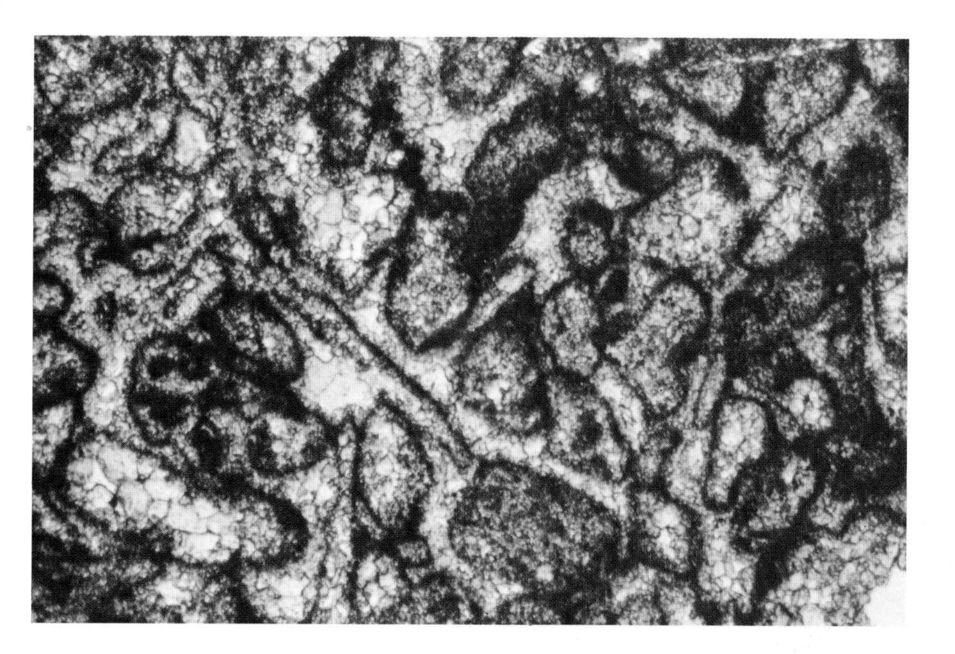

Up. Cretaceous Pilot Knob reef facies of Austin Group
Texas

Leached grains with micrite envelopes. Most grains show no internal structure and a single dark micritic rim. This is produced by algal boring of original aragonitic skeletal grains (producing dark envelopes) and subsequent dissolution of the unstable aragonite.

0.38 mm

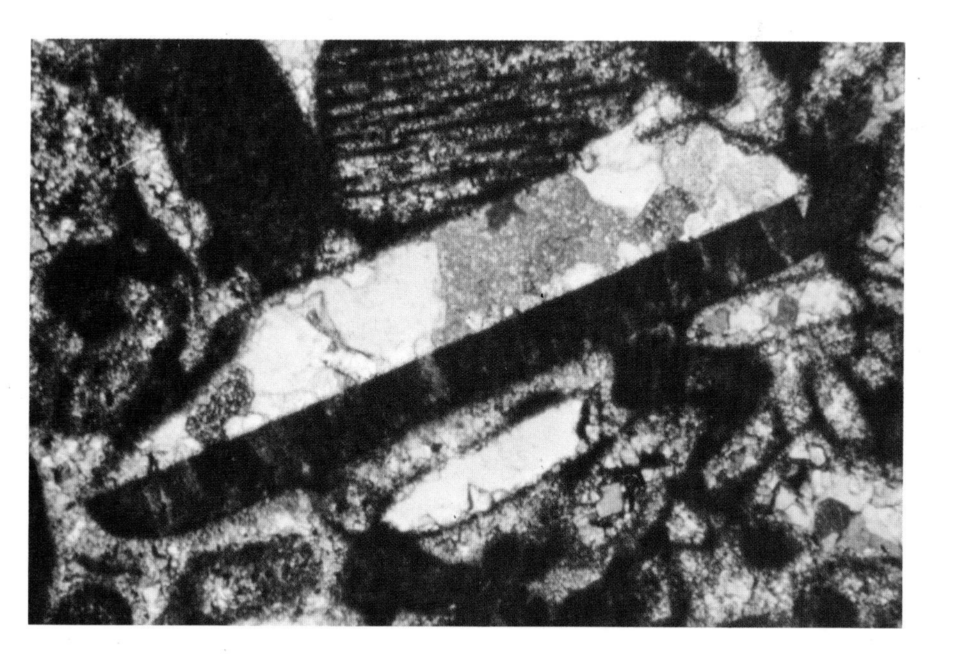

Up. Cretaceous Pilot Knob reef facies of Austin Group
Texas

Large mollusk grain which originally had two-layer structure (one aragonite, one calcite). Calcitic layer preserved while the aragonitic layer was dissolved and replaced by sparry calcite. Note micrite envelope defining former aragonitic zone.

X.N. 0.30 mm

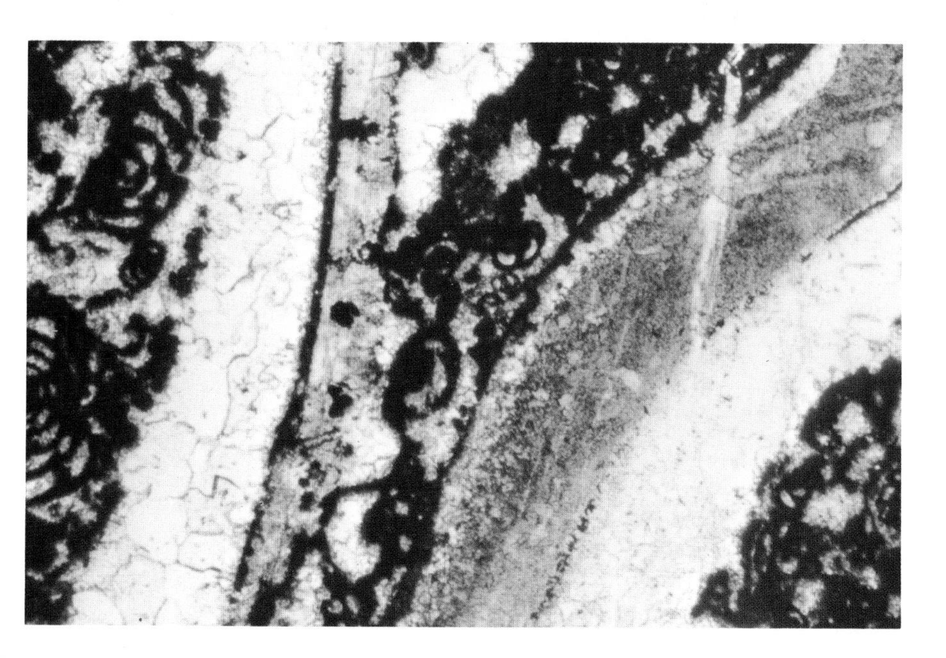

Up. Cretaceous El Abra Ls.
Mexico

The odd texture, shown here, of miliolids essentially "floating" in sparry calcite is the product of dissolution of the aragonitic shell layer of a ***Toucasia*** rudistid. The calcitic wall layer is still preserved in brownish original material while the aragonitic layer was leached after partial cementation of the matrix sediment. The leach void was later filled with clear, sparry calcite.

0.64 mm

Pleistocene aeolianite
Abu Dhabi

Fossils, especially originally aragonitic ones, can be leached, leaving a series of unfilled micrite envelopes. In this example, several free-standing micrite envelopes are acting as a template for the internal and external cementation by sparry low-Mg calcite cement. If continued, this process will yield fabrics such as those seen in the previous three photos.

X.N. 0.07 mm

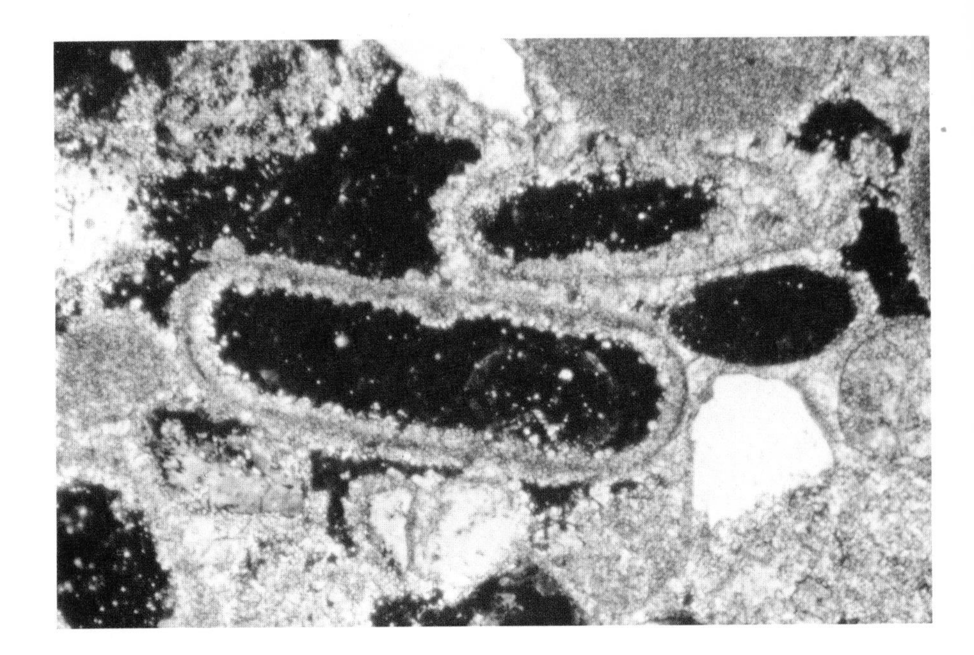

Mid. Eocene Avon Park Ls.
Florida

Leached fossils in a dolomitic rock. Either the matrix was dolomitized and the fossils leached, or fossils were leached and rock matrix was then dolomitized (dolomite apparently replaces more readily than it cements). Thus, determining the paragenetic sequence in such rocks can be very difficult.

G.P. 0.38 mm

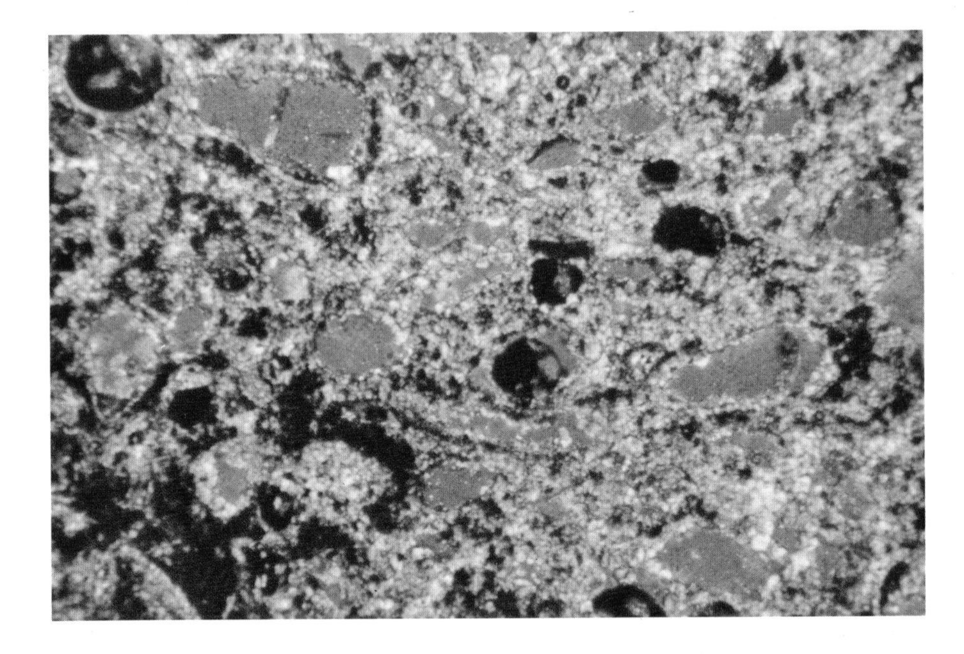

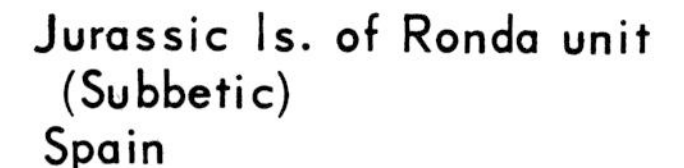

Jurassic ls. of Ronda unit
(Subbetic)
Spain

Selective leaching of dolomite (de-dolomitization) has taken place in this rock. This is a common near-surface process and is aided by the presence of evaporites (sulphate minerals).

X.N. 0.38 mm

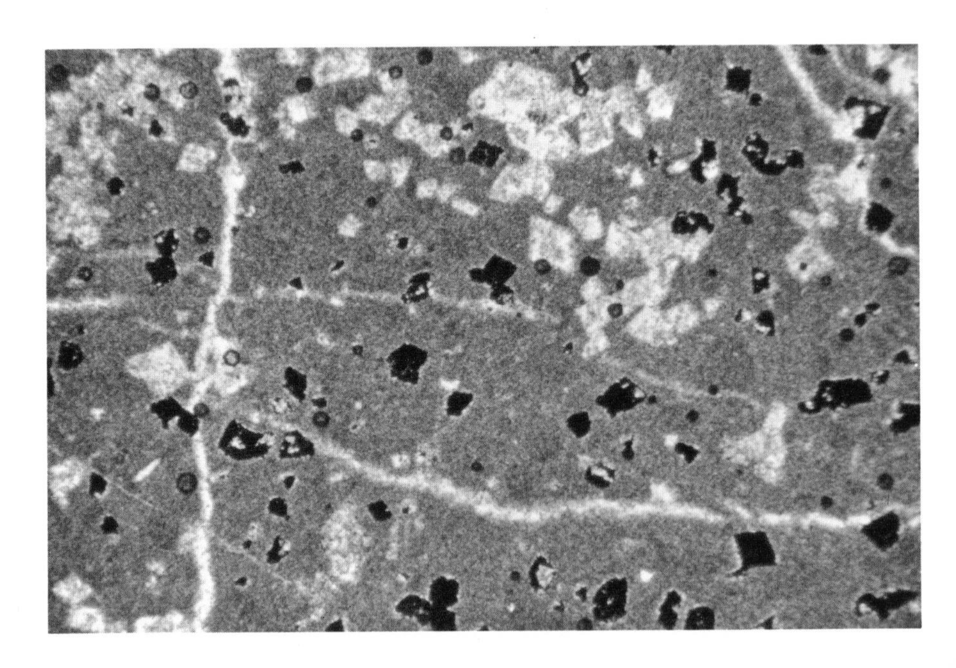

Cretaceous Tamabra Ls.
Mexico

Occasionally the timing of leaching can be closely dated. Here, in the lower center of the photo, one can see sediment infilling a fossil. This clearly implies very early dissolution of the fossil (probably vadose diagenesis) and infilling with internal sediment or vadose silt.

0.64 mm

Replacement and Neomorphism

(diagram after Folk, 1965)

PROCESS	EXAMPLE	METAMORPHIC OR IGNEOUS TERM	METALLURGICAL TERM	BATHURST (1958) VOLL (1960) ETC.	THIS PAPER: LOOSE	THIS PAPER: STRICT USAGE	
ONE MINERAL REPLACES ANOTHER OF A DIFFERENT COMPOSITION	CALCITE → DOLOMITE, PYRITE, ETC.	REPLACEMENT OR METASOMATISM	———	REPLACEMENT	REPLACEMENT		
A MINERAL IS REPLACED BY ITS POLYMORPH	ARAGONITE MUD OR SKELETON → CALCITE MOSAIC	INVERSION OR TRANSFORMATION	ALLOTROPIC RECRYSTALLIZATION	NOT SPECIFIED	REPLACEMENT (LOOSELY); RECRYSTALLIZATION (LOOSELY)	INVERSION	NEOMORPHISM
A DEFORMED MINERAL CHANGES TO A MOSAIC OF UNDEFORMED CRYSTALS OF THE SAME MINERAL	STRAINED CALCITE → UNSTRAINED CALCITE	RECRYSTALLIZATION	(STRAIN) RECRYSTALLIZATION	RECRYSTALLIZATION		STRAIN RECRYSTALLIZATION	
AN UNDEFORMED MINERAL CHANGES ITS FORM, GRAIN SIZE OR ORIENTATION	CALCITE MUD OR FIBERS → CALCITE MOSAIC, ETC.	RECRYSTALLIZATION	GRAIN GROWTH	GRAIN GROWTH (OR GRAIN DIMINUTION)		RECRYSTALLIZATION (AND DEGRADING RECRYSTALLIZATION)	
A MINERAL DISSOLVES LEAVING A CAVITY; CAVITY IS FILLED LATER	SKELETON → CAVITY → CALCITE	RARE NO SPECIAL TERM	———	NO SPECIAL TERM	SOLUTION–CAVITY FILL		

Comparison of terminology for mineral alterations as observed with petrographic microscope, observe particularly the varying status of the term *recrystallization.* It is proposed that grain growth not be used for sedimentary carbonate rocks, because the process of grain growth in metals does not resemble any diagenetic process in normal limestones.

A Code for Directly Precipitated Calcite (after Folk, 1965)

Formative Mechanism	Foundation	Grain Shape	Term	Symbol	Terms of Bathurst (1958)	
Directly Precipitated (P)	Syntaxial Overgrowth (O)	Equant; on monocrystalline nucleus	x-Monocrystalline[1] Overgrowth	$\mathrm{P.E{*}O_m}$	Rim-Cement	
		Equant; on polycrystalline nucleus	x-Equant Overgrowth	P.E*O	Granular Cement (if between allochems)	Drusy Mosaic (inside fossils, etc.)
		Bladed-on polycrystalline nucleus	x-Bladed Overgrowth	P.B*O		
		Fibrous-on polycrystalline nucleus	x-Fibrous Overgrowth	P.F*O		
	Crusts-(C)-oriented physically on surface	Equant	x-Equant Crust	$\mathrm{P.E{*}C_{(w)}}$		
		Bladed	x-Bladed Crust	$\mathrm{P.B{*}C_{(w)}}$		
		Fibrous	x-Fibrous Crust	$\mathrm{P.F{*}C_{(w)}}$		
	Randomly-oriented	Equant	x-Mosaic	P.E*		
		Bladed	Very rare or non-existent "Bladed Mosaic" may perhaps exist; "Fibrous Mosaic" probably non-existent.			
		Fibrous				

* Add number for crystal size, e.g. $\mathrm{P.E_4O_m}$, $\mathrm{P.F_2C}$.
x Add crystal size term, e.g. Finely Bladed overgrowth (or finely crystalline Bladed overgrowth); Medium Fibrous crust (medium crystalline Fibrous crust).
[1] Monocrystals may be Bladed or rarely Fibrous; use "Bladed Monocrystalline overgrowth" if desired, $\mathrm{P.BO_m}$, etc.

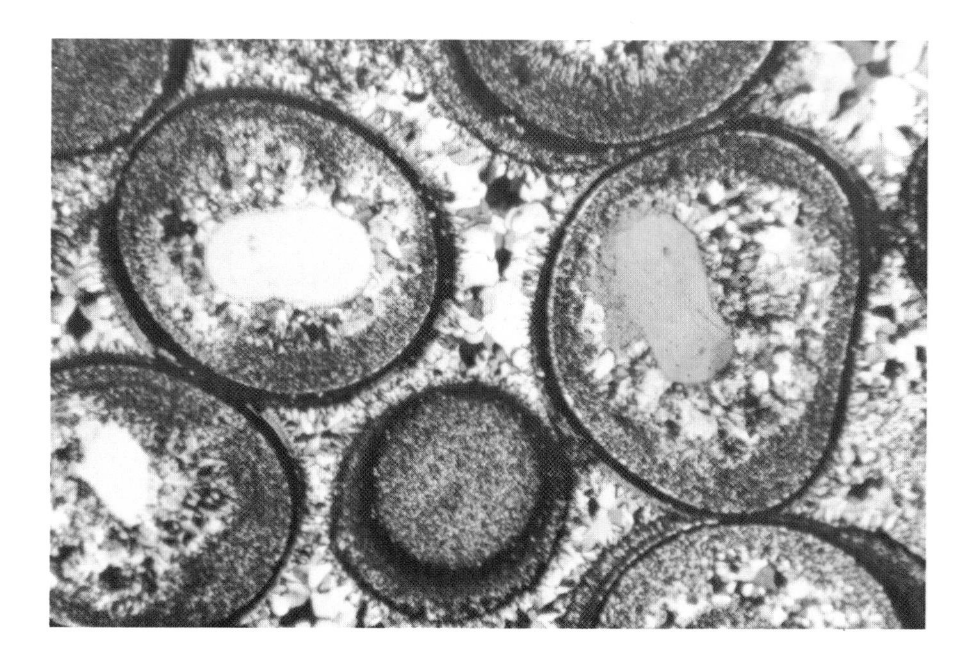

Up. Cambrian Mines Dolomite Mbr. of Gatesburg Fm. Pennsylvania

Replacement. Original oolitic limestone has been completely replaced by quartz in the form of chert, chalcedony, and mega-quartz. Although this example shows remarkable preservation of original detail, many replacement textures simply obliterate original fabrics.

X.N. 0.38 mm

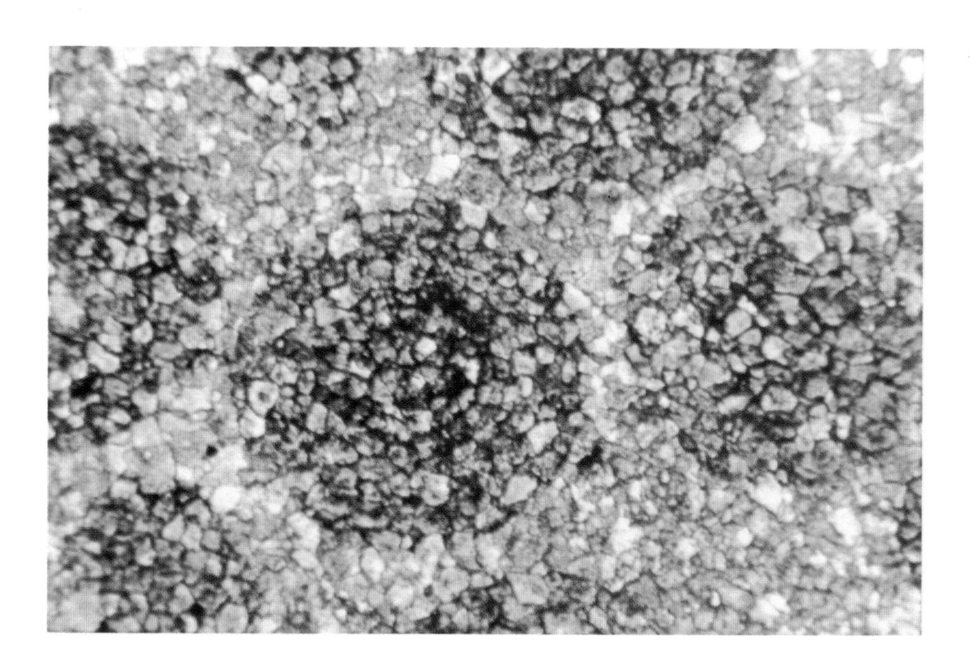

Up. Cambrian Kittatinny Ls. New Jersey

Dolomite replacement with retention of relict or "ghost" texture. Original sediment was a porous oolite, probably later spar cemented, and finally completely replaced by dolomite. Ooids can be recognized because organic matter and other impurities have been retained in the dolomite replacement.

0.38 mm

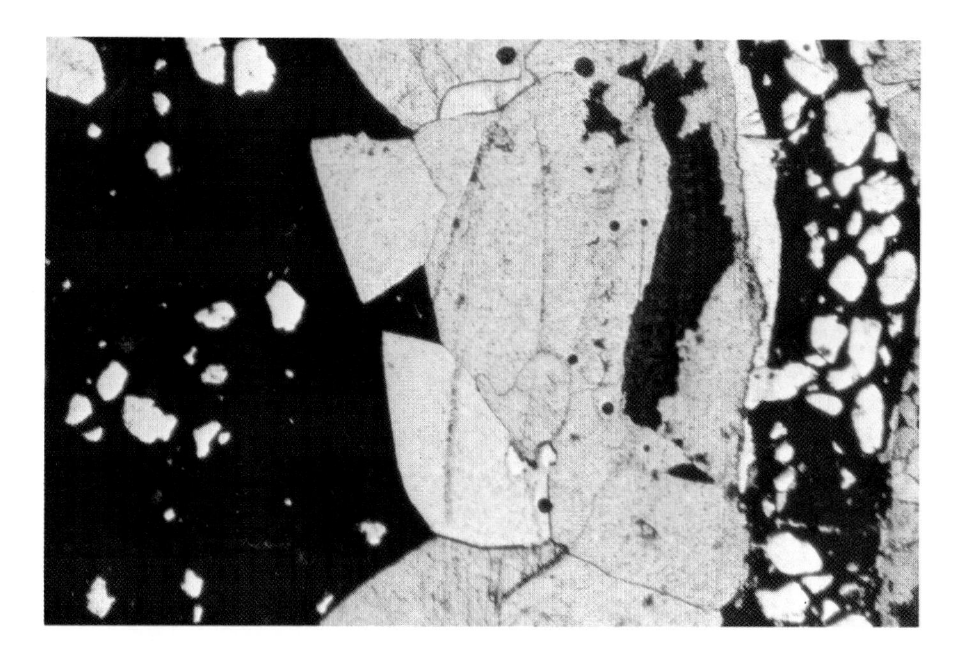

Pliocene and Pleistocene Caloosahatchee Fm. Florida

Inversion. Originally aragonitic shell of a vemetid gastropod is here partly inverted to calcite. The coarse spar is calcite and the darker patches are still aragonite. Note total loss of primary texture within the inverted areas and the partial overlap of the crystals (by displacement or inversion) into micritic matrix.

0.64 mm

Mid. Ordovician Chazy and Black River Ls.
Pennsylvania

Recrystallization of micrite to microspar and pseudospar (see Folk, 1965). Virtually complete diagenetic alteration of primary texture.

0.38 mm

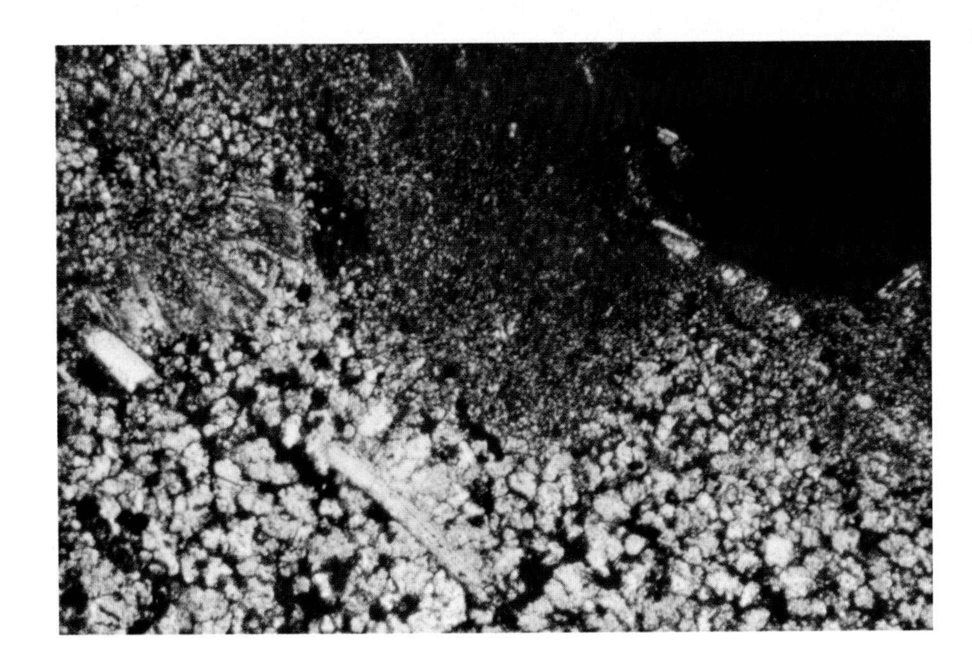

Mid. Ordovician Chazy and Black River Ls.
Pennsylvania

Recrystallization of micrite to microspar and pseudospar. Note irregular grain shapes of pseudospar crystals. Complete diagenetic alteration of primary depositional texture.

0.38 mm

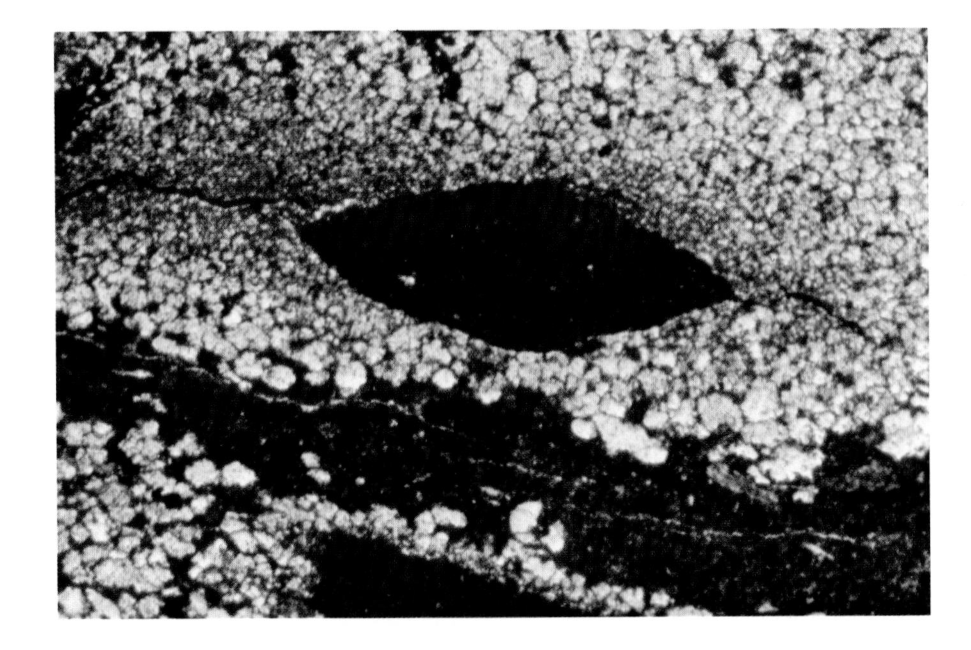

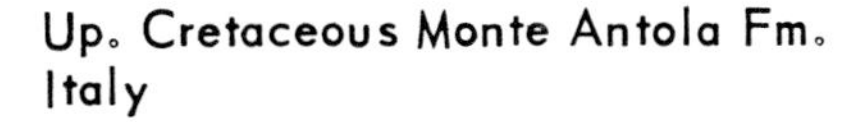
Up. Cretaceous Monte Antola Fm.
Italy

Strain recrystallization. These veins are thin zones of microspar which have formed along virtually invisible microfractures or zones of shear. Although these veins can grow to look very similar to fracture veins, there never was any open void space associated with these, as evidenced by fossils and other grains going across them without breakage.

0.11 mm

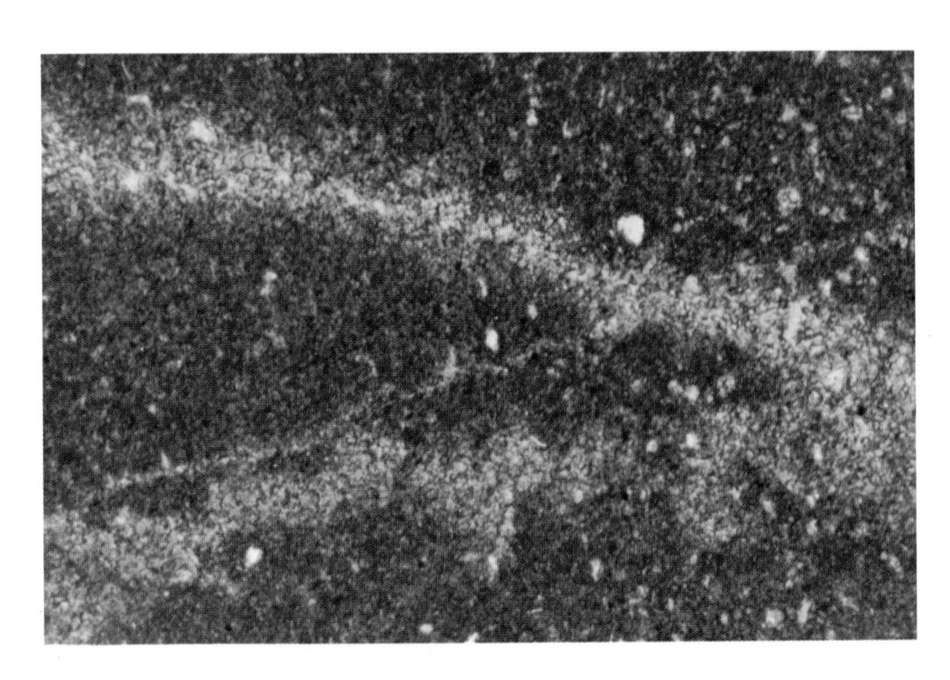

Jurassic(?) Carrara Marble
Italy

The end product of extensive strain recrystallization. The entire rock has gone to a uniform mosaic of medium-crystalline, equant crystals with rather straight intercrystalline boundaries. The driving mechanisms were temperature as well as stress during metamorphism. This unusually uniform marble is used for sculpture and building stone.

X.N. 0.64 mm

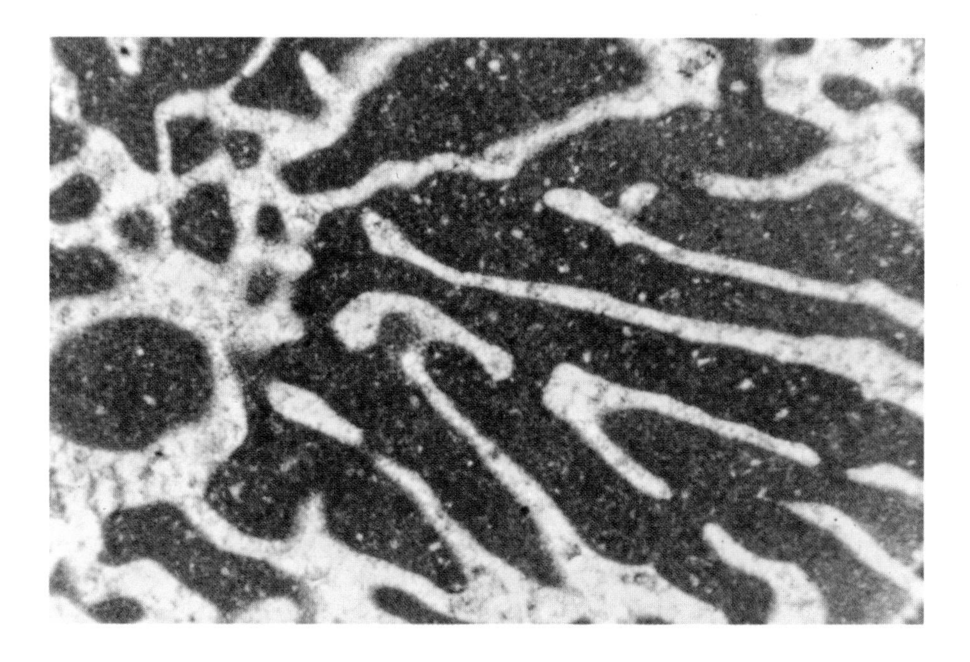

Paleocene (Danian) Faxekalk
Denmark

Solution-cavity fill. This originally aragonitic skeleton of a coral *(Dendrophyllia candelabra)* has been altered to calcite by going through a phase of complete dissolution and later infilling of the leached void. This has led to complete obliteration of the internal wall structure of the coral; only skeletal outline remains.

0.40 mm

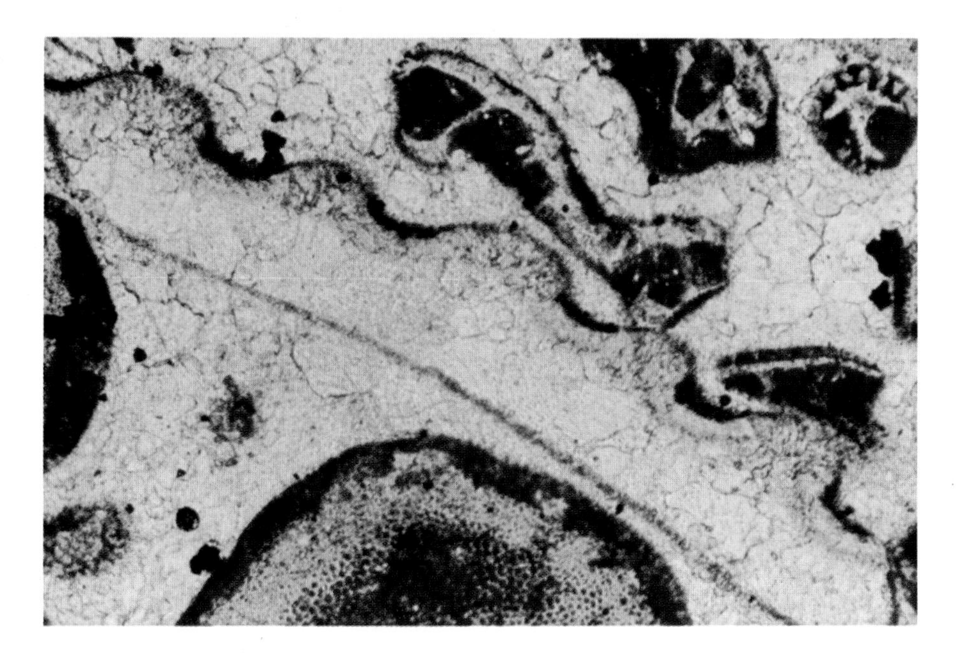

Up. Mississippian Hindsville Ls.
Oklahoma

Solution-cavity fill. A pelecypod shell with two-layer structure has undergone alteration from aragonite to calcite by either solution-cavity fill or inversion. Original shell structure has been largely obliterated and coarse, sparry calcite now fills shell. Distinguishing between solution-cavity fill and inversion can be impossible in many cases.

0.19 mm

Compaction and Deformation

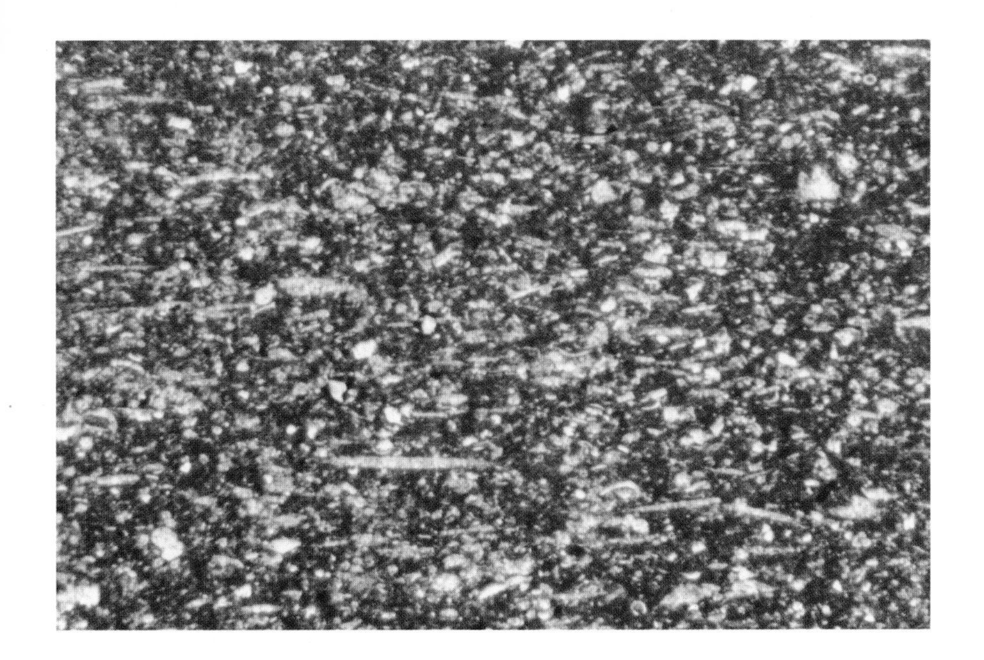

Cretaceous Gmunden Flysch
Austria

Excellent orientation of grains (especially sponge spicules). This is produced, in part, by primary sedimentary processess but is greatly enhanced by compaction which produces something akin to fissility by grain rotation into bedding-plane parallelism. Although compaction has not yet been widely recognized in shallow-water carbonates, it is common in deeper-water carbonates.

X.N. 0.38 mm

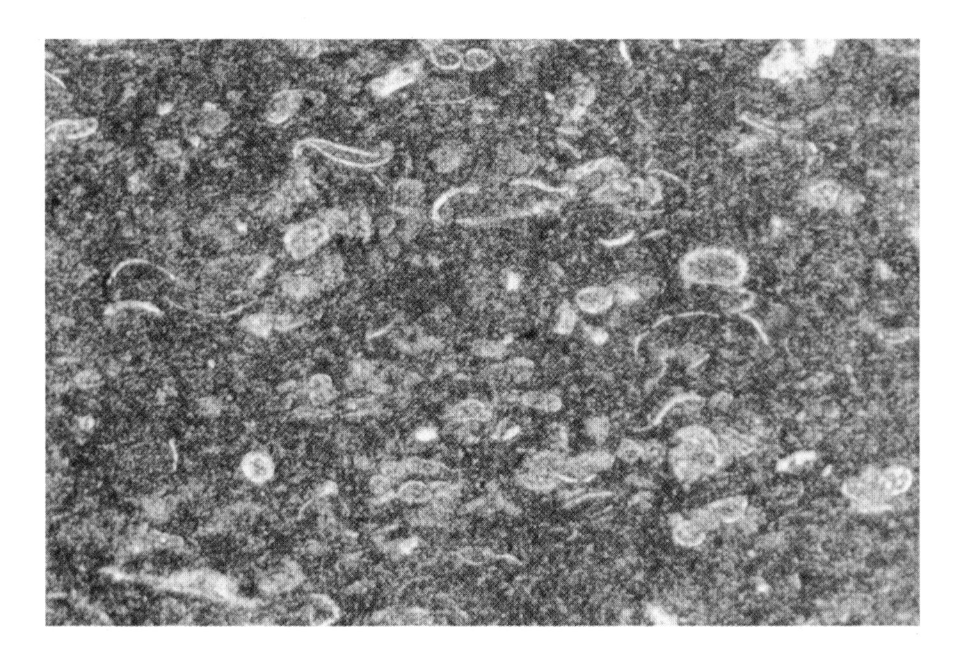

Up. Cretaceous White Ls.
Northern Ireland

Intense, bedding-parallel compaction and crushing of thin-walled organisms in a chalk. Such compaction is especially common in areas where there has been extremely rapid loading of overburden. In this area, thick basalt flows overlie the chalk.

0.11 mm

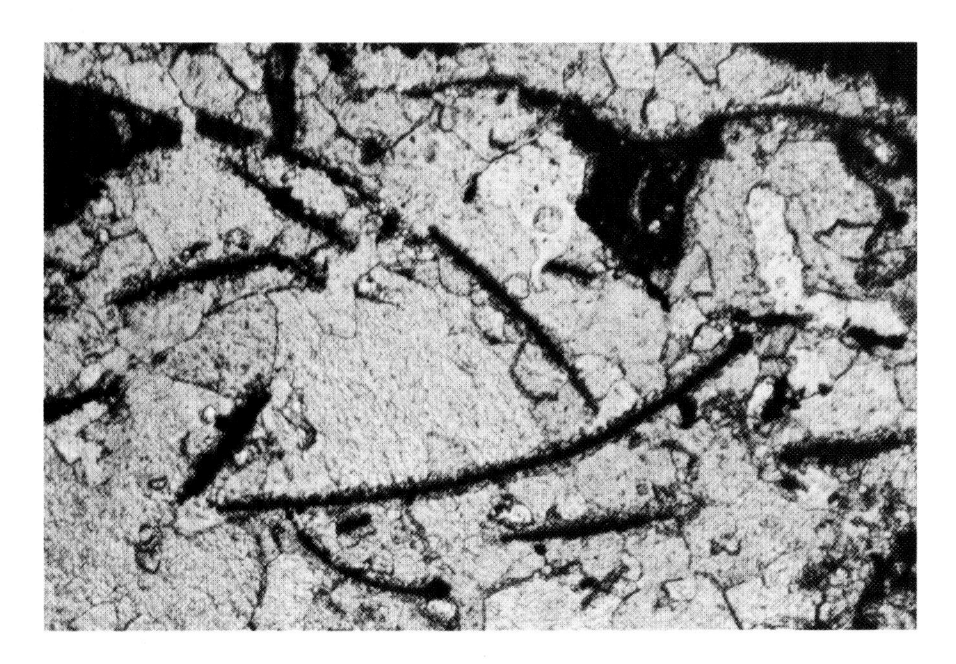

Lo. Cretaceous Edwards Ls.
Texas

Shattered micrite envelopes are an indicator of compaction or deformation of the sediment before major cementation. Because of the fragile nature of these envelopes, however, it does not take much compaction to destroy them.

0.25 mm

Mid. Ordovician Black River(?) Ls.
Pennsylvania

Stylolite solution zone. Dark lines are solution interfaces with concentration of insoluble minerals (especially clays and iron minerals). Dolomite is abundant along these stylolites. Stylolites are often associated with strong deformation or compaction and provide late diagenetic cement for many units.

0.38 mm

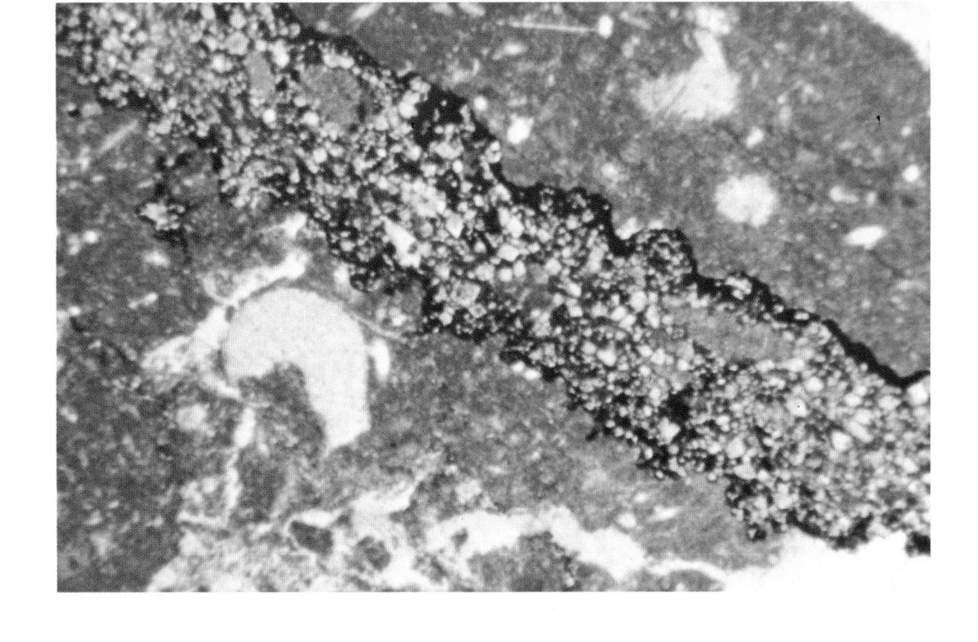

Lo. and Mid. Pennsylvanian Marble Falls Ls.
Texas

Solution zone marked by stylolitic insoluble residue shown cutting a crinoid fragment. Loss of a large volume of carbonate is apparent from the truncation of the crinoid.

X.N. 0.38 mm

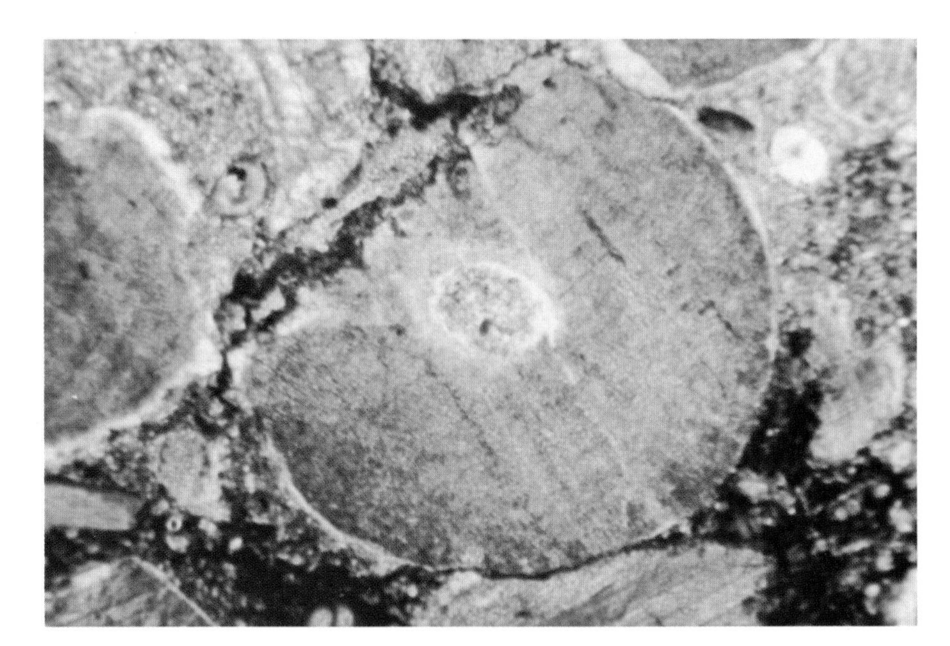

Lo. Devonian Becraft Ls.
New York

Extreme case of stylolitization in which a large brachiopod shell and numerous echinoderm fragments (with syntaxial overgrowths) have mutually dissolved and interpenetrated each other.

0.38 mm

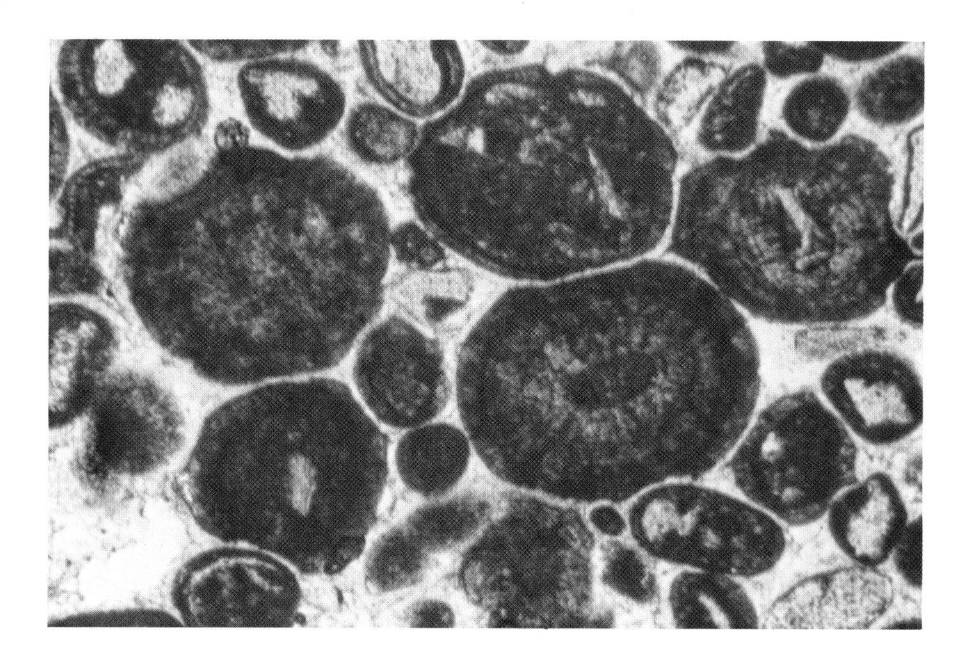

Up. Mississippian Ste. Genevieve Ls.
Tennessee

An interesting example of compaction postdating cementation. Although ooids show extensive grain interpenetration there are virtually no grain-to-grain contacts. In each case, a cement rim intervenes. Thus, thin rinds of submarine cement must have formed early in the history and compaction and grain interpenetration followed later.

X.N. 0.38 mm

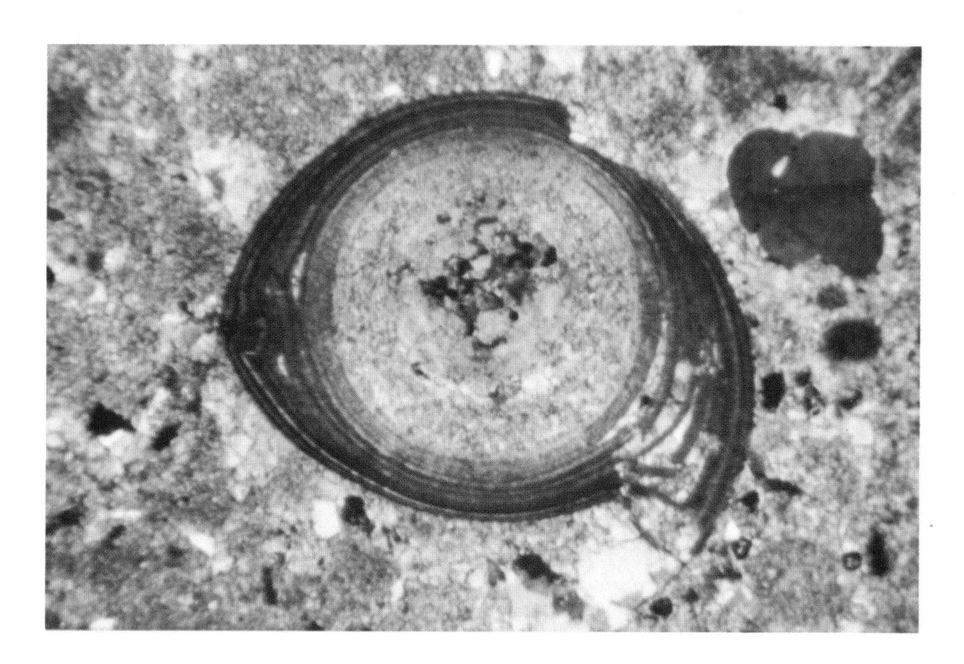

Up. Cambrian Gatesburg Fm.
Pennsylvania

Moderately deformed ooid showing how concentric laminations are sheared from the nucleus. Such textures may indicate that cementation was late relative to deformation.

X.N. 0.30 mm

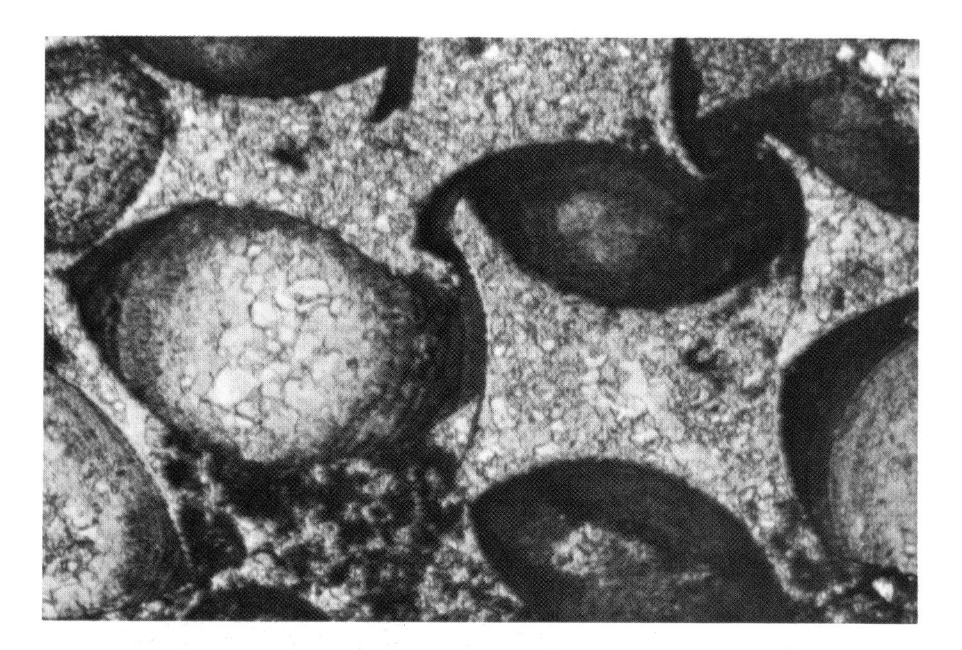

Up. Cambrian Gatesburg Fm.
Pennsylvania

Strongly deformed ooids (spastoliths) showing how strong compression will shear off outer concentric coatings of the ooids and produce shapes often difficult to identify as ooids.

X.N. 0.38 mm

Lower Jurassic Hierlatzkalk
Germany

Multiple fractures (now filled with sparry calcite) cutting a sparse biomicrite. This rock comes from a thrust slice in the northern Alps.

0.38 mm

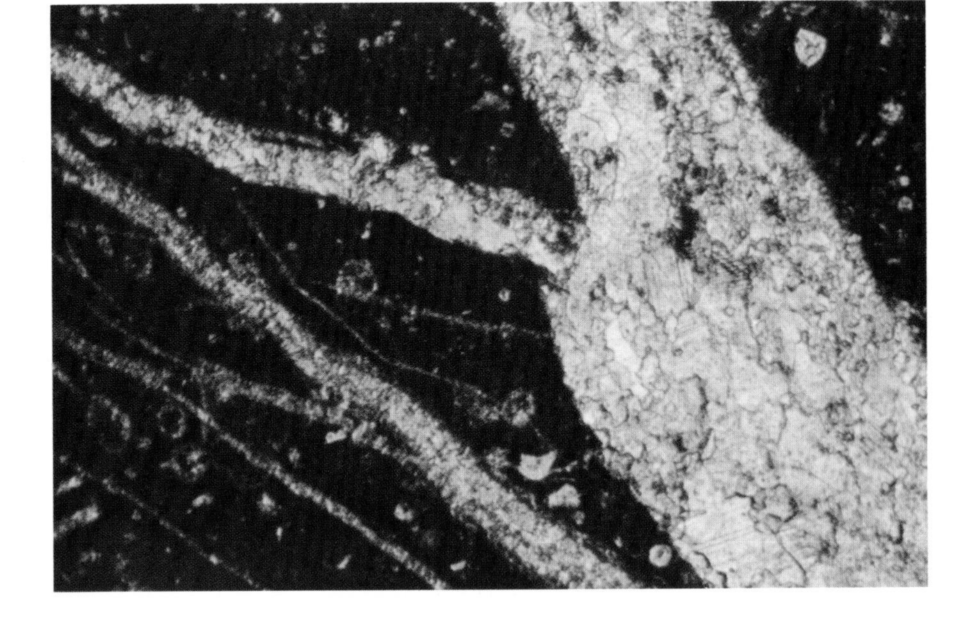

Up. Cretaceous Chalk
England

A strongly fractured chalk from an area of only mild deformation. Such fractures are commonly late diagenetic, and postdate most other diagenetic features in the rock. They can occasionally be used to predict the orientation and proximity of fault zones.

0.40 mm

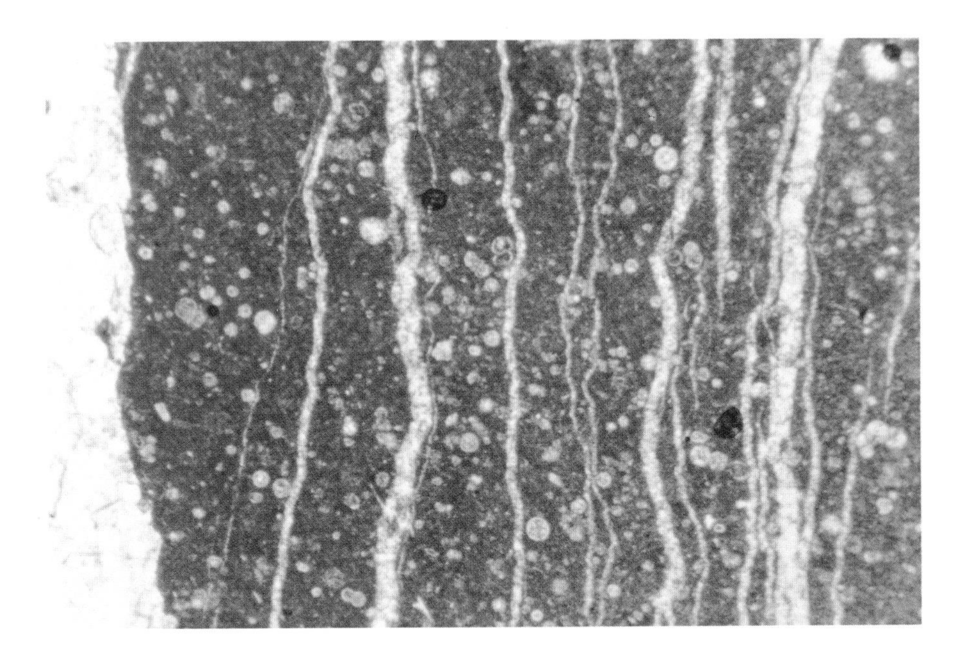

Jurassic Pennine Bündnerschiefer
Switzerland

Extreme deformation and alteration of an impure limestone. Crenulate folding, development of new micas, and strain recrystallization of calcite are all present.

X.N. 0.64 mm

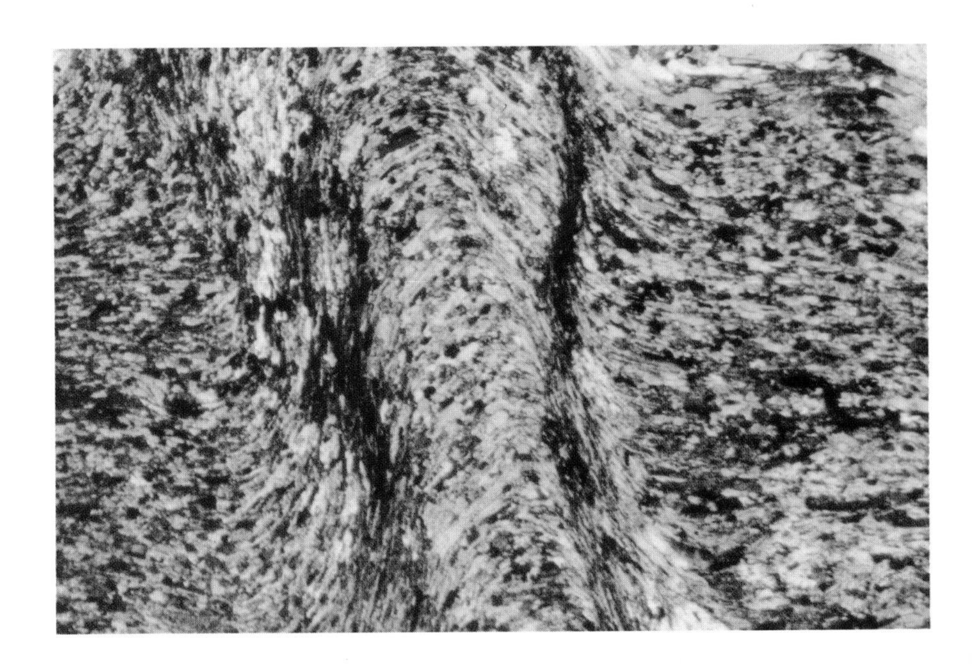

Geopetal Fabrics

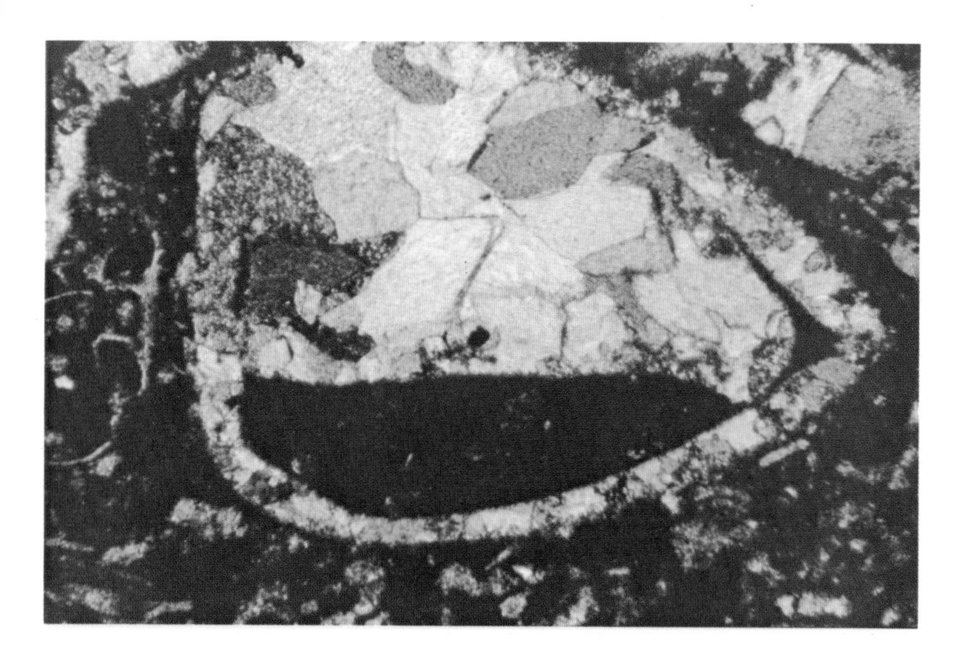

Up. Mississippian Hindsville Ls.
Oklahoma

Geopetal sediment infilling of a pelecypod shell. Note the alteration of the shell wall without the displacement of the sediment filling--this implies inversion rather than complete dissolution and subsequent void fill. Not all geopetal fills are exactly horizontal so a statistical sampling should be used.

P.X.N. 0.20 mm

Pliocene and Pleistocene Caloosahatchee Marl
Florida

Geopetal filling of a gastropod chamber. Mud (micrite) fills the lower part of the cavity while sparry calcite fills the upper part--the contact indicates an approximately level surface at the time of deposition. Such structures are very useful in determining original dips of units.

0.38 mm

Up. Cambrian Allentown Dolomite
Pennsylvania

Geopetal ooids. Original circular outline of ooids can be seen. Dissolution of part of the ooids allowed the remainder to fall to bottom of cavity. Upper part then filled with spar. Indicates level surface at time of dissolution.

0.38 mm

Up. Cambrian Morgan Creek Ls. Mbr. of Wilberns Fm.
Texas

"Umbrella" void. Presence of the large fossil fragments prevented the complete infiltration of lime mud and maintained shelter voids which were later filled with sparry calcite. This can be used as a geopetal indicator.

0.38 mm

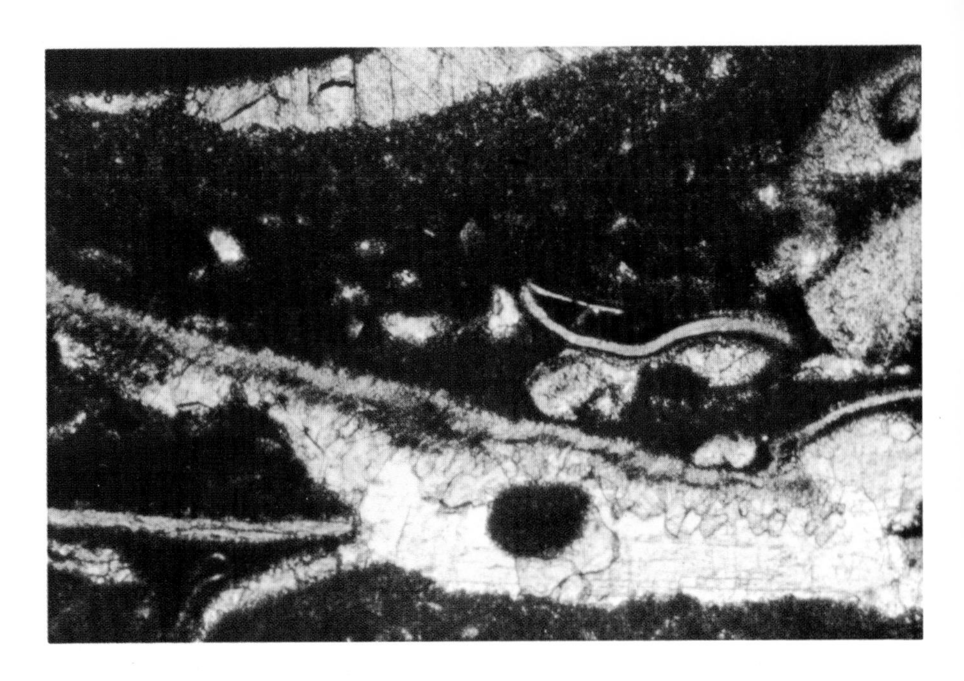

Pliocene and Pleistocene Caloosahatchee Marl
Florida

Umbrella void--large mollusk fragment prevented infiltration of micrite and maintained cavity filled by later crust of bladed spar and equant spar.

0.38 mm

Cretaceous Tamabra Ls.
Mexico

Geopetal sediment or vadose-silt infilling of voids of fossils (aragonitic) which were leached. These geopetal features may reflect postdepositional but prediagenetic structural or compactional deformation, and must thus be distinguished from synsedimentary ones. 0.64 mm

Porosity Classification

Philip W. Choquette and Lloyd C. Pray

BASIC POROSITY TYPES

FABRIC SELECTIVE

Type	Abbreviation
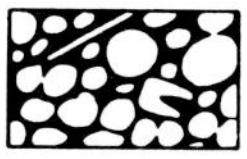 INTERPARTICLE	BP
 INTRAPARTICLE	WP
 INTERCRYSTAL	BC
 MOLDIC	MO
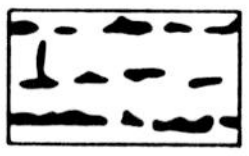 FENESTRAL	FE
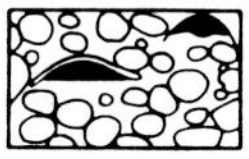 SHELTER	SH
 GROWTH-FRAMEWORK	GF

NOT FABRIC SELECTIVE

Type	Abbreviation
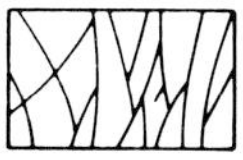 FRACTURE	FR
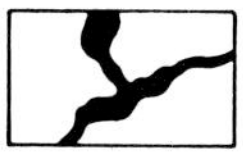 CHANNEL*	CH
 VUG*	VUG
 CAVERN*	CV

*Cavern applies to man-sized or larger pores of channel or vug shapes.

FABRIC SELECTIVE OR NOT

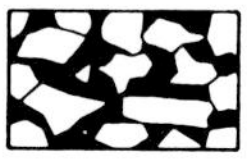

Type	Abbreviation
BRECCIA	BR
 BORING	BO
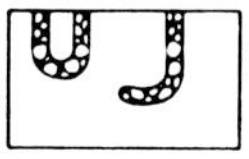 BURROW	BU
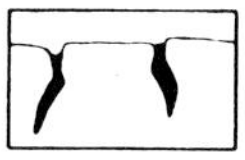 SHRINKAGE	SK

MODIFYING TERMS

GENETIC MODIFIERS

PROCESS

SOLUTION	s
CEMENTATION	c
INTERNAL SEDIMENT	i

DIRECTION OR STAGE

ENLARGED	x
REDUCED	r
FILLED	f

TIME OF FORMATION

PRIMARY	P
pre-depositional	Pp
depositional	Pd
SECONDARY	S
eogenetic	Se
mesogenetic	Sm
telogenetic	St

Genetic modifiers are combined as follows:

PROCESS + DIRECTION + TIME

EXAMPLES:

solution-enlarged	sx
cement-reduced primary	crP
sediment-filled eogenetic	ifSe

SIZE* MODIFIERS

CLASSES				mm†
				256
MEGAPORE	mg	large	lmg	32
		small	smg	4
MESOPORE	ms	large	lms	1/2
		small	sms	1/16
MICROPORE	mc			

Use size prefixes with basic porosity types:

mesovug	msVUG
small mesomold	smsMO
microinterparticle	mcBP

*For regular-shaped pores smaller than cavern size.

†Measures refer to average pore diameter of a single pore or the range in size of a pore assemblage. For tubular pores use average cross-section. For platy pores use width and note shape.

ABUNDANCE MODIFIERS

percent porosity (15%)

or

ratio of porosity types (1:2)

or

ratio and percent (1:2) (15%)

Classification of Porosity Types (Choquette and Pray, 1970)

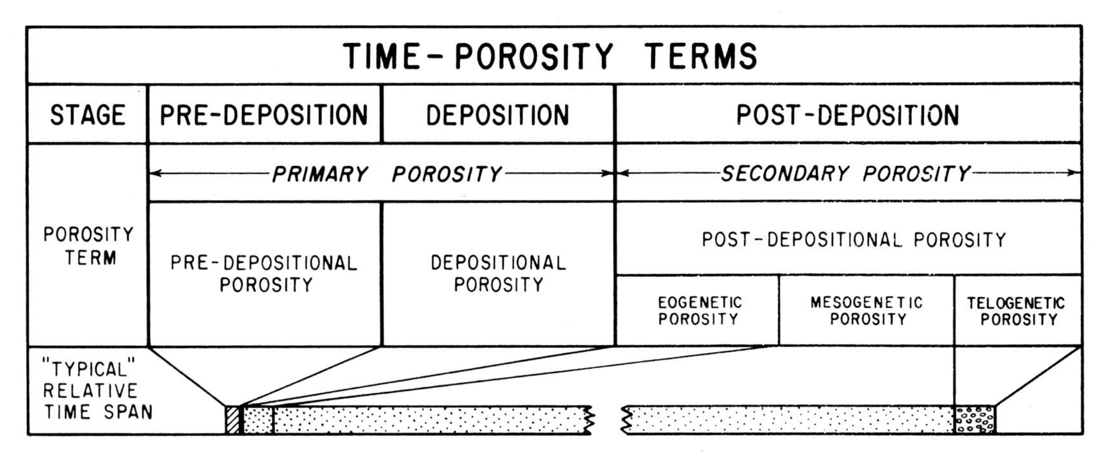

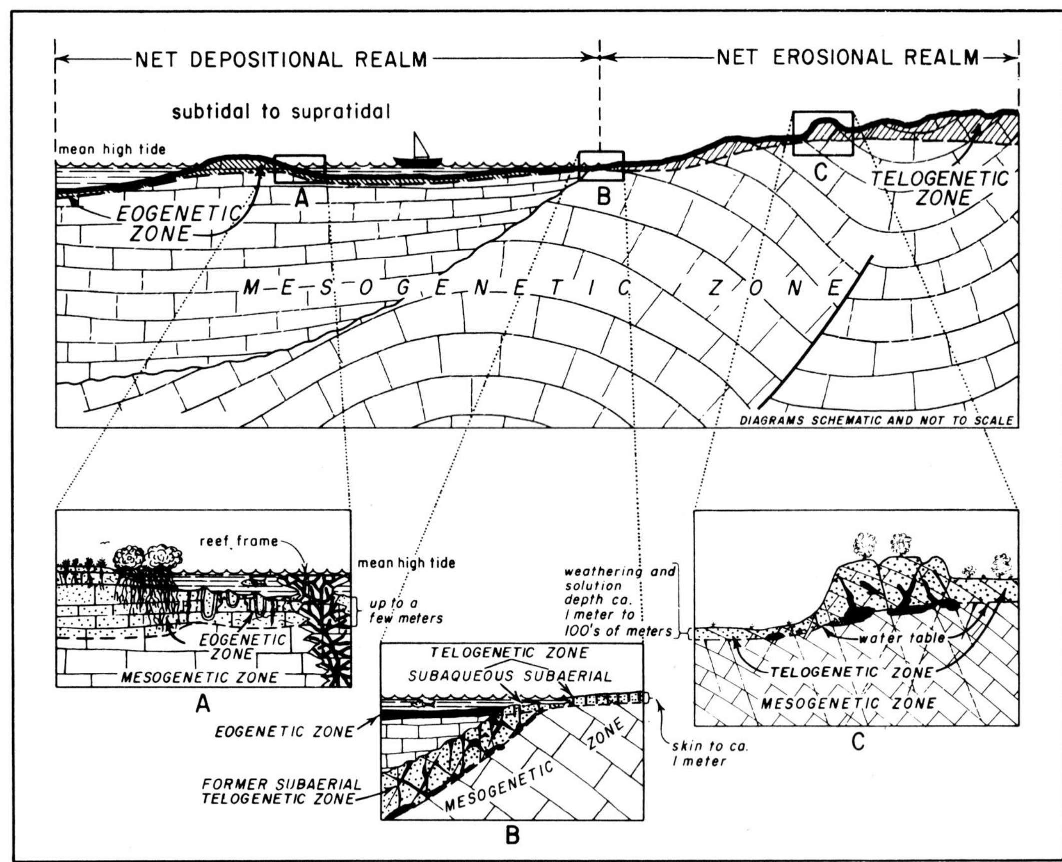

Time-porosity terms and zones of creation and modification of porosity in sedimentary carbonates.

Upper diagram: Interrelation of major time-porosity terms. Primary porosity either originates at time of deposition (depositional porosity) or was present in particles before their final deposition (predepositional porosity). Secondary or postdepositional porosity originates after final deposition and is subdivided into eogenetic, mesogenetic, or telogenetic porosity depending on stage or burial zone in which it develops (see lower diagram). Bar diagram depicts our concept of "typical" relative durations of stages.

Lower diagram: Schematic representation of major surface and burial zones in which porosity is created or modified. Two major surface realms are those of net deposition and net erosion. Upper cross section and enlarged diagrams **A, B,** and **C** depict three major post-depositional zones. Eogenetic zone extends from surface of newly deposited carbonate to depths where processes genetically related to surface become ineffective. Telogenetic zone extends from erosion surface to depths at which major surface-related erosional processes become ineffective. Below a subaerial erosion surface, practical lower limit of telogenesis is at or near water table. Mesogenetic zone lies below major influences of processes operating at surface. The three terms also apply to time, processes, or features developed in respective zones. (from Choquette & Pray, 1970)

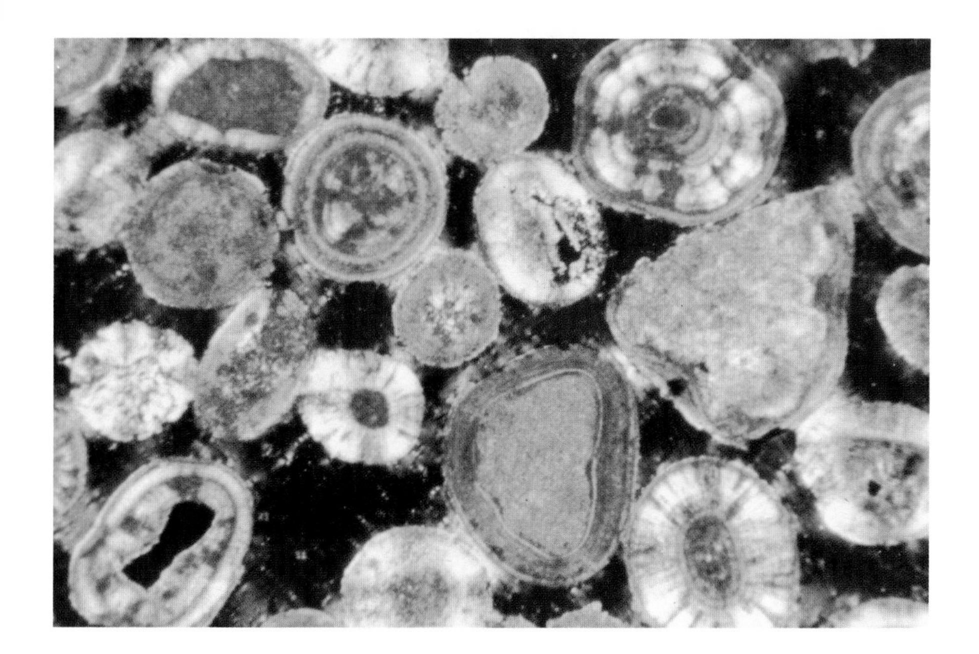

Holocene oolite
Great Salt Lake, Utah

Unfilled interparticle porosity (in oolite). Porosity in black. Classification used in these and following photos is that of Choquette and Pray (1970).

X.N. 0.30 mm

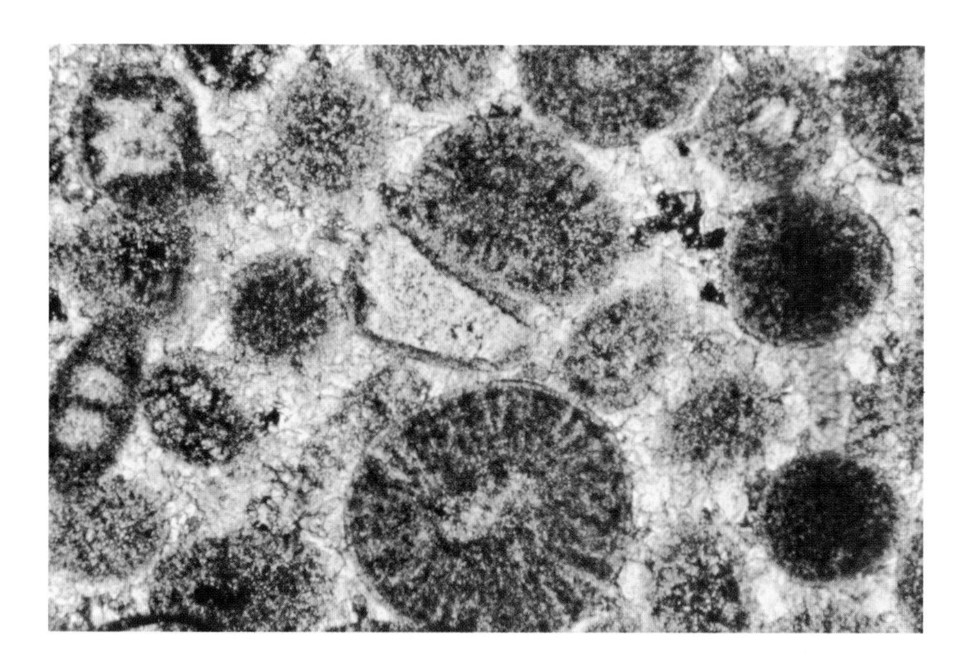

Up. Mississippian Pitkin Ls.
Oklahoma

Sparry calcite-filled interparticle porosity (in oolite).

X.N. 0.10 mm

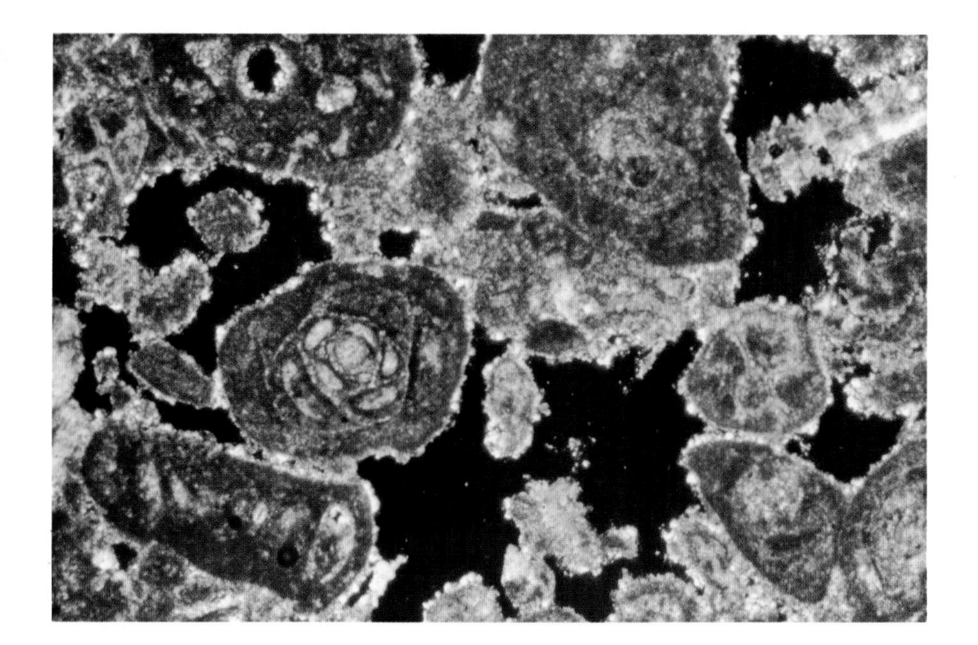

Up. Eocene Ocala Fm.
Florida

Enlarged intergranular (interparticle) porosity (probably enlarged through solution). Porosity in black. Grains are mostly miliolid Foraminifera.

X.N. 0.38 mm

Holocene back-reef beach sediment
Belize (British Honduras)

Unfilled intraparticle porosity (within a large coral fragment). Porosity in black.

X.N. 0.38 mm

Pleistocene Key Largo Ls.
Florida

Reduced interparticle and intraparticle porosity (in Foraminifera and mollusks). Porosity in black.

X.N. 0.38 mm

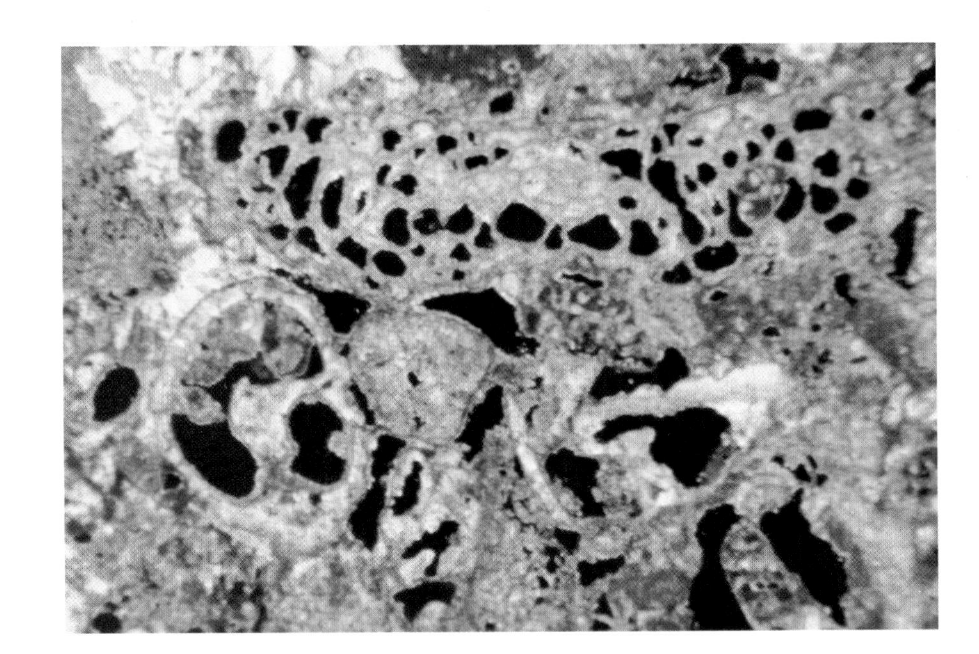

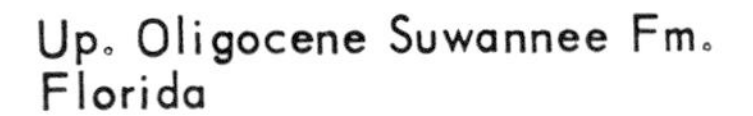

Up. Oligocene Suwannee Fm.
Florida

Reduced intraparticle porosity (especially within miliolid Foraminifera) and reduced interparticle porosity (especially adjacent to echinoid fragments). Porosity in purple.

G.P. 0.38 mm

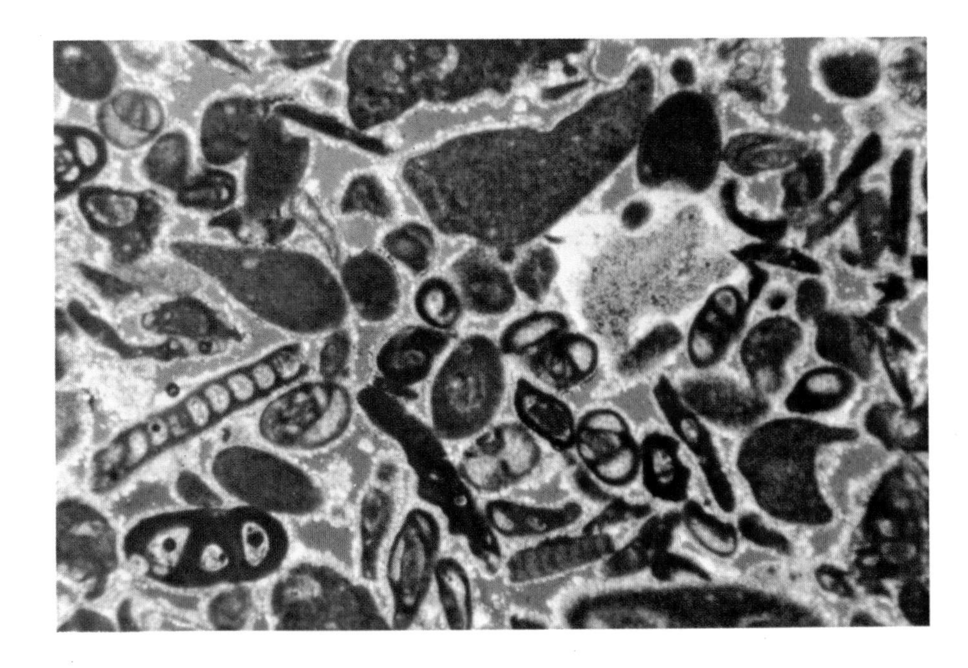

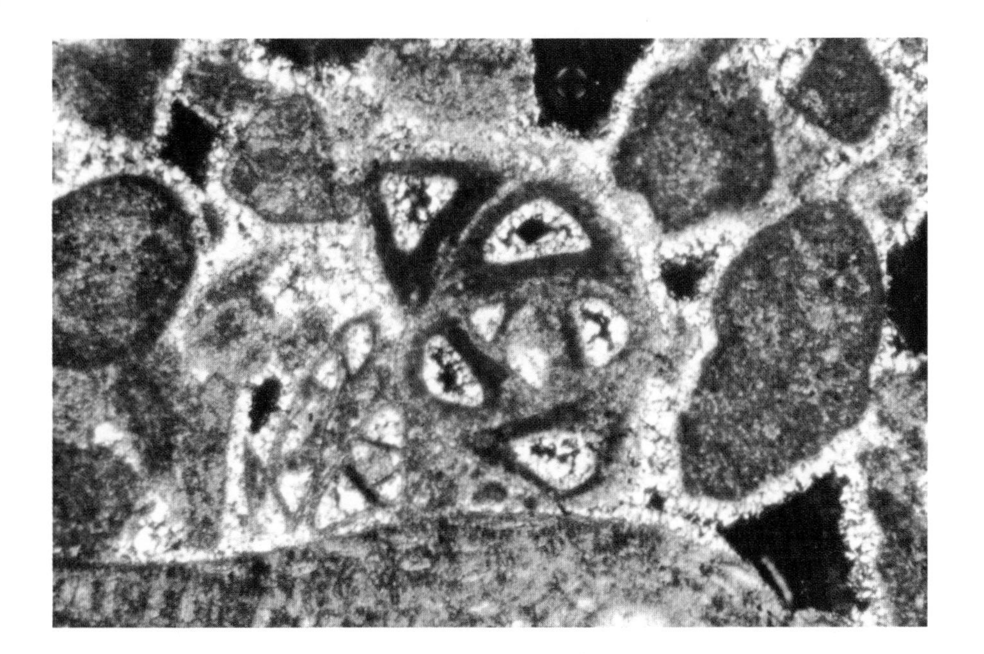

Up. Eocene Ocala Fm.
Florida

Reduced intraparticle (within miliolid Foraminifera) and interparticle porosity. Cement here is chert. Porosity in black.

X.N. 0.38 mm

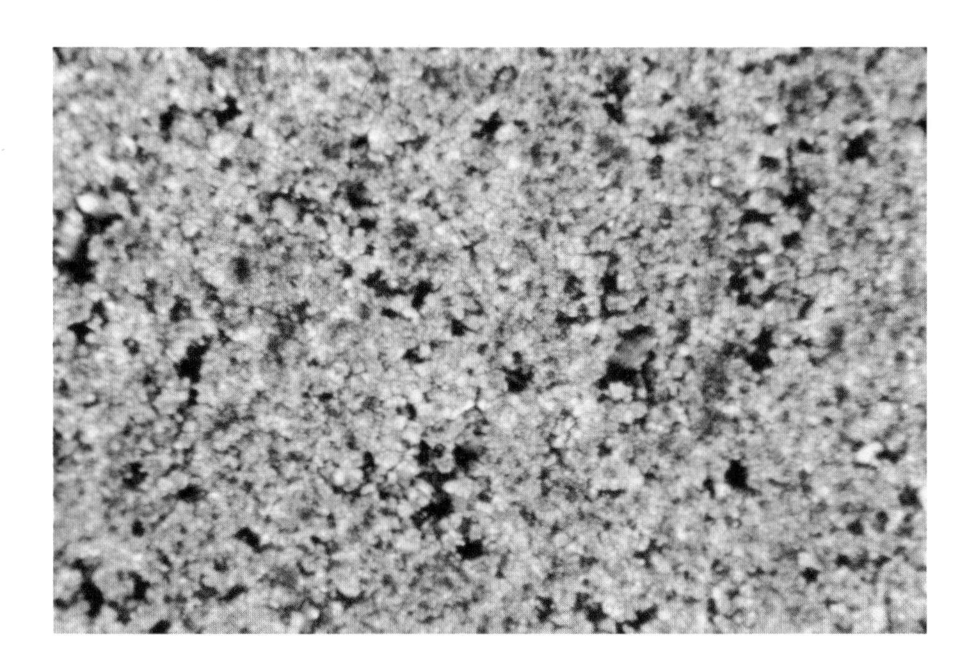

Mid. Eocene Avon Park Ls.
Florida

Intercrystal porosity in a fine- to medium-crystalline dolomite (replacement). Porosity in black (with strain patterns).

X.N. 0.38 mm

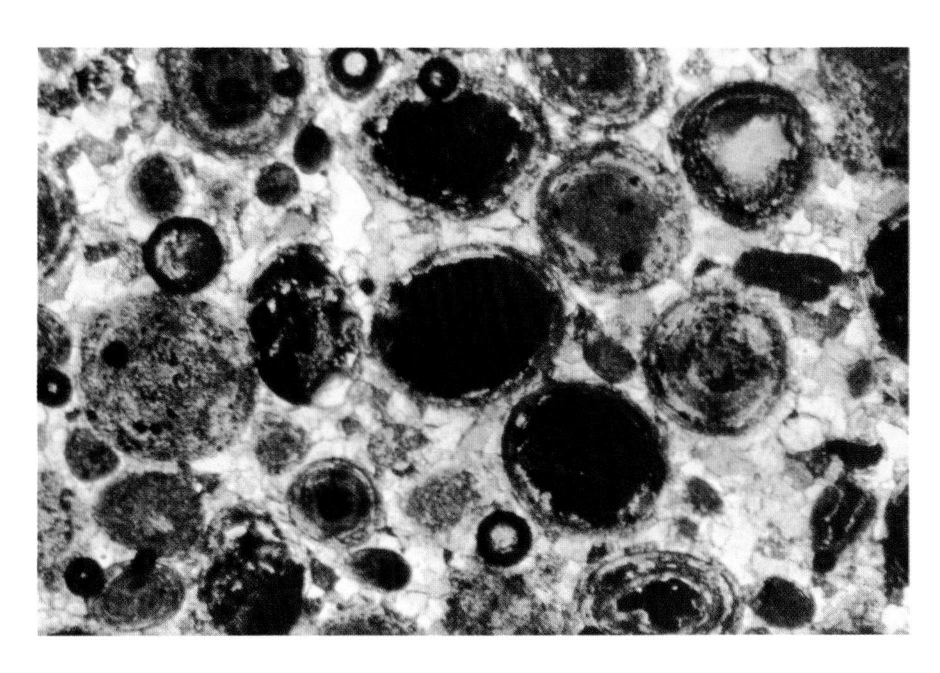

Pleistocene Miami oolite
Florida

Moldic (oomoldic) porosity (black).

X.N. 0.38 mm

Up. Eocene Ocala Ls.
Florida

Moldic porosity. Formerly a fossiliferous micrite, the micrite has been replaced by dolomite and the more soluble calcitic fossils have been dissolved to leave a moldic porosity. Porosity in purple.

G.P. 0.38 mm

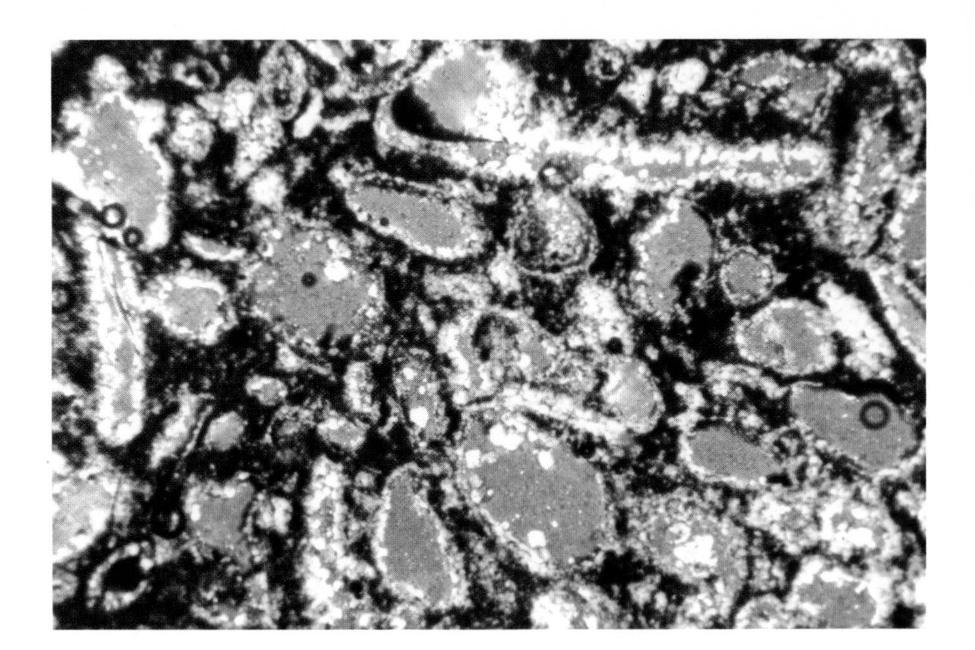

Up. Eocene Ocala Ls.
Florida

Moldic porosity. As above, dolomite replaced the micrite matrix, then allowed dissolution of more soluble fossils to yield moldic porosity. Porosity in purple.

G.P. 0.38 mm

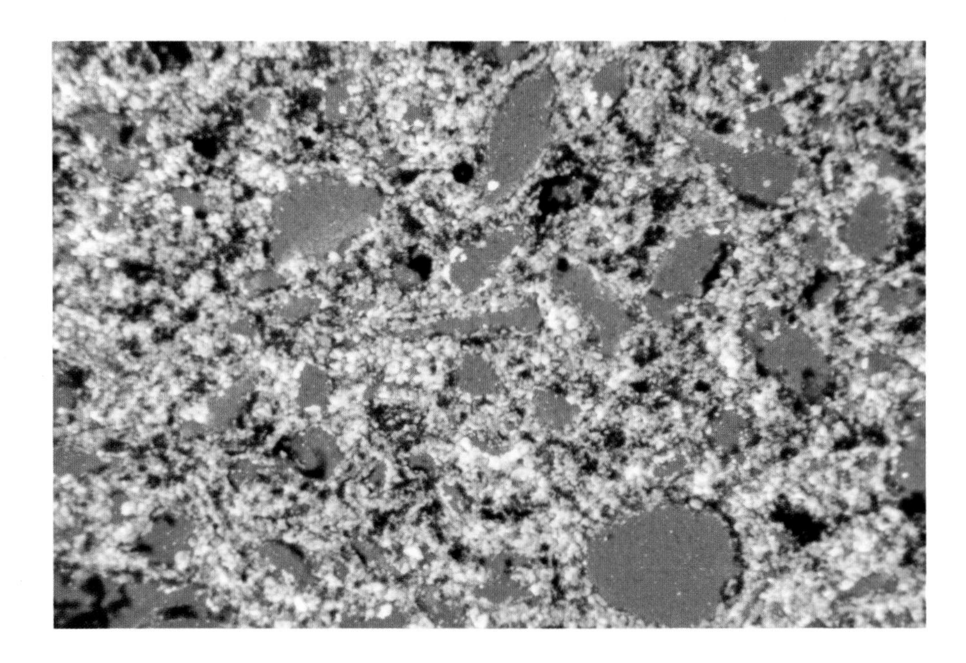

Up. Cretaceous Pilot Knob reef facies of Austin Group
Texas

Filled moldic porosity. Rock was cemented by calcite, then aragonitic fossils were dissolved, leaving voids. Subsequently these voids were filled with sparry calcite. Former fossils marked by micrite envelopes.

0.30 mm

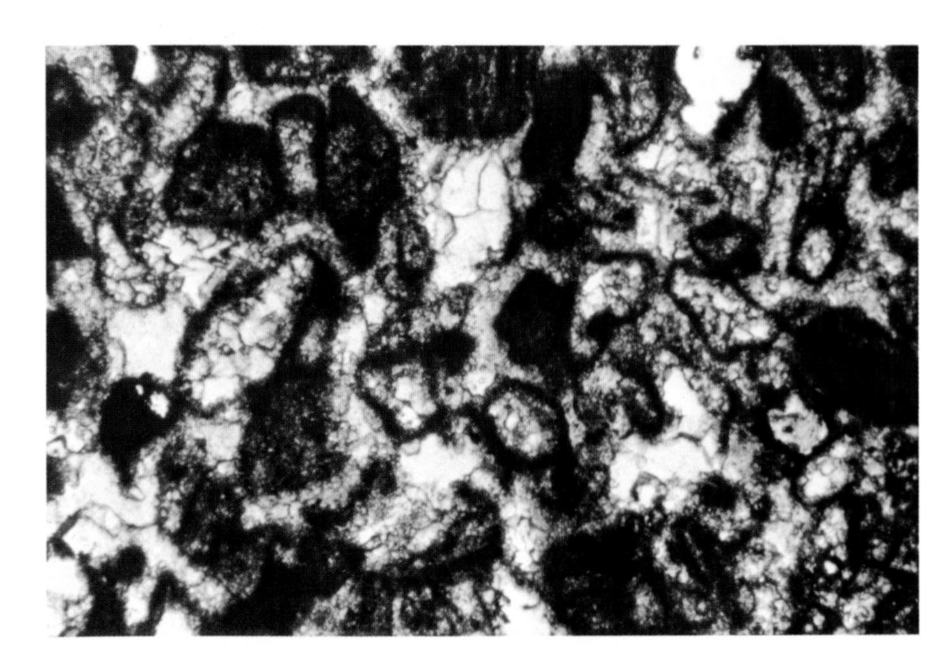

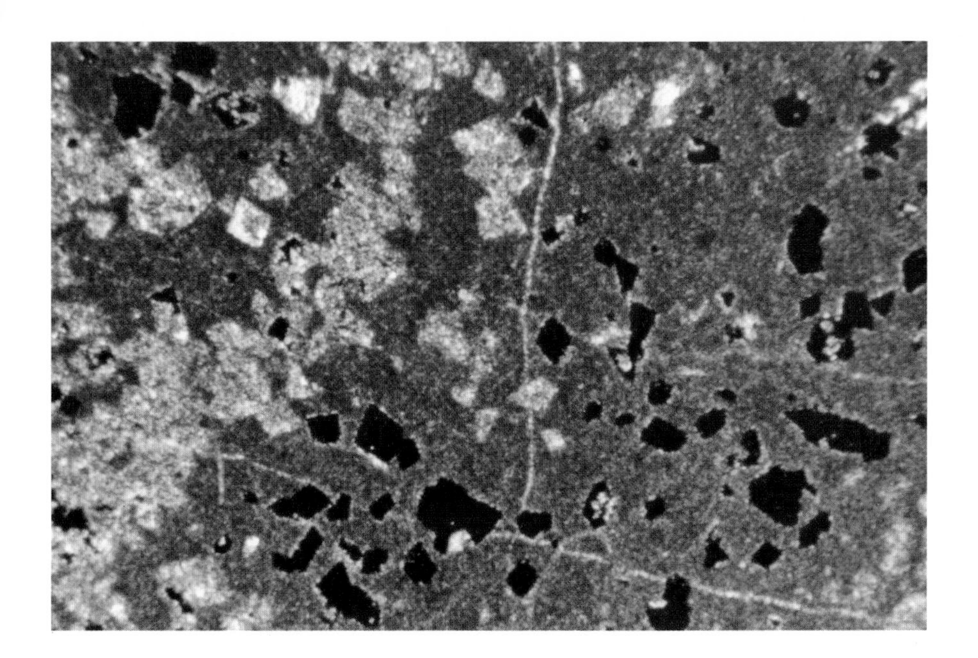

Jurassic Ls. of Ronda unit (Subbetic)
Spain

Moldic porosity produced by selective leaching of dolomite crystals (dedolomitization). This is probably a telogenetic (outcrop) process which is enhanced by the presence of evaporite (sulphate) minerals in the section.

X.N. 0.38 mm

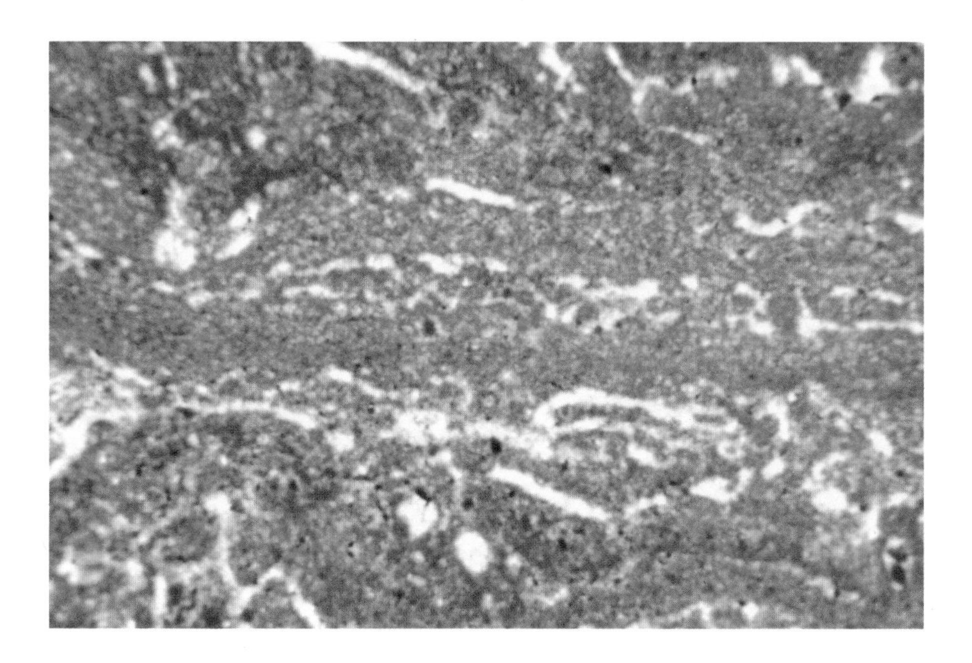

Up. Silurian part of Tonoloway-Keyser Ls.
Pennsylvania

Filled fenestral porosity in a blue-green algal biolithite. Porosity may be due to air spaces in crinkled mat sediments.

0.38 mm

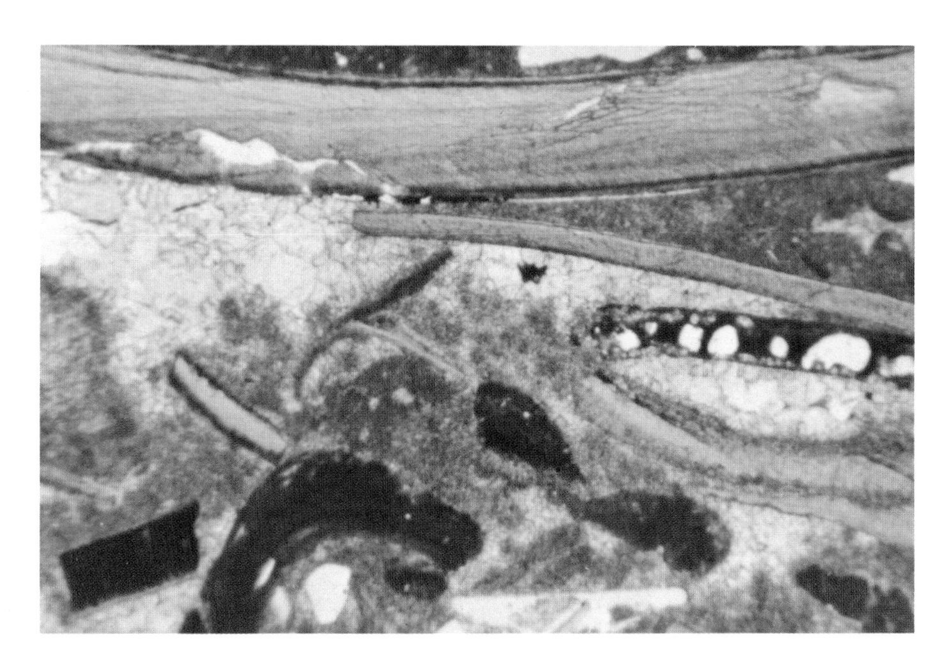

Pliocene and Pleistocene Caloosahatchee Marl
Florida

Filled shelter porosity beneath a large mollusk fragment.

0.38 mm

Up. Jurassic Solnhofen Ls. (Kelheim facies)
Germany

Reduced fracture porosity. Porosity in black.

X.N. 0.38 mm

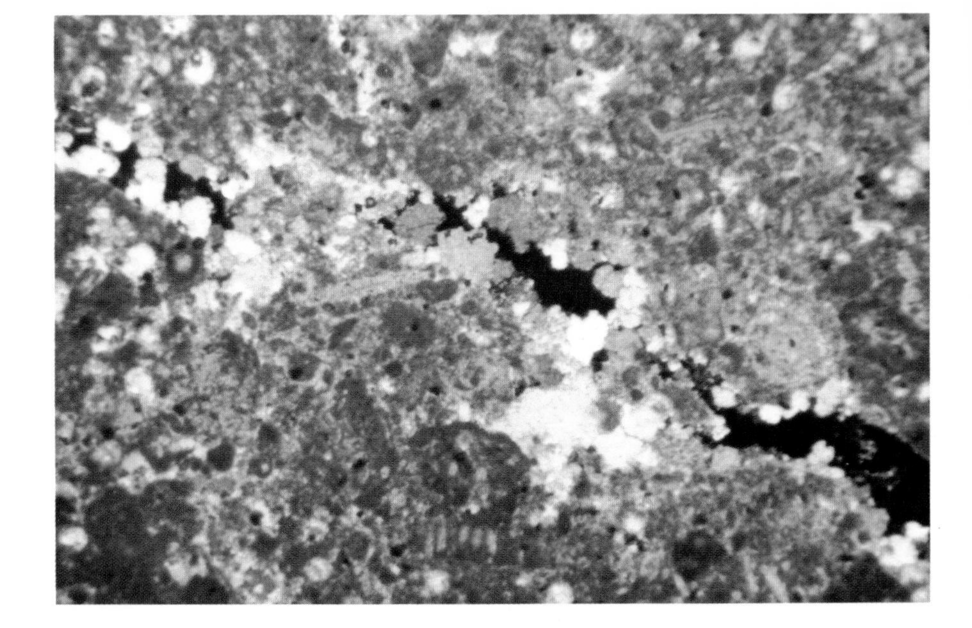

Mid. Ordovician Black River(?) Ls.
Pennsylvania

Filled fracture porosity. Conjugate set of large fracture veins.

0.38 mm

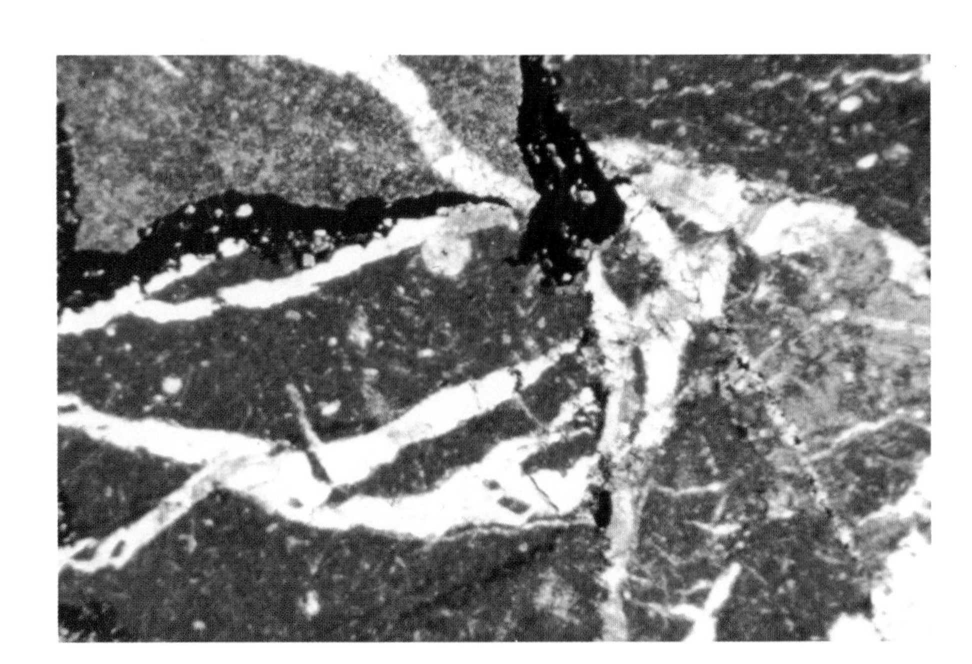

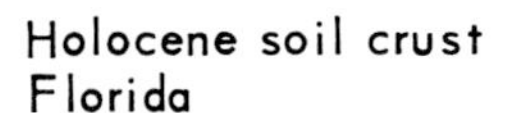
Holocene soil crust
Florida

Secondary enlarged porosity yielding small solution vugs. Porosity in purple. Dark grain in upper left-hand corner is a wood fragment.

G.P. 0.38 mm

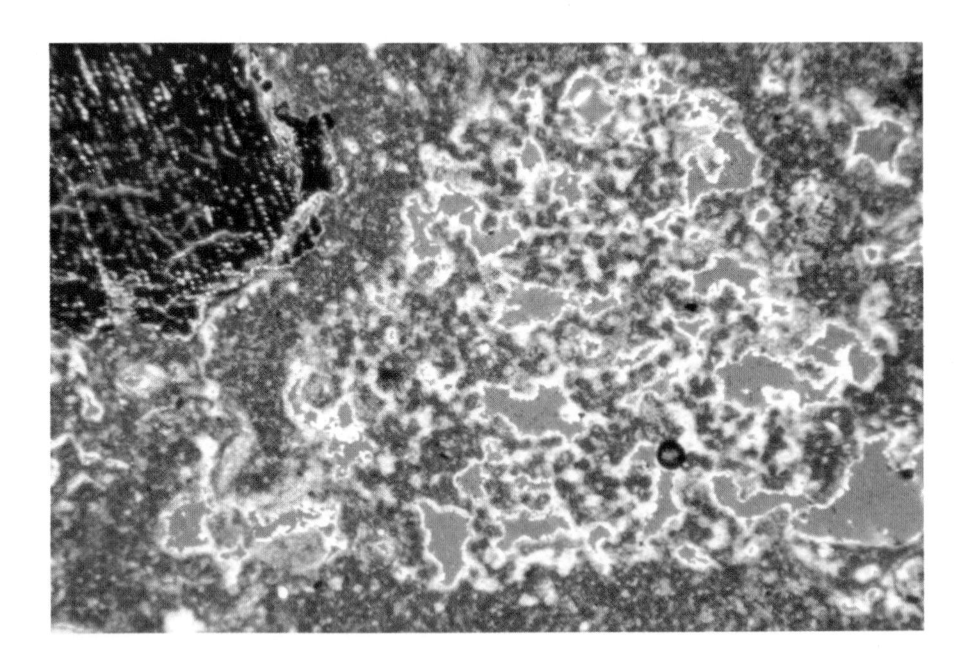

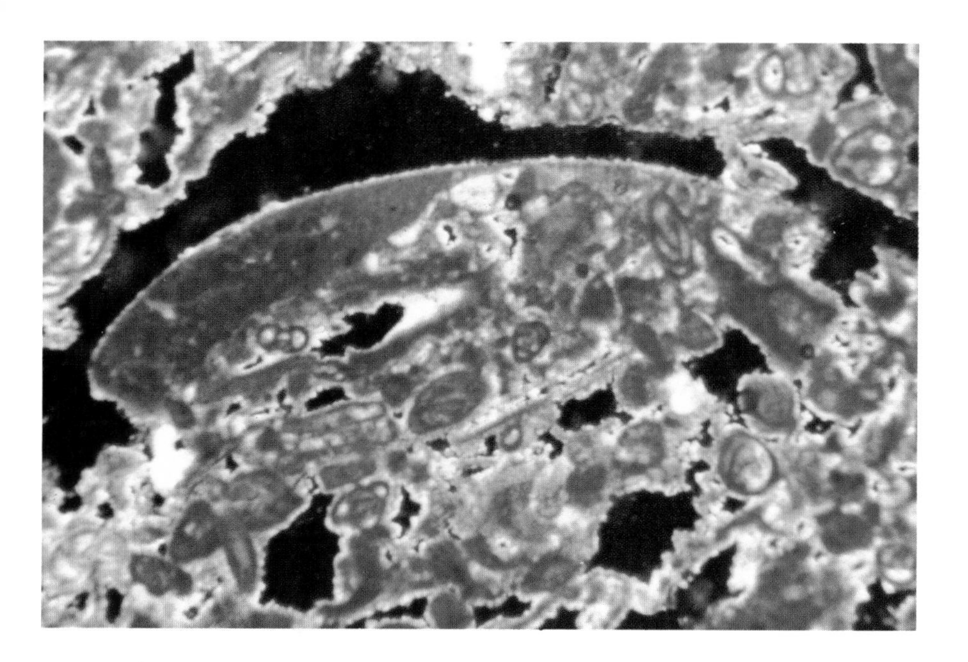

Up. Oligocene Suwannee Ls.
Florida

Enlarged moldic porosity. Note large pelecypod mold with upper edge enlarged. Porosity in black. Other grains include mainly miliolid Foraminifera.

X.N. 0.38 mm

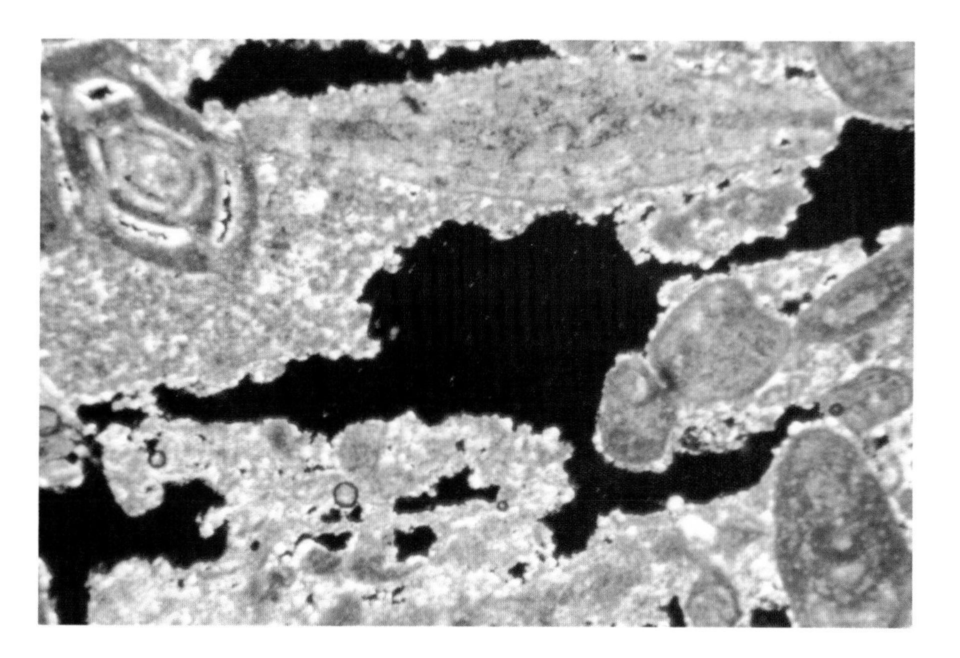

Up. Eocene Ocala Fm.
Florida

Vuggy porosity. Probably solution enlarged. Porosity in black. Grains include miliolids and other Foraminifera.

X.N. 0.38 mm

Up. Cretaceous Pilot Knob reef facies of Austin Group
Texas

Vuggy porosity, clearly enlarged and not fabric selective. Porosity in black.

X.N. 0.38 mm

Pleistocene Key Largo Ls.
Florida

Vuggy porosity. Probably solution enlargement of original interparticle porosity. Porosity in black.

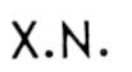

X.N. 0.38 mm

Carbonate Techniques

Techniques

Although petrography is an extremely valuable tool for the identification of minerals and their textural interrelationships, it is best used (in many cases) in conjunction with other techniques.

Precise mineral determinations are often aided by staining of thin sections or rock slabs, by X-ray analysis, or by microprobe examination. Where noncarbonate constituents are present in carbonate rocks they often are better analyzed in acid-insoluble residues than in thin section. Where detailed understanding of the trace element chemistry of the sediments is essential, X-ray fluorescence, microprobe, atomic absorption, or cathodoluminescence techniques may be applicable.

Commonly, also, sediments may be too fine-grained for adequate examination with the light microscope. The practical limit of resolution of the best light microscopes is in the one to two micrometer (μm) range. Many carbonate and noncarbonate grains fall within or below this size range. Furthermore, because most standard thin sections are about 30 μm thick, a researcher examines 10 or 20 of these small grains stacked on top of one another, with obvious loss of resolution. Smear mounts or strew mounts (slides with individual, disaggregated grains smeared or settled out onto the slide surface) are an aid in examining small grains where the material can be disaggregated into individual components. However, in most cases, scanning and transmission electron microscopy have proved to be the most effective techniques for the detailed examination of fine-grained sediments.

The bibliography of this book provides a number of references to techniques useful in supplementing standard petrographic analysis. Although many of the techniques require sophisticated and expensive equipment, others, such as staining, acetate peels, and insoluble residues can be performed in any laboratory.

Because of the potential desirability of these techniques, it is often useful to prepare epoxy-cemented thin sections without coverslips. These sections can be examined under a light microscope either by placing a drop of water and a coverslip on the sample during viewing, or by using mineral oil or index of refraction oils with or without coverslips. These methods involve some loss of resolution but do allow the cleaning and drying of the surface of the section and subsequent staining, luminescence, or microprobe examination. One can even partially or completely immerse the thin section in acetic or hydrochloric acid and decalcify the section thereby sometimes enhancing organic structures or insoluble-mineral fabrics. Finally, uncovered thin sections can also be gound thinner in cases where examination of very fine-grained sediments is needed.

Clearly, one can spend a lifetime analyzing a single sample using all possible techniques. Efficient study requires a thorough understanding of all available tools and proper application of the most useful and productive of these.

Staining

Staining techniques are among the fastest, simplest, and cheapest methods for getting reliable chemical data on carbonate phases. The following list of minerals and their diagnostic stains is derived from the work of Friedman (1959), Dickson (1966), Milliman (1974), and others. The original papers, listed in the bibliography, will provide details about the exact application and methods.

Aragonite Distinguished from calcite by use of Feigel's Solution. Aragonite turns black whereas calcite remains colorless for some time. Mixing Feigel's Solution requires 7.1 g of $MnSO_4 \cdot H_2O$; 2 to 3 g of Ag_2SO_4; 100 cc distilled water and a 1% NaOH solution. Difficult to prepare and store.

Calcite Can be distinguished from dolomite with a simple stain of Alizarin Red-S in a 0.2% HCl solution (cold). Calcite and aragonite turn red, whereas dolomite remains colorless.

Dolomite Can be distinguished from calcite by the above method or one can stain specifically for dolomite with a number of organic stains including Titan Yellow, Trypan Blue, and Safranine O. All these stains require boiling the sample in a concentrated NaOH solution.

(Mg) Calcite Can be distinguished from aragonite and low-Mg calcite by use of Clayton Yellow stain. This is made by adding 0.5 g of Titan Yellow, 0.8 g of NaOH, and 2 g of EDTA to 500 ml of distilled water. The grain or section is etched in dilute acetic acid for 30 seconds, and is then put in Clayton Yellow solution for 30 minutes. Mg-calcite turns red while low-Mg calcite remains colorless. Mg-calcite can also be stained with Alizarin Red-S in 30% NaOH; calcite remains colorless whereas Mg-calcite turns purple.

(Fe) Calcite Can be distinguished (along with ferroan dolomite) from normal calcite by the use of a potassium ferricyanide stain in a weak HCl solution (details in Dickson, 1966). Ferroan minerals turn pale to deep turquoise, and nonferroan ones remain colorless.

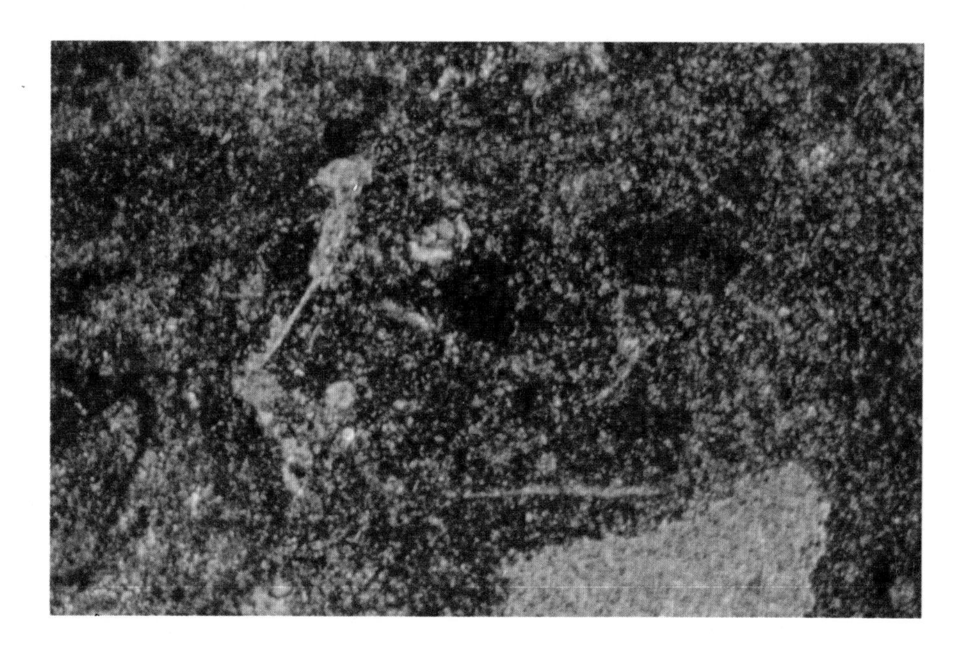

Upper Permian Tansill Formation
New Mexico

An example of Alizarin Red-S staining of a thin section. Large echinoderm fragment and other skeletal grains have stained red (calcite) whereas matrix is largely unstained (dolomite). Stains can be preserved with covering of mineral oil or glass cover slip.

0.25 mm

X-Ray Diffraction of Carbonate Rocks

In addition to being a fast and reliable method of determing the bulk mineralogy of carbonate rocks (aragonite, calcite, dolomite, siderite, etc.), X-ray diffraction allows the moderately accurate determination of the amount of magnesium (Mg) substitution in the calcite or dolomite lattice. For this, one needs to do careful scans at low chart speeds. By including an internal standard (galena for ancient carbonates and NaCl or CaF_2 for modern sediments) one can determine very accurately (in terms of the angle, 2θ) the reflection peak positions representing the (112) plane of the calcite crystal lattice. By matching that data against the chart given below (modified from Goldsmith, Graf, and Heard, 1961) one can approximate the magnesium content of the lattice to about 0.5 mol%. However, other cations besides magnesium can cause lattice spacing shifts and this data should thus be checked occasionally with microprobe or atomic absorption.

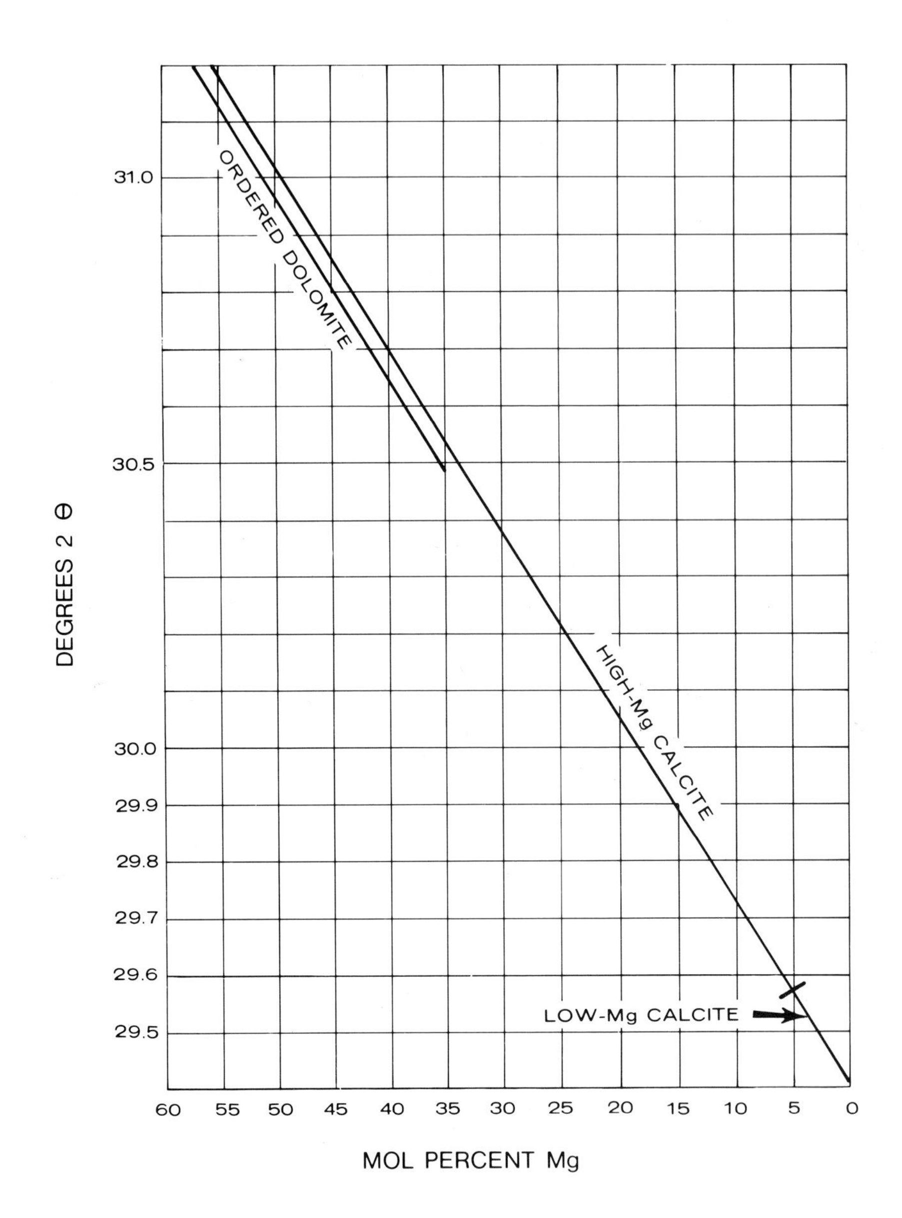

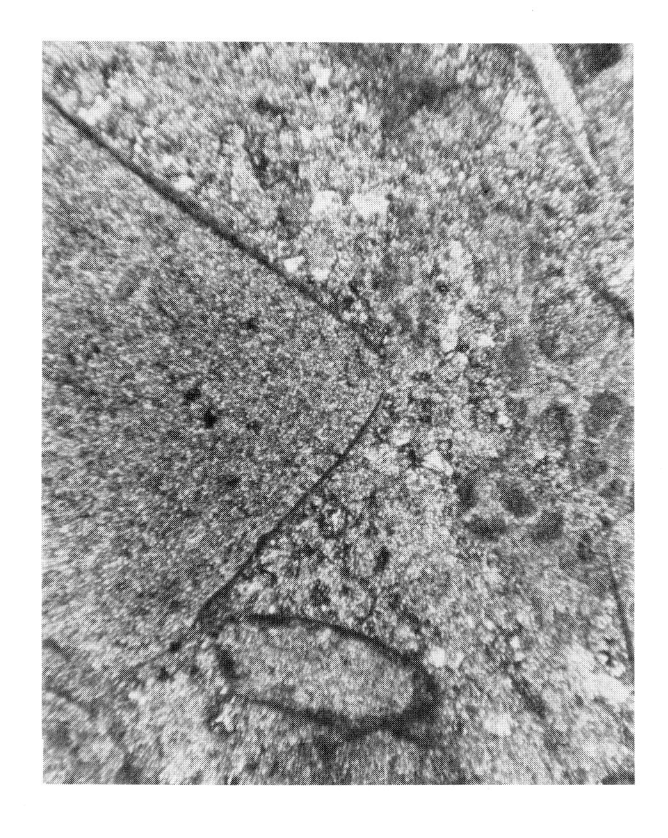

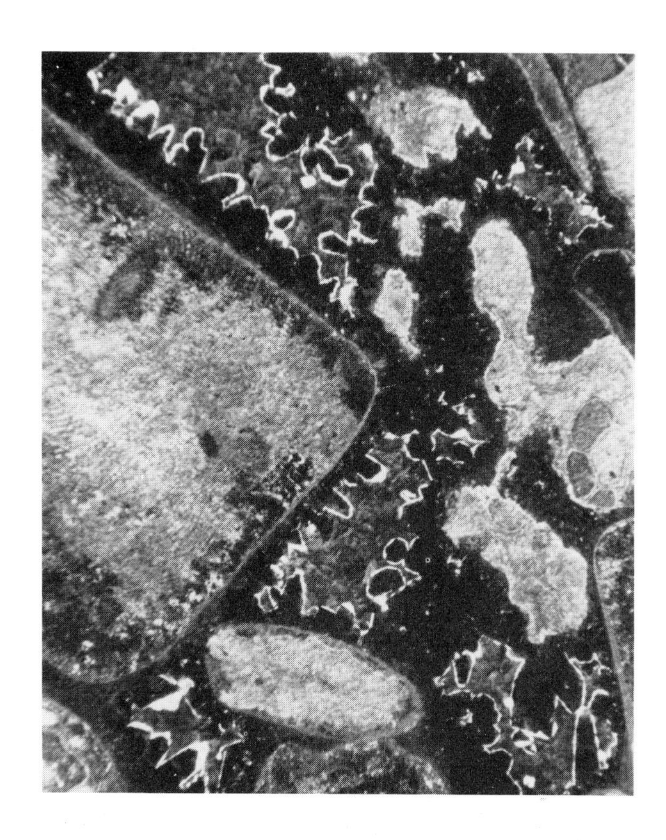

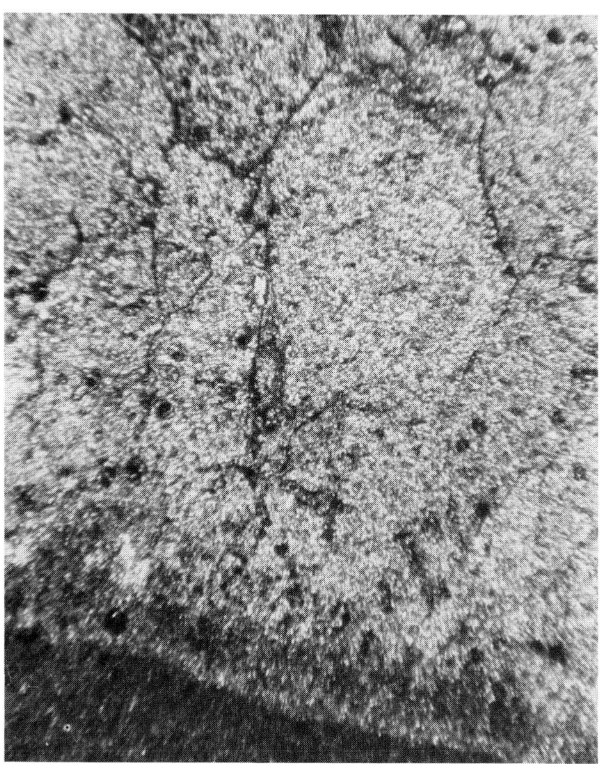

Cathodoluminescence can provide useful information about the spatial distribution of trace elements in carbonates although it does determine which elements are involved. These photos show comparison of the same field of view under plain light microscopy and cathodoluminescence. Note distinction of two or more generations of cement with complex boundaries in cathodoluminescence photos.

Upper samples are from Lower Pennsylvanian Bloyd Formation, Oklahoma; a 1.25 cm scale bar would be 0.25 mm. Lower samples are from Triassic of Austria; a 1.25 cm scale bar would be 0.20 mm.

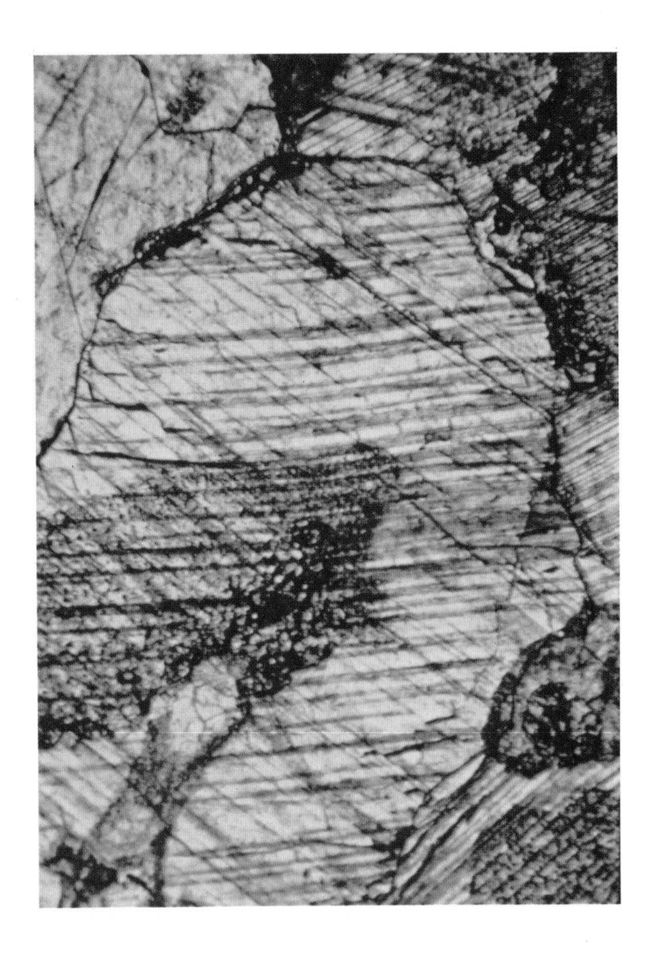

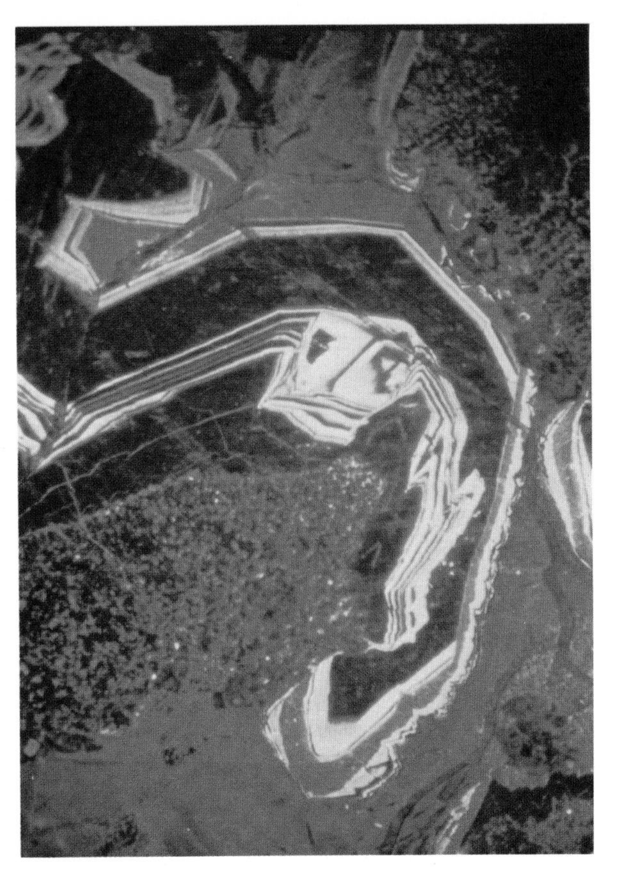

Color photography of cathodoluminescence (although sometimes difficult because of the low light intensities involved) can often yield spectacular results.

This example is from the Lower Mississippian Lake Valley Limestone of New Mexico. The upper photograph shows a cement zone in a crinoidal biosparite. From this transmitted light view only one generation of overgrowth cementation would be inferred.

The lower photograph illustrates approximately the same field of view, however, and shows that with cathodoluminescence at least five generations of cementation are visible.

These cement generations can be correlated from sample to sample and can be related to a variety of tectonic and erosional events (see Meyers, 1974).

A 1.25 cm scale bar would be equal to 0.30 mm on original sample. Photographs courtesy of W. J. Meyers.

Scanning electron microscopy (SEM) offers enormous advantages in the 60- to 15,000X-magnification range; easy preparation, excellent resolution, and a much greater depth of focus than with light microscopy. First photo shows details of coccolith morphology and crystal texture in this chalk (Upper Cretaceous Austin Group, 1.3 μm). The great depth of focus of the SEM allows the observation of whole specimens, such as this Foraminifera *Spirillina decoratal(?)* shown in the second photo.

A 1.25 cm scale bar would be 55 μm. The main difficulties are in correlation of SEM observations with those made with the light microscope.

Transmission electron microscopy allows very high resolution and magnification. Disadvantages include working with polished, etched, and replicated surfaces which may introduce artifacts, and difficulty of getting low magnification views. Two coccoliths seen here from Upper Cretaceous Monte Antola Formation, Italy. Scale bars are each 2 μm.

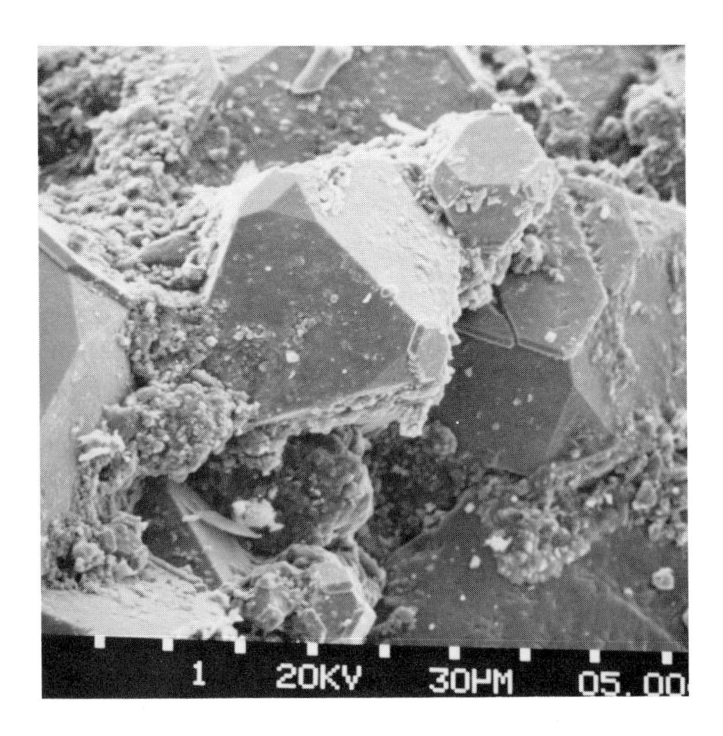

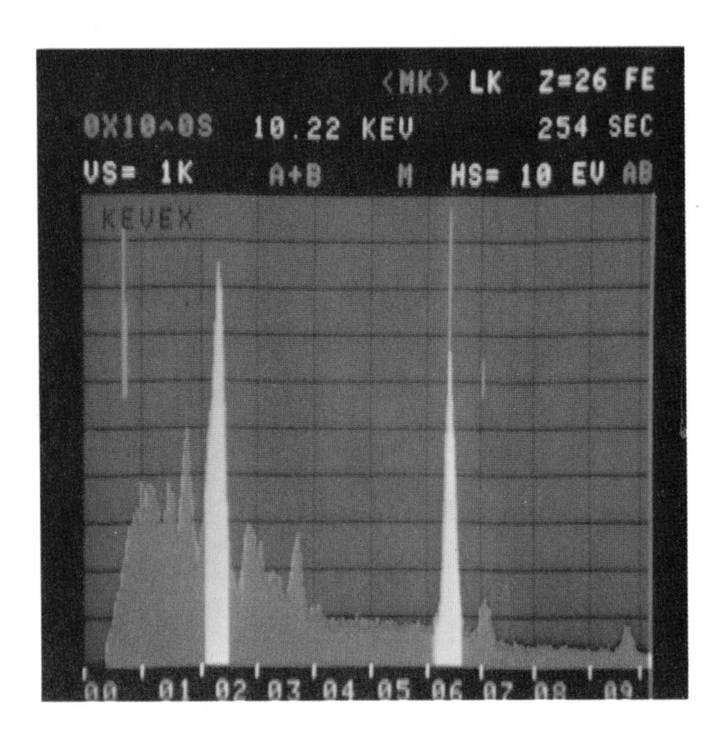

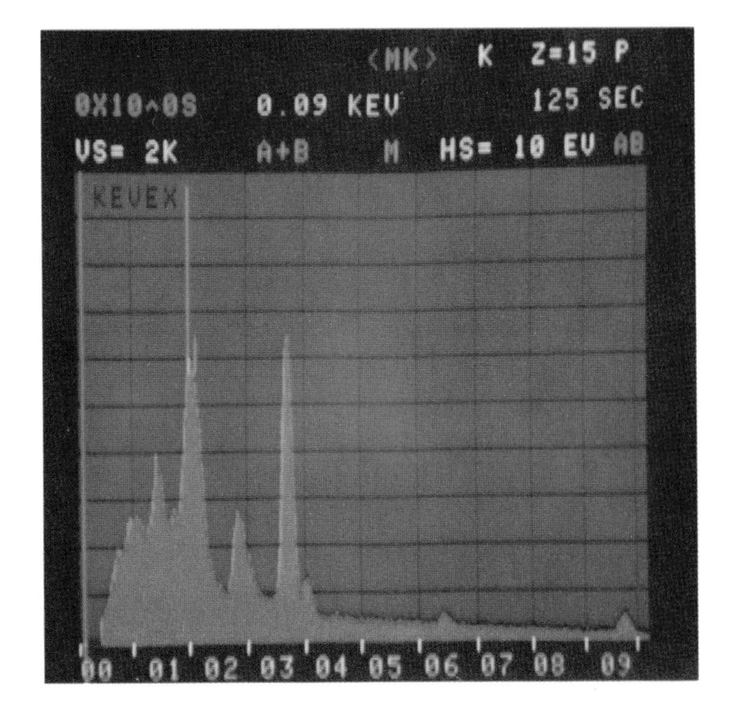

Scanning electron microscopy is useful not only for examining sediment textures, but when equipped with an energy dispersive analyzer it can be used effectively for mineral identification and semi-quantitative chemical analysis.

The above examples from the Upper Cretaceous Atco Formation (Austin Group) of Texas show accessory minerals in the chalk.

The pair of photographs on the left (top and bottom) illustrate pyrite crystals (about 60 μm in diameter) and the chemical analysis of the same area. Note prominent light-colored peaks for Fe and S.

The pair of photographs on the right show a fragment of phosphatic material (circular depressions are about 100 μm across) with major P and Ca peaks. These analyses can be performed in seconds and are both convenient and accurate ways of doing elemental analyses.

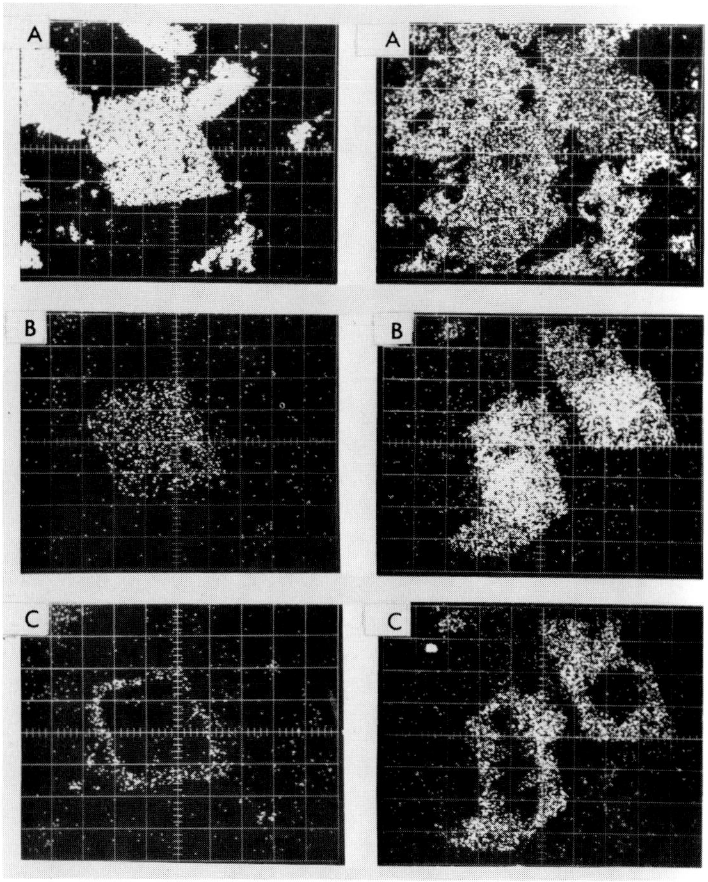

A A
B B
C C

A A
B B
C C

Microprobe

The microprobe offers the advantage of getting chemical analyses of areas as small as a few microns or scans of much larger regions. These photographs illustrate one application. The rock was filled with rhombs which were all presumed, on petrographic grounds, to be dolomite.

When placed under the microprobe, they were analyzed for calcium (top photos, A, in each of the four series), magnesium (middle photos, B, in each of the four series), and iron (bottom photos, C). Note that the crystals in the two upper series all contain calcium and magnesium (A and B) while having outer, iron-rich zones (C). The lower two series have calcium and iron-rich zones (A and C) but are completely lacking the magnesium (B). Thus the two upper crystals are dolomite and ferroan dolomite whereas the bottom two crystals have been dedolomitized (but without losing zonation) and are now calcite, and ferroan calcite. Simple petrographic distinction was impossible, although staining or cathodoluminescence might have revealed the same features.

The main disadvantages of the technique are cost, time of sample preparation, need for accurate standards, and inability to distinguish many elements in the trace amounts in which they are found in carbonate rocks.

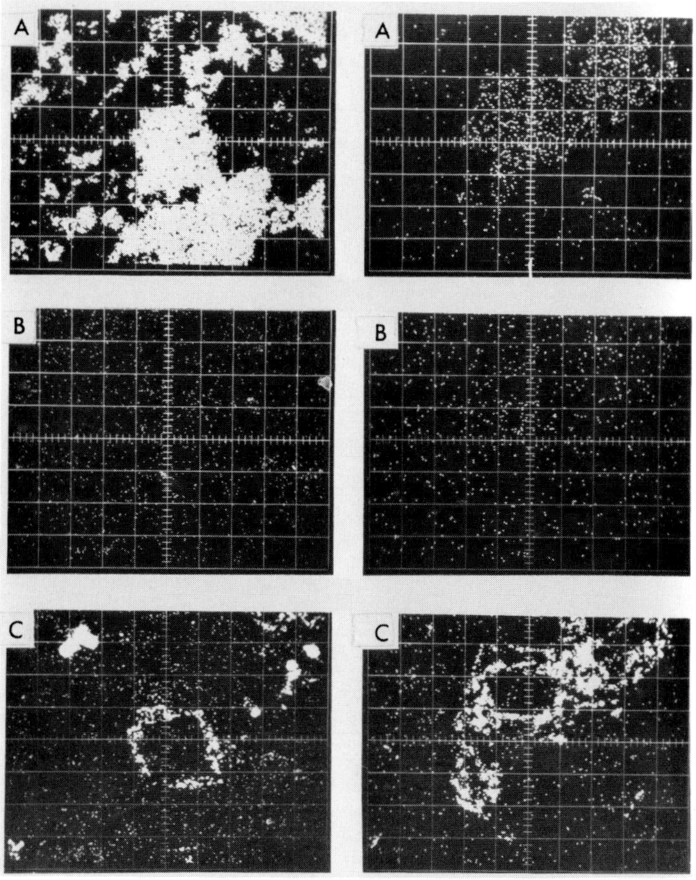

Index

A reference is indexed according to its important, or "key" words.

Three columns are to the left of the keyword entries. The first column, a letter entry, represents the AAPG book series from which the reference originated. In this case, M stands for Memoir Series. Every five years, AAPG merges all its indexes together, and the letter M will differentiate this reference from those of the AAPG Studies in Geology Series (S) or from the AAPG Bulletin (B).

The following number is the series number. In this case, 27 represents a reference from Memoir 27.

The last column entry is the page number in this volume where the reference will be found.

Note: This index is set up for single-line entry. Where entries exceed one line of type, the line is terminated. (This is especially evident with manuscript titles, which tend to be long and descriptive.) The reader sometimes must be able to realize keywords, although commonly taken out of context.